America's
National Parks
and Their Keepers

America's National Parks and Their Keepers

RONALD A. FORESTA

Published by Resources for the Future, Inc., Washington, D.C.

Library of Congress Cataloging in Publication Data

Foresta, Ronald A., 1944–
America's national parks and their keepers.

Bibliography: p.
Includes index.
1. National parks and reserves—United States.
2. United States. National Park Service. I. Resources
for the Future. II. Title.
SB482.A4F67 1984   333.78'3'0973   83-43262
ISBN 0-915707-02-0
ISBN 0-915707-03-9 (pbk.)

Published by Resources for the Future, Inc., 1755 Massachusetts Avenue, N.W.,
Washington, D.C. 20036
Resources for the Future books are distributed worldwide by The Johns Hopkins University
Press.

Resources for the Future is a nonprofit organization for research and education in the development, conservation, and use of natural resources, including the quality of the environment. It was established in 1952 with the cooperation of the Ford Foundation. Grants for research are accepted from government and private sources only on the condition that RFF shall be solely responsible for the conduct of the research and free to make its results available to the public. Most of the work of Resources for the Future is carried out by its resident staff; part is supported by grants to universities and other nonprofit organizations. Unless otherwise stated, interpretations and conclusions in RFF publications are those of the authors; the organization takes responsibility for the selection of significant subjects for study, the competence of the researchers, and their freedom of inquiry.

This book was edited by Ruth B. Haas and designed by Elsa Williams. The maps were drawn by Julie Cocks. The photographs were kindly supplied by the U.S. Park Service Photo Library. The index was prepared by Florence Robinson.

Contents

Foreword

Outdoor recreation policy in the United States has many unique features. We recognize that outdoor recreation can add significantly to the quality of life, and our magnificent natural resources permit us to provide outstanding outdoor experiences for many citizens. Our system of national parks reflects this policy and our rich natural endowment of scenic resources.

Yet, we have entered a new era. Most of our unique natural areas have either passed into public ownership, or already have been commercialized. Increasing affluence and mobility, combined with a new awareness of the importance of natural environments, have led to levels of use that Steven Mather and Horace Albright could not have dreamed of when they took over the reins of the newly founded Park System in 1916. In 1982, 245 million people visited lands in the National Park System, an increase of 152 percent during the past twenty years. The result has been crowded facilities, damaged vegetation in some parks, noise, crime, and in many cases a loss of experiencing what John Muir described as a landscape in repose. At the same time, the Park System has been expanded to include areas far removed from what might be described as the unique "crown jewels." Old homesteads, battlefields, even old railway sites are a part of the System. As the System has changed, so have the expectations of those who use the parks. The demands of backpackers for more wilderness and the cries of biologists and historians for more preservation are joined by equally intense requests for more roads, more accessible parks, more services for the public.

It is not surprising the National Park Service should sometimes feel at sea when appropriate policies are not obvious. Should additional acres of land be added, or should funds be used instead for maintenance or other means of increasing supply? Should policies, such as the imposition of user fees, be adopted that would influence the quantity taken?

This book is a study of an expanding National Park System; it is also a study of the bureaucracy that shaped it, how it grew, and the stresses it faces. As Dr. Foresta shows, the Park Service is no stranger to controversy and change. One of the Service's strengths has been its pragmatism and adaptability, but current guidelines are hard to come by and the decisions of the future will be neither simple nor easy. This study is valuable not only for its analysis but for the informed, revealing picture it presents of an agency and a system that have enriched the lives of countless citizens and visitors to this country. Dr. Foresta does what is so rarely accomplished— he reduces a difficult and complex policy issue to an interesting treatise without sacrificing completeness and rigor.

This is the first book to emerge from the Gilbert White Fellowship program at Resources for the Future and it sets a high standard for those that will follow. Established in honor of Gilbert White, distinguished geographer and former chairman of the RFF Board of Directors, the fellowships offer young scholars an opportunity to devote an uninterrupted year at RFF to a project in their field of interest and to interact with our staff formally and informally. Gilbert White will no doubt be pleased that the first published product of the program is from a geographer.

October 1983   Emery N. Castle
                                                          President

Acknowledgments

The debts I have accumulated in researching and writing this book are many and although I will never be able to pay them off, I can at least acknowledge them.

First I would like to thank the men and women of the National Park Service. They gave generously of their time in explaining how their agency works and in guiding me to its key documents and data sources. They frequently took me into their confidence, invited me into their homes, and treated me as a friend. The mix of professionalism and easy egalitarianism in the Park Service men and women I met make them the very best in the American tradition of public service. Their faith in the possibility of improving the future through an understanding of the present is inspiring.

The generosity of Resources for the Future in the form of a Gilbert White Fellowship allowed me to devote a full, undistracted year in Washington to this project. RFF's contribution extended far beyond financial support, however. RFF's reputation is such that its sponsorship opened many doors which otherwise might have remained closed to me. It also put me in touch with the many RFF staff members who offered advice and support. Herb Morton was a source of guidance and unfailingly sound advice from the incipient stages of the research to the finishing touches. In addition to trying to teach me some economics, Irving Hoch and Roger Sedjo provided insights, moral support, and friendship. Marion Clawson made me the beneficiary of his great wisdom and knowledge of public land policy. Ruth Haas was the ideal editor—supportive, tactful, supremely competent, and more.

Thanks also to those who read all or part of the manuscript along the way and whose comments improved the final version. Susan Fainstein's comments were the bedrock on which the final draft was based. Jerry Fish, as usual, rose to the occasion with his truly original acumen. Robert Nelson provided a perceptive and reasoned critique of the first draft and Jim Doncaster's close reading of the second draft improved it greatly. Richard Liroff and Frank Popper brought their own experience to the manuscript and left it better than they found it. The comments of the four anonymous reviewers were helpful and appreciated.

I would like to particularly thank Sid Jumper, my department chairman, who encouraged me in this project, who arranged for the cartography, and who, when he saw my phone and photocopying bills, swallowed hard and didn't say anything. Julie Cocks did a fine job on the maps. Pam Sharpe managed the manuscript production with awesome competence, unflagging energy, and a sense of humor that made her a pleasure to work with. Thanks also to Pam's assistant, Starr Reilly, who always came through. Finally, the Southern Regional Education Board provided a grant that allowed me to put the finishing touches on the research. If there is anyone whom I have omitted from this list, I hope they will forgive me.

Now that I have revealed the extent of my debts to others, I would like to add that I am solely responsible for all the errors, lapses of judgment, moments of mean spirit, and canards found herein.

October 1983   R.A.F.
                                                               Knoxville,
                                                               Tennessee

1 Introduction

In the public's mind, the images of the National Park System and its keeper, the National Park Service, are clear ones. The public sees the System as "big trees, tall mountains, beautiful lakes" in the words of one park ranger.[1] The image of the Park Service is that of the ranger on horseback—part naturalist, part policeman, part resource manager, and even part educator. The images of the Park System are of remote places and past times. They are tied up with American memory and mythology; they are collages of Bierstadt paintings, heroic pioneers, log cabins, and Indian tepees. However, things are not what they appear to be. The reality beneath the image is that neither the national parks nor their keepers stand apart from our times; they are very much subject to the problems and dilemmas of modern American life. We are given occasional hints of this. A television program focuses on pollution and crime in Yellowstone National Park. A newspaper article discusses vandalism in a national park close to Cleveland or Atlanta or New York. Not only are the parks and their keepers very much part of the present, but many of those close to them believe that the modern era has posed questions and created problems that have overwhelmed the System and its keepers.

It is felt that as a result, the Park Service no longer knows what its purpose is nor that of the Park System it manages. An administrator in a sister agency said, "In a deep sense, the Service has no idea of what it ought to be doing; there is only confusion. I look for principles and consistent criteria [for decisions] but I don't see them." A member of Washington's environmental community conveyed a similar thought when he said that "[the National Park Service] doesn't seem to have a purpose, or at least you'd never be able to define one from its actions." Robert

Cahn, a journalist who has followed the Park Service for years and who perhaps knows it as well as anyone, writes of "accumulated contradictions."[2] Others see "a confusion of purpose" in the National Park Service and "a crisis in dilution of purpose."[3] A student of the agency told the author that "the past ten years have been years of transition for the Service but exactly where it's going is something the Service doesn't know." And as one observer summed it up, "Mostly the Service is confused by America's changing values and policies."[4]

This confusion is not merely a matter of academic interest; rather, it is a serious practical concern. Criticisms of the agency's behavior indicate that there is basic uncertainty about its decisions. For example, one of its students told the author, "Many of the Service's problems would be solved if only it would say 'yes' and 'no' with some sort of predictability." Administrators in other federal agencies and environmental activists also perceive this inconsistency and blame it for the Service's unsettled and erratic relationship with its supporters. The respected management consultants, Coopers and Lybrand, in a recent study of the agency's management problems, wrote that "perhaps the largest constraint on [the] National Park Service in fulfilling its mission is the mission itself."[5] As clarified later, this meant that there was no clear, commonly agreed-upon sense of mission within the agency; instead, there was "confusion about the purpose, goals and objectives of the Park Service." Coopers and Lybrand concluded that failure to arrive at and project a clear sense of mission would "ultimately lead to inability to manage the Park System."[6]

If the agency is in fact confused about its mission, it is for good reason. From the 1960s onward there have been profound changes in attitudes toward just about everything the Park Service does. New ways of looking at things have emerged, old certainties which served as trusted policy guides for the park management have disappeared; and old solutions to problems are no longer acceptable. New problems have emerged for which there do not seem to be obvious solutions.

Through its management of the National Park System, the Park Service has considerable responsibility for the preservation of nature in America, yet the past two decades have seen a large change in the way Americans view the natural world around them. With the increased environmental awareness of the 1960s and 1970s has come a view of nature as the fragile base on which our civilization rests. The idea that civilization would, of its own accord, come to a comfortable equilibrium with nature has given way to a view of American material culture as a juggernaut which is greatly damaging the natural world in many obvious as well as subtle ways. These changes in attitude, which have occurred on both the popular and scientific level, have raised new questions about park management. For one, they have called into question the Park Service's traditional policy of intensely developing some parts of the national parks and leaving others wild. The Sierra Club and other environmental groups claim that the park visitor is a threat to the parks, and that development to accommodate visitors degrades the natural environments of the Park System in the same way that civilization has degraded the wider natural world. If this is the case, then perhaps past efforts to accommodate visitors have been too zealous.

The new, more embattled, view of nature has also raised the issue of the criteria for selecting parks. Traditional criteria emphasized human appeal; a national park should be sublime above all. Nature in the national parks was anthropocentric; it was scenery and its value was based on direct human appreciation. The new view of nature has forced the Park Service to ask about the value of natural areas with less direct human appeal. Should endangered ecosystems be preserved in the National Park System even if their aesthetic appeal is minimal? What about traditional selection criteria which stress the inclusion of only the finest examples of natural objects or areas in the Park System? If nature is truly under siege, does not an agency whose past is associated with the protection of nature have a responsibility to defend natural areas wherever they are threatened by civilization's inroads? Should it stand aside and let areas be destroyed simply because they are less than the greatest remaining examples of their type?

While our view of nature was undergoing this deep change, the construction of the interstate highway system was destroying whole districts of historical interest. Public urban redevelopment projects and the private building boom were obliterating countless more. As the cost of maintaining the buildings of past eras rose, their owners replaced them with modern, efficient structures, or sometimes they simply walked away from them. One result of these assaults on the past was a great fear of losing touch with our national history, with the texture and scale of past times, and with the relics and sites of great deeds. This fear aroused concern at the highest levels of government. For example, the Department of Interior's major policy statements in the early 1960s and the speeches on quality of life by the decade's presidents stressed the parallels between preserving nature from the excesses of material progress and protecting the important remnants of the past from like excesses. The Historic Preservation Act of 1966 cast this concern into law.[7] Given that the Park Service was already involved in historic preservation through its management of several score historic sites, it was only logical that increased concern for the past at the highest levels of government would affect the agency. First, the Historic Preservation Act created an important new role for the agency as the keeper of the National Register of Historic Places. Second, historic preservation was raised to formal parity with the protection of the nation's great natural areas in the agency's range of responsibilities.

With this new emphasis on history, there arose unanticipated questions for which there were no easy answers. For example, what parts of the past were worth preserving in the System? Surely the sites of national triumphs such as Gettysburg and Valley Forge belonged in the Park System. But what about the darker side of our past? Did Andersonville belong? Should all our history be represented, or only those events with special, clear associations with physical sites and material remnants? Should the System be primarily a repository of patriotic symbols or should the System's approach to the past be one of detachment and balanced explanation? Were historical sites shrines or places of education? Finally, should the Park System be used to adjust our view of the past to present attitudes and values? For example, should the agency seek out sites associated with the early days of the movements for women's or minority rights so as to be in a position to honor or at least teach about these movements?

In the 1960s and 1970s the question of the government's responsibility for rectifying social and economic inequality rose to prominence on the federal agenda. Should the federal government make a major effort to improve the lot of those left behind by prosperity? If so, how? The plight of the nation's cities raised related and similar questions about the nature and limits of federal responsibility. Should the federal government attempt to maintain the cities and the quality of life for city inhabitants when the city governments were incapable of doing so? If so, what policies should be adopted and what agencies should have a role in carrying them out?

The emergence of these concerns affected the Park System by raising proposals for national parks on the periphery and sometimes even in the center of the nation's largest cities. It was argued that such an expansion would bring the System close in time and distance to those who had little chance to visit the great national parks of the West. It would also relieve the cities of part of their responsibility for providing recreation facilities for their citizens and thus serve as direct federal aid to them. These proposals opened debate on some elementary questions about national parks. Was a national park something that could be established anywhere? In a trivial sense, yes, of course; a piece of land could be set off and called a national park. But in a deeper sense, didn't a national park have to have spectacular nature or important history? If so, national parks were where you found them, not where you thought they should be. The appropriateness of the National Park System as an instrument of national urban policy depended on the answer to this basic question of definition.

When the Park System was drawn into cities with the establishment of several urban national parks in the 1970s, tough new issues arose over how these urban parks were to be developed. Should the popular access side of the System's mandate be emphasized by the development of mass recreation facilities or should the dominant theme in park development be the preservation of the natural and historical resources found in them?

The modern era's increasing rejection of the clear demarcation between public and private interest in land development also had a strong impact on thinking about the Park System. Spreading awareness of the economic externalities of land development and an increasing sensitivity to the fact that ecological processes did not respect property lines brought many to view public and private interest in land use as intimately intertwined. In fact, interest in land came to be seen as even more complicated; private owners, the local public, the state, and the nation as a whole all might have definite and distinct interests in a single land-use decision. Since the public or even several publics might have a stake in how private lands were developed, it followed that they should, through the political process, have a voice in those decisions. At the same time there was an increased sensitivity to the impact that public projects or development on public lands could have on neighboring private lands and a willingness to give neighboring landowners a say in public land-use decisions.

This new sense of intertwined responsibility forced those close to the System to wrestle with the problem of defining the limits of agency responsibility in protecting the parks from external threats. Should the agency adopt an aggressive policy of protecting the parks from visual intrusions or pollution threats? If so, how far from the parks should the agency's concern extend? Traditionally it stopped at park boundaries. Should it now extend to neighboring lands? To the entire region in which the park was set? These attitude changes in wider society also forced those close to the Park System to reexamine traditional emphasis on fee simple acquisition to protect natural and historic sites of national importance. Perhaps in many cases just the acquisition of development rights or scenic easements might be the best way of protecting both the rights of private landowners and the public interests. In some cases, perhaps the existence of a local or regional planning process would be sufficient to protect the national interest in a particular region, and no land purchases at all would be necessary. Perhaps a national park should not even be thought of as a discrete, publicly owned entity but rather as an area in which the national interest, be it in access or preservation, was protected by whatever means were appropriate. Here too, the agency was forced to reexamine not only the purpose and goals of the Park System but the very definition of a national park as well.

These then are some of the questions which the National Park System and those responsible for it have faced in recent years. They are many and varied. Each in its own way raises the matter of standards to be maintained, people to be served, and goals to be pursued. In the ideas and values they reflect, they strike deeply into prior assumptions about what national parks are supposed to be. They shake traditional policy to its foundations, and they raise fundamental questions about the proper mission of the System's keepers, the National Park Service. It is little wonder that defining the role of the System has become an obsession with the agency and a great concern to its critics.

ORGANIZATION OF THE BOOK

These questions are the subject of this book. Subsequent chapters examine the circumstances which gave rise to these questions, gauge their effect on the agency's sense of mission, and determine the impact that attempts to grapple with them have had on the National Park System. On a more general level, this is a case study of mission and purpose in public agencies, of how the broadest of organizational goals are formulated, and how a sense of purpose is sustained. It is hoped that in treating the Park Service's sense of mission and the challenges to it that recent times have brought, this study will cast light on these general questions.

There are two contexts in which the current problems of the System and Service must be understood. Each is discussed in the following two chapters. One context is that of tradition and precedence. The first national park was founded over a hundred years ago, and the National Park System was brought together as an administrative entity under the National Park Service more than sixty years ago. A century of past attitudes weighs heavily on the System and the Service. Past policies form a baseline of established practice against which all change is measured. Traditional responsibilities are the core of the public image of the national parks and to a large extent they are the core of the Park Service's image of itself. To understand the System and Service today one must know what baggage they carry from former eras.

The other context is that of political power. Here we find Congress, the White House, and other public agencies arrayed around the System. Many private interest groups also have their own goals for the System, and each has its own sources of power and its ways of exercising that power. The wishes of these interests and institutions combine to give park policy its parameters and to define the limits of the possible and the realistic. Political forces, like tradition, must be understood if today's Park System and Service are to be understood.

After a discussion of these contexts of tradition and power, each of the following four chapters deals with one of the major areas of national park policy into which the doubts and contradictions of recent American life have been funneled. Chapter 4 looks at park policy with regard to the preservation of nature in the Park System. Chapter 5 examines the Park Service's interpretation of its historic preservation mission. In chapter 6, the Park System's urban involvements are explored, and in chapter 7 the Park Service's interest in land beyond park boundaries is discussed.

The overall approach to the subject is analytic; my primary aim is to develop an understanding of what the System and Service are today and how they have come to their present states. Mere understanding, however, is not in itself a sufficient goal of public policy analysis, especially since understanding is too often an excuse for acceptance. Judgment, prediction, and evaluation should also flow from good analysis of policy. Accordingly, in the concluding chapter, some suggestions are set forth for the management and growth of the Park System, and some points of policy orientation offered for its keepers.

USE OF INTERVIEWS

The research for this book included extensive interviews and a word about the use of quotes is in order here. When I interviewed a subject I first established the degree of confidentiality with which the information elicited would be treated. Some informants did not want to be quoted directly. Some did, and wanted to be identified as the source of an idea or opinion. These wishes were of course respected. A number of informants—about half a dozen—did not even want to be identified as having been interviewed by the author. Their wishes too were respected and their names do not appear on the list of interviews presented in appendix B. Treating the information gathered in these restricted interviews presented little problem; it was a simple matter of honoring agreed-upon confidentiality.

The majority of informants, however, were quite willing to have their words attributed to them, even on what might have been rather sensitive subjects. It was with this group that a problem arose. I conducted most of my interviews during the last months of the Carter administration and the first months of President Reagan's term. It was a halcyon period of "organizational democracy," when the atmosphere within Interior was relatively relaxed and policy researchers were welcomed. Shortly after this research was concluded, however, the atmosphere changed and remarks or opinions that could have been shared with outsiders a few months before were now indiscretions or worse. In light of this change of atmosphere, I adopted a cautious approach to attribution, even if it had to be at the expense of scholarly convention. If a quote is completely innocuous, its source is identified (if, of course, that was part of the original understanding). If the matter appears sensitive it is not, even if the source agreed to be quoted when he was originally interviewed. If I was unsure, I erred on the side of caution and have identified the source only in a general way. The exception to this rule is the informant who insisted that he be identified. This happened infrequently and only with people well insulated from retribution, so these informants are identified.

2 The First Fifty Years

THE BEGINNINGS

When the National Park Service was established in 1916, Steven T. Mather was appointed its first director, and he in turn appointed Horace Albright as his assistant. The two could not have been more different in personality.[1] Mather was flamboyant, imaginative, given to improvisation, and possessed of a flair for public relations that approached genius (figure 2-1). All of these characteristics combined to make him, in the words of a student of the Park Service, "the very model of an American salesman."[2] Albright was his complement; he was a phlegmatic, diligent lawyer fresh out of law school who was strong on organization, administration, and planning. Together they would see the Service through its first thirteen years, and with Mather gone, Albright himself would guide it for another four (figure 2-2). By the time Albright retired in 1933, the Park Service was a secure agency, managing a well-established system of national parks. It exuded a strong sense of purpose; it had public support and powerful allies in and out of government.

To understand Mather's and Albright's accomplishment, one must first be aware of the ambiguities associated with the agency's birth. Although the Park Service was founded to unify and bring order to the management of the national parks, it was unclear exactly what the national parks were to be. It seems that everyone knew generally and no one knew specifically.[3]

The parks that Mather and Albright found themselves managing in those first days were truly a motley collection. They included such magnificent places as Yellowstone, Glacier, Sequoia, and Yosemite national parks, places whose preservation had been a national cause and whose beauties had been made widely known to the American public through popular

9

Figure 2-1.
Steven Mather (foreground), first director of the National Park Service.

Figure 2-2.
Horace Albright as superintendent
of Yellowstone National Park.

naturalists like John Muir. But these parks with their beauty and high-minded origins were not the only ones the Service inherited in 1916. There was Mesa Verde in Colorado, established largely to protect an archaeological site of scientific value from vandalism.[4] There was Platt, 900 acres of hot springs in Oklahoma, whose waters were almost perpetually contaminated by sewage from an adjoining town. Platt's establishment as a park had been contrived by Oklahoma's congressional delegation, who could be counted on in session after session to call for increased federal spending in connection with the park.[5] The delegation wanted fish hatcheries, dams, and disabled veterans' homes in the park, and railroads, highways, and telephone lines through it. Wind Cave in South Dakota was also in the System. It had been established for reasons which no one could remember in 1916. Then there was Sully's Hill; no one was sure it was supposed to have been a national park in the first place. In 1904 as part of an Indian treaty, Congress reserved a small part of the public domain in the Sully's Hill area of North Dakota and declared that the area should be a park. It was no more specific than that. Thus for lack of a better idea of what to do with it, Sully's Hill was included in the National Park System.[6]

While these parks were not in the class of Yellowstone, they were not always obscure, little-visited ones. Table 2-1 shows park visits for 1916. The entire system had 356,000 visitors, with 36,000 and 33,000 at Yellowstone and Yosemite, respectively. Platt, with 30,000 visits that year, came close to these in popularity. Hot Springs, Arkansas, a medicinal spa, enjoyed 119,000 visitors. This was more than one-third of the total visits to the National Park System in its first year under the Park Service. It made Hot Springs by far the most visited park in the System, a distinction it retained for years to come.

Nash suggests that the American "invention" of the national park depended on four things: our unique experience with nature in North America, our democratic ideals, the vastness of our public domain, and the affluence of our society.[7] But the system Mather and Albright inherited in 1916 was too disparate in origins to have been invented; it accumulated. And as much as its origins may have depended on Nash's four factors, it also depended on pork barrel politics, on the need for a catch-all category for miscellaneous withdrawals of public land, and on purely idiosyncratic circumstances. Given this heritage and the fact that they had little guidance from the National Park Service's founding legislation in sharpening the purposes and goals of the new agency, Mather and Albright's initial confusion about its function was understandable. Nevertheless, there were guides to success for those who could find them in the interplay of the era's social and economic forces, in the examples set by other agencies, and above all in the spirit of the times. If Mather and Albright knew nothing else, they knew where to look.

Table 2-1. Visits to the National Park System in 1916

Visits 1916	Year of establishment	Park	State
118,740	1832	Hot Springs	Arkansas[a]
35,849	1872	Yellowstone	Montana–Wyoming–Idaho
—	1875	Mackinac Island	Michigan[b]
10,780	1890	Sequoia	California
33,390	1890	Yosemite	California
15,360	1890	General Grant	California
23,989	1899	Mount Rainier	Washington
12,265	1902	Crater Lake	Oregon
9,000	1903	Wind Cave	South Dakota
1,500	1904	Sully's Hill	North Dakota[c]
30,000	1906	Platt	Oklahoma
1,385	1906	Mesa Verde	Colorado
12,839	1910	Glacier	Montana
51,000	1915	Rocky Mountain	Colorado
N.A.	1916	Hawaii	Hawaii
N.A.	1916	Lassen Volcanic	California

Sources: Lee, Family Tree, p. 10; National Park Service, *Public Use of the National Parks: A Statistical Report (1904–1940),* p. 3.
N.A. = Not available.
[a]Hot Springs, Arkansas was withdrawn from claims in 1832 and was run as a park by the federal government thereafter. It was part of the original charge of the NPS although it did not become a national park until 1921.
[b]Ceded to Michigan, 1895.
[c]Converted to game preserve, 1931.

Conservation and Progressivism

The three decades between 1891 and 1920 were called the conservation era because the period was one of extraordinary concern for the fate of the nation's natural resources.[8] During those decades, the wise management of American resources was a major cause that captured much media attention and united "men of good will from nearly all stations, all skills and all types of education," under its banner.[9] It drew support from organizations based on profession—associations of biologists, geographers, geologists, and so on. It also drew support from outdoor clubs such as the Appalachian Mountain Club and the Sierra Club, and from scenic preservation organizations, garden clubs, and civic betterment associations. The conservation cause was also supported by those businesses which had economic stakes in provident natural resource management—some timber companies, many railroads, and sections of the mineral extraction industry. This breadth of concern inspired and was reflected in the high level of political attention conservation received. Both major parties hewed large platform planks out of the conservation issue, while Gifford Pinchot, one of conservation's most forceful partisans, became one of Roosevelt's chief lieutenants and ultimately one of the most powerful men in Washington.

As a movement and as a cause, conservation did not stand alone; it had strong ties to its contemporary, the progressive movement. The term "progressive conservation" is an appropriate descriptor for the conservation movement of the late nineteenth and early twentieth century.[10] McConnell sees a similar connection between the movements, and he writes that the ideals and objectives of the conservation movement "were deeply imbued with the doctrine of progressivism."[11]

Progressivism was intimately tied to the emergence of an industrial economy and a modern, middle-class society in the decades following 1890, and one of its many articulated goals was to make government responsive to the rapidly broadening and increasingly powerful middle class. Another of its central aims was to bring government in line with the latest rational scientific principles.[12] Out of this concern for rationality came a demand for efficiency in government. In its abhorrence of "politics" and "special interest," it reflected a homogeneous notion of the public good that made little allowance for class or region-based differences in needs.

Like progressivism, the conservation movement was directed toward accommodating the changes associated with the emergence of modern industrial society. Indeed, Hays suggested that the wider significance of the movement stemmed from the role it played in transforming a decentralized, nontechnical and loosely organized society into a highly organized and centrally directed society that could meet a complex world with efficiency and purpose.[13]

The Public Lands

The progressive conservation agenda for land resources, especially what remained of public domain, reflected some of the movement's dominant themes. The nineteenth century's answer to the question of what to do with the public domain that the federal government inherited as large blocks of western land were acquired had been to sell this land for almost nothing or, later, to give it outright to homesteaders. But the conservation movement no longer considered this appropriate. The frontier had closed, and the post-frontier homesteader was viewed, not as a sturdy pioneer like his predecessor, but as "a dubious character not interested in acquiring a farm and a home, but making a short term investment in cheap land with an eye to quick and profitable sale."[14] To men who thought this way, further alienation of the public domain could only lead to waste and, as such, was unacceptable to a cause in which the idea of the public good was so bound up with the notion of efficiency. Instead of alienation, the conservation movement advocated continued public ownership of much of the public domain. Public ownership would give scientists and resource management professionals the greatest latitude in working toward the public good. Moreover, public land ownership fit into a homogeneous, universalist notion of the public good; all Americans, regardless of class and region would become the beneficiaries of its bounty.

Although the idea of perpetual public ownership of much of the public domain was not a new one, and isolated areas had been exempted from private claims beginning in the early nineteenth century, it was not until the era of progressive conservation that a policy of massive, systematic exemption gained broad popular support and made political headway.[15] Before the progressive era closed after World War I, more than 150 million acres of the public domain had been placed in national forests, parks, or monuments and for all practical purposes could not be transferred to private ownership.[16]

While settling some questions, these withdrawals raised a new one: What should be done with the millions of acres of land that the federal government was to manage in perpetuity? When this question did arise, the conservation movement was not capable of answering with one voice. Although all conservationists could agree on the most general goals of the movement, such as "the wise use of the nation's resources," these were inadequate for guiding land management in and of themselves. In fact, the broad goals on which there was agreement covered many notions of the public good, each with its roots in very different soil. Schmaltz writes that the *zeitgeist* of the conservation era was broad enough to include "a reform Darwinist mastery over nature, a more efficient management of nature by experts and a solemn call for reverence toward the ecological system of which man was but a part."[17] Although all these themes could and sometimes did appear concurrently in the writings of individual conservationists, once actual decisions concerning land management practices had to be made, a split emerged between those for whom ideas of the public good were based on a utilitarian, materialist humanism and those whose ideas were grounded more in a sense of aesthetics or a mystical reverence for nature.

Utilitarian vs. Recreational Uses

There was a cleavage . . . between humanism and something more mystical."[18] The writings of Gifford Pinchot and John Muir, close friends in 1896, bitter enemies fifteen years later, illuminate the poles of thought which the movement included. According to Pinchot, "the first principle of conservation is development, the use of natural resources not existing on this continent for the benefit of people who live here now."[19] The spirit of Muir's conservation, which could not have been farther from this blunt utilitarianism, speaks for itself in his description of California's Sierra Nevada range:[20]

> benevolent, solemn, fateful, pervaded with divine light, every landscape glows like a countenance hallowed in eternal repose; and every one of its living creatures, clad in flesh and leaves, and every crystal of its rocks, whether on the surface shining in the sun or buried miles deep in what we call darkness, is throbbing and pulsing with the heartbeats of God.

Although on first glance Muir's writings appear to be the very antithesis of modernism, with their seemingly primitive strands of animism and nature worship, Muir was as much a man of his era as was his eventual nemesis, Pinchot. While Pinchot offered the material means of a satisfying modern life, what Muir offered in his writing, and what he hoped to create through his preservation efforts, were psychic and physical antidotes for what he saw as the dehumanizing and anti-spiritual aspects of modern industrial and urban society. Untrammeled nature could restore what modern society deadened: the primitive, natural side of man. According to Muir, those who went to the Sierras were bound "to find that going to the mountains is going home." He held that "wilderness is a necessity . . . which can arouse man from [the] stupefying effects of the vice of over industry and the deadly apathy of luxury."[21]

Although later in the century, criticism of industrialism and urbanism would grow into unrestrained indictments of modern civilization, in early twentieth-century America, faith in progress was too strong and the optimism bred of material advances was too robust to permit criticisms of the spiritual costs of modern life to become a condemnation of modern life itself. What the likes of Muir wanted was a balance of the spiritual and the material, and out of that would come total well-being. The preservation of the nation's great scenic places from material exploitation would be an important means to this end.

Mere preservation was not sufficient, however. In keeping with progressivism's broad-based, democratic notions of public interest, it was also important to ensure access to these unspoiled places. An approving Muir wrote that "all the western mountains are still rich in wilderness, and by means of good roads are being brought nearer civilization every year."[22] Horace McFarland, who was to be influential in the founding of the National Park Service, insisted that access to nature must be made as wide as possible. He believed that the right to take refuge in nature belonged to all men, and furthermore that periodic refuge in nature was "a plain necessity for good citizenship."[23] McFarland saw great injustice in the fact that while the rich and powerful could get away from the ugliness industrialism had created, working men were condemned to live in unhealthy, depressing cities surrounded by the "all too common unnatural scenery of Man's commercial filth." Once he made this point at the expense of President Taft (and the state of New Jersey).[24] Addressing an audience which included the president, he noted that "The President has but just returned from a weekend to his castle of rest in the Virginia Hills." Then he asked, "Could he have had equal pleasure in Hoboken?"

It would be a mistake, however, to read much class antagonism into McFarland's demands or to see in them any fundamental disillusion with material progress or American life as it was evolving. The cause of preserved and accessible natural areas under public ownership was like progressive conservation itself in the broad range of people it brought together under its banner. In the early Sierra Club there was no polarization of the preservationist and the entrepreneur.[25] Schrepfer suggests that both saw the establishment of great natural reserves under public ownership as a final conquest of the West, the supreme expression of civilization and progress, and as something that, in the vision of all concerned, "would complement a rapidly industrialized society in a universe that tended toward the good." On the ideal landscape of the benevolent future, much of the land could be put to its most productive material and extractive uses, but some of the greatest natural areas were to be preserved for the enjoyment of all.

Thus the preservation wing of the conservation movement, like its utilitarian complement, dipped into the well of progressivism and drew from it many shared values—a faith in the federal government as an able manager of public land, a broad, classless, and regionless view of national interest; an optimism about the future; and a desire to bring the modern world in its best possible form into being. Unfortunately, these common elements were not enough to prevent acrimonious differences from developing on issues of substance, and it was out of these differences and out of the intragovernment rivalries to which they gave rise that the National Park Service emerged.

The Park Service Is Established

Through the late nineteenth and early twentieth centuries, a dozen or so national parks had accumulated. In 1872 Yellowstone, usually considered the first national park, was established. Three years later Mackinac Island was accorded national park status (only to be decommissioned and ceded to Michigan twenty years later). In 1890 Yosemite, which the federal government had given to California as a park in 1864, was returned to federal ownership and established as a national park. That same year saw Sequoia and General Grant national parks created. The year 1899 brought national park status to Washington's Mt. Rainier, and by the turn of the century, the national military parks had been established on the great battlefields of the Civil War. The next ten years saw more national parks created. They also saw the first reserves established by presidential order rather than by congressional statute—the national monuments.

These preserved places were under a variety of management arrangements in the early twentieth century, and there was no coordination from one park to another. The army had been involved in running Yosemite and Yellowstone, and it managed the Civil War battlefield parks.[26] There were national monuments assigned to the army and to the departments of Agriculture and Interior. Those under the Department of Agriculture were the responsibility of the Forest Service. Those under Interior were usually the charge of the General Land Office. Some of those under Interior were on Indian lands, however, so their management involved the Bureau of Indian Affairs as well as the Public Lands Office. Although proposals for a new government agency to consolidate the management of these places had been made as early as 1900, they made little headway during the first decade of the century, in large part because no real harm seemed to come of this anarchy of management. After all, parks and monuments were being established and protected, so consolidation seemed likely to confer little benefit beyond, perhaps, convenience.

By 1910, however, the situation had changed. The Forest Service had become aggressively utilitarian in its outlook and aims under the directorship of Gifford Pinchot. Although the Forest Service was authorized by its organic act of 1897 to manage its land for a mix of ends, including recreation and preservation, it showed little interest in activities other than those which involved resource extraction or harvesting.[27] This utilitarianism, coupled with the introduction of legislation supported by Pinchot to bring all the national parks under the jurisdiction of the Forest Service, alarmed preservationists, who feared for the integrity of the parks under Forest Service management. In the meantime the controversy over whether to build a dam in the Hetch Hetchy Valley of Yosemite National Park had erupted, and it convinced those who opposed the dam that unless the national parks were under the unified administration of a single agency willing to fight for their integrity, the parks could fall to development schemes one at a time.[28]

In 1910, the Sierra Club took up the cause of a bureau to run the national parks and appointed a special promotion committee to advance the idea. In 1911 and 1912 national parks conferences were convened by an alliance of public interest groups. They recommended a parks bureau within the Interior Department, where there would be organizational distance between it and the Forest Service. By 1912 the proposal had won the support of President Taft, who in a special message told Congress that he recommended the establishment of a bureau of national parks. He insisted that such legislation was essential to the proper management of "those wonderful manifestations of nature, so startling and so beautiful that everyone recognizes the obligations of government to preserve them."[29]

Not surprisingly, the Forest Service objected to the idea, claiming that the efficiency of public land management would be furthered by unifying the management of the national parks, monuments, and forests—under its supervision, naturally. Although it was headed by the strongly prodevelopment Franklin Lane after 1913, the Department of Interior supported the idea of a bureau of national parks within Interior. Interior had been slowly transforming itself from a catchall department of internal affairs into a natural resource management agency.[30] Consolidating, and therefore solidifying, its jurisdiction over the national parks would be in keeping with this transformation and perhaps might even be a step toward consolidating management of all public lands under its authority. Finally, in 1916, with the support of Interior, the Wilson administration, and a large number of civic groups, legislation directing the Department of Interior to set up the National Park Service was approved by Congress and signed by the president.[31]

SECURING THE BASE OF SUPPORT

Threats to Agency Survival

Mere establishment of an agency carries no guarantee of success. The authority to act vested in an agency is a necessary, though hardly a sufficient, condition of success.[32] Nor does establishment guarantee longevity. Agency disbandings are not a rare occurrence in the federal government, and new and old bureaus alike face the possibility of extinction.[33] In the United States, more than in most countries, it is up to the agency administrators to ensure agency survival and gather sufficient power to act. Here, the public administrator is expected to be an "administrative politician" and his style is expected to be one of "managerial activism."[34] In short, the agency, under the guidance of its leadership, must make its own way in the world of bureaucracies. Under the American political system, the bureaucracy has a large share of responsibility for promoting policy and organizing the political basis for its survival and growth.[35] Consequently, federal agencies usually devote much time and effort to building and maintaining a constituency.

The agency for which Mather and Albright took responsibility had some of the advantages common to new bureauracies. It was not ossified and rendered inflexible by an accumulation of past commitments and entrenched procedures. It was supported by the coalition which brought it into existence, a coalition still active and fresh from success. It rode the crest of good will which frequently greets the creation of a new agency not charged with an unpopular task.[36]

The Park Service also had to overcome some inherent weaknesses, however, and it faced some formidable external threats to its survival. The Service was a small agency with a limited charge. It ran a mere fifteen parks. Its field staff numbered a few hundred and its Washington staff, consisting of Mather, Albright, a draftsman, a few clerks, messengers, and secretaries, could be housed in a couple of offices. It had little in the way of precedents to guide it. But perhaps most dangerously, it was born in an unsettled bureaucratic environment of ideological conflict and interdepartmental rivalry. It was a small ship on a rough sea, with few clear bearings.

These circumstances gave rise to one immediate and pressing task for the agency's leadership. It had to consolidate and expand the first and fundamental source of power for administrative agencies—outside support.[37] This meant several things had to be done. First, the agency had to keep together the coalition which brought it into being and keep it convinced that it made the right move in creating the agency in the first place. The agency also would have to get its supporters used to working with it. This was not as easy as it may seem; a government bureau has, perforce, a different modus operandi and a different set of restraints than does a nongovernment group. These differences can cause friction between a public agency and private group even if they completely agree on policy goals.[38] The agency also had to expand its original base of support until it was sufficient to ensure its effectiveness as well as its existence, and until the agency had more power arrayed in support of it than against it, or arrayed in support of its rivals.[39] Finally, through this outside support (and through direct entreatment), it had to make Congress favor it at budget time and do its bidding when it needed legislation or special authorization. Part of the task of securing and stabilizing the agency involved developing an ideology and coherent sense of mission: in short, developing its own view of how it fit into the world around it.[40]

The Forest Service

Among the specific threats to its existence, one agency eclipsed all others in the way it forced the fledgling Park Service to form alliances, cultivate clients, and work out a sense of agency purpose. A great irony which attended the creation of the Park Service was the degree to which the behavior of its chief rival, the Forest Service under Pinchot, permitted and even encouraged that creation. Indeed, it almost seems as if Pinchot singlehandedly forced the Park Service into being with his eagerness to preach and aggressively apply his gospel of materialist utilitarianism. Perhaps Pinchot's intolerance of any but commodity uses of the forests under his charge was necessary in establishing an unambiguous sense of mission for the Forest Service, itself a relatively new agency, and in guaranteeing the Forest Service's immediate survival. It meant forfeiting its chances of capturing responsibility for forest preservation and recreation, however, since Pinchot's attitude convinced others that the Forest Service could not be trusted with such tasks as the preservation of great scenic areas or the provision of public access to them.

Pinchot left the Forest Service in 1910 and was replaced as Chief Forester by Henry Graves, who, although he would remain a friend of Pinchot, was not as aggressively utilitarian in his views of the proper management of the national forests.[41] But even before Graves became Chief Forester, some administrators were convinced that the most picturesque sites in the national forests ought to be preserved for their aesthetic qualities, and that the recreation demands that the public was making on those forests should be accommodated.[42] In the forests themselves, rangers had initiated tentative policies to accommodate recreation while Pinchot was the agency head. For example, campsites were laid out and livestock grazing was restricted in recreation areas. Under Graves, this policy of accommodating recreation was encouraged at the highest level within the agency, and in 1915 Congress, with Forest Service prompting, passed legislation which allowed the leasing of Forest Service land for the construction of "summer homes, hotels, stores or other structures needed for recreation or public convenience."[43]

The Forest Service also began issuing internal bulletins on management of forest land for recreation, and it hired a landscape architect who, as one of his assignments, designed a tourist center for the south rim of Grand Canyon, then under Forest Service management. By taking these actions, the Forest Service was closing the potential niche in the federal bureaucracy for a preservation and recreation agency which Pinchot had opened with his aggressive utilitarianism, and it was doing so even while the Park Service was being established.

As students of public administration recognize, the most dangerous organizations are those with closely related goals, "functional competitors."[44] The Forest Service, under the pragmatic Graves, was far more dangerous to the National Park Service than it would have been under his ideologue predecessor because Graves chose to abandon the utilitarian ideology and thereby make the Forest Service a functional competitor with the new Park Service.

For a new bureau, opposition from its functional competitors will be fierce, and its antagonists will often seek to capture the new bureau's responsibility for themselves.[45] There can be little doubt that capture of the Park Service's charge and its extinction as an independent bureau were important goals of Forest Service leadership during the first years of the Park Service. Graves publicly decried the inefficiency of separate management for national parks, which were frequently surrounded by national forests. In an article published in 1920,[46] he insisted that the Forest Service was interested in more than timber harvesting and argued that some parts of the national forests ought to be managed for preservation and others for intensive recreation. But he did not stop there; he insisted that the national parks be kept free from resource exploitation and by doing so, he asserted that his agency's concern for preservation was ecumenical. Having staked this claim, he went on to argue that first, national recreation policy ought to be unified, with responsibility for it consolidated under one agency—his own—and second, the national parks should be transferred to the Department of Agriculture, the Forest Service's parent department.

Thus, to Mather and Albright the early years of the Park Service must have seemed like a period of great and increasing danger. Support had to be shored up and expanded, and a secure niche and sense of purpose had to be quickly established.

Enlisting Allies

Mather and Albright spent much time building outside sources of support for their new agency, and they shaped many of their policies and much of the agency's sense of purpose toward this end. Their success was impressive; they gained both the passive support of the American public and the active support of many special societal groups. By background and inclination, Mather could not have been better qualified for the task. He was a man of considerable self-made wealth who came from a solid New England family, and he commanded respect on both counts. He was a superb, imaginative salesman for his ideas, and his energy and enthusiasm, once engaged, seemed boundless. His most important asset to his agency, however, was probably his ability to establish close, easy peer relationships with some of the nation's most influential men, relationships which turned many of them into fast friends of the Park Service.

Perhaps nothing illustrates his mix of skills and the free-wheeling style of leadership to which it gave rise better than the backwoods Yosemite expedition he organized for a select group of influential men during the campaign to establish the Park Service.[47] His guests included a vice-president of the Southern Pacific Railroad, the editor of *National Geographic*, the president of the American Museum of Natural History, and a congressman who was later to become the Speaker of the House. The aim of the expedition was to give the men a good time, impress them with the beauties of the national park, and, not incidentally, persuade them to use their influence on behalf of the upcoming legislation to establish the Park Service. Augmented by cooks, guides, and aides of every sort, the group traveled in rustic splendor and was feted with backwoods haute cuisine which included venison and gravy, fried chicken, hot rolls, salad, and freshly made pies. The expedition was a great success; many of Mather's guests put their support behind the legislation and some became lifelong friends of the Park Service.

The elite backwoods trips became a common way of building support. On them Mather concocted a winning mix of elitism, male bonding, and aggressive, if superficial, egalitarianism (figure 2-3). Later Albright took Calvin Coolidge on a fishing trip in Yosemite and on it the simple pleasures prevailed; Albright persuaded Coolidge that he would have better luck if he gave up worms and baited his hooks with Albright's special hand-tied fly instead. Even stiff-mannered Herbert Hoover was reportedly turned into one of the boys on an agency-sponsored fishing trip to one of the western parks.[48]

Cultivation of this elite support paid large dividends for the Service. Rich friends gave large sums of money for projects such as the Tioga Road through Yosemite, which had failed to get congressional appropriations. (A good part of the road was paid for by Mather himself out of his private fortune.) John D. Rockefeller, Jr., and several others (again including Mather) helped defray the cost of special study commissions such as that which investigated the possibility of a large park in the southern Appalachians. Rockefeller and some of his peers were often as generous with land as they were with money. For example, Acadia National Park was established on land donated by Rockefeller. These men could also be counted on

Figure 2-3.
An elite camping expedition typical of the trips organized by Mather and Albright to enlist support for the National Park System. Albright is at the far right.

to put in a good word when political decisions affecting the National Park Service were being made. For example, in 1920 the Service's influential friends convinced the newly elected Harding administration that the agency's leadership posts (which were occupied by their original Wilson appointees) should not be considered political posts and therefore the proper spoils of victory.[49]

Mather established a fine relationship with that epitome of progressive era organizations, the General Federation of Women's Clubs, whose conservation committee had actively supported the establishment of the new agency. He also courted the support of conservation organizations such as the National Federation of Audubon Clubs, the Sierra Club, and the Save-the-Redwoods League. These organizations tended to be most interested in specific types of conservation and in specific geographical regions, so each had a special concern for particular parks. For example, the Sierra Club focused its interests on California's national parks in the Sierras: Yosemite, Kings Canyon, Sequoia. The Audubon Society, on the other hand, was very interested in the establishment of a national park in south Florida to serve as a great bird sanctuary. Nevertheless, these organizations could be counted on to offer more general support for the entire Park System when it was needed.

The public interest group which was most closely aligned with the agency during the Mather–Albright era was the National Parks Association (NPA), an organization whose primary concern was for the national parks. In fact, the group was established by Mather and his associates shortly after the National Park Service was founded to act as a fifth column for agency leadership, and as "a non-governmental voice that spoke words unwise in official statements."[50] Robert Sterling Yard, a friend of Mather's and for a while an associate within the Interior Department, left his government post at Mather's request to head the NPA. What Mather had in mind for the organization can be seen in what he told the departing Yard: "With you working outside the government and with me working inside, we ought to make the National Park System very useful to the country."[51] Thus, Mather and Albright were able to build a platform of elite support under the System and Service, and to secure the backing of a coalition of organized public interest groups. They did not stop there, however.

Both Mather and Albright considered themselves businessmen at heart. In fact, Albright thought of his stint in the agency as a prelude to a proper career in business. Mather had made a fortune in business, and he frequently stressed his goal of putting park management on a businesslike basis. Mather loved to give the impression he was slightly at sea in the world of public bureaucracy. His casual request to Secretary Lane for an assistant who knew the ways of government illustrates this pose: "Say, Frank, . . . they do things so differently in government than in business that I'll probably get myself in jail in a week unless you give me a guardian."[52]

Given these attitudes and attributes, plus the example of Pinchot's skillful cultivation of timber industry support, it is not surprising that Mather and Albright sought out allies in business by portraying national parks as good opportunities for private enterprise and especially for the tourist business. Mather and Albright realized that the agriculture, timber, and power opportunities which the establishment of the parks had precluded could be offset by tourism, and they knew that those whose livelihoods came to depend on the existence of the parks could be counted on to defend them. Mather pointed out to businessmen the great profits to be made in expanding facilities in national park concessions. He formed close working relationships with western tourism organizations and with western railroads. At the same time he coordinated the publicity campaigns of private industry with those of the National Park Service. He even approved a tire company's billboard advertising, which linked the beauties of Yellowstone with the virtues of their tires. Albright had a speech with the title (and theme) of "Parks Are Good Business" which he made to chambers of commerce. Yard of the NPA wrote a how-to-do-it article entitled "Making a Business of Scenery" in *Nation's Business*.[53] The point of Yard's article was that scenery was more than a luxury, it was a hard-headed business opportunity for those astute enough to recognize it as such.

For their part, those in the travel and tourist business were not slow to recognize the opportunities for profit that the parks offered. Louis Hill of the Great Northern Railroad saw the parks as a chance to increase passenger volume at little extra expense, since tourists going to the parks would largely be using rail lines already in place. "Every passenger to the national parks represents practically a net earning," Hill was quoted as saying.[54] Seventeen western railroads contributed to the publication and wide distribution of the National Parks Portfolio, a glossy publicity portfolio that Mather sponsored and promoted. The western tourist industry, largely through their National Park Highway Association, worked with the Park Service to improve access to the parks, mostly by lobbying for the construction and upgrading of roads connecting the parks to major highways.

While support outside government helped to put the agency in congressional favor,[55] Mather and Albright also courted legislators directly by respecting congressional prerogatives and by giving key legislators the V.I.P. treatment in the parks. Eventually they succeeded in building a coterie of active congressional supporters led by Representative Cramton, chairman of the House Appropriations Subcommittee for Interior. With such friends in Congress, appropriations were usually adequate during the Mather and Albright era. In fact, there were years in which appropriations were even greater than the agency requested.[56] Its good relations with key congressmen, moreover, allowed the Park Service to retain a large measure of control over its relationship with Congress, and most of the initiatives for legislation related to the national parks came from the agency itself. If they did not, they at least had the agency's active support.[57] The congressional supporters of the agency respected and trusted Mather and Albright and were content to allow the initiative on park matters to rest with them.

Mather and Albright realized, however, that the active or even enthusiastic support of a small body of congressmen by no means assured long-term success for the agency or security for the System. The National Park Service had its detractors among the legislators, especially among those close to the Forest Service. And beyond these two groups of supporters and detractors was the majority of Congress, to whom park issues may have been important occasionally, but who had little ongoing interest in the System or, for that matter, in the bureaucratic fortunes of its managing agency. More needed to be done.

Courting the Public

There are two basic strategies by which an agency can gain the support of the public.[58] The first is to cultivate a strongly favorable image, to convince the public that what the agency does is in keeping with the highest of popular values. The second is to build strength with the public, or sections of it, by providing specific tangible benefits. Although distinct, the two are by no means mutually exclusive strategies.[59] Early agency leadership was aware of and willing to use both.

When Mather first assumed responsibility for the national parks, he undertook to create a public awareness of them by an ambitious publication campaign which included articles strategically placed in mass circulation magazines like *National Geographic* and *The Saturday Evening Post*, as well as by the publication and wide distribution of the National Parks Portfolio mentioned above. With his prompting, the national parks became a media fad, and there arose a tidal wave of newspaper and magazine publicity.[60] Not incidentally, Mather took great pains to identify his agency with the national parks he was so busy publicizing. He appeared to know what Robert Moses would soon discover in New York: if the public sees you as being on the side of parks, you can do no wrong in its eyes.[61]

Mather did not stop with publicizing national parks. He campaigned for the preservation of great natural areas outside the National Park System and over which neither he nor his agency had any jurisdiction. Mather used his influence, for example, to place an article entitled "The Last Stand of the Giants" in *National Geographic*.[62] The article, about the cutting of the coastal redwoods, tripped off a wave of publicity about the trees and a deluge of contributions to "save the giants" (including, predictably, those highly publicized ones that involved little children breaking open their piggybanks). Although Mather had no official business being involved in this campaign since there was no national park within hundreds of miles, his involvement identified him and his agency in the public mind not only with the national parks but with the cause of preserving great natural areas in general.

Although he was successful in identifying the Park System with popular values, Mather knew that it would be wise to deliver substantive as well as ideological benefits to the public. While public support based on image can be a potent force, such support is also, as political scientists have noted, diffused, inconsistent, and unreliable.[63] There had been the lesson of Hetch Hetchy. The defeat of those who opposed the dam was due in large measure to the fact that, while they convinced much of the public that the valley about to be destroyed had sublime qualities, they did not prove that solid, tangible benefits would result from not building the dam.

Given that the parks had to be made useful, several things compelled Mather and Albright to strive to make them so by opening them up to as broad-based a national client group as possible. First, although Mather and Albright attempted to recruit allies among park neighbors by pointing out the business opportunities that tourism created, there were times when these neighbors saw the refusal of the Park Service to permit resource exploitation in the parks—especially in connection with irrigation projects—to be positively injurious to local well-being. Hence, taken as a class, park neighbors were never likely to become full allies of the Park Service. According to Stratton and Sirotkin, "This lack of appeal meant that the service was compelled to find its support in a dispersed national constituency."[64] Second, there was the tradition of progressive conservation to consider. The movement out of which the Park Service was born put a high premium on, in Pinchot's words, "the greatest good" and it saw a socially and geographically undifferentiated "greatest number" as the proper recipient of this good. Although Pinchot's stark commodity utilitarianism offered the agency little that was useful in developing its sense of mission, the agency could not—and did not—reject the broader notions of public service which both wings of the conservation movement shared.

Moreover, the emergence of modern society was making it possible to consider the bulk of the American public as potential clients of the National Park System. Increasingly, there was time to visit the parks as industrialism broke down the agrarian work cycle in which summer was the period of heaviest work and greatest commitments. The length of the work week was being shortened, and the paid annual vacation was spreading through the industrial and clerical work force. A new world of mass leisure was unfolding.[65] Furthermore, there was now a means of getting to the parks. Although the railroad had served some of the national parks, the auto could make all of them accessible. In 1900 there were 8,000 autos; in 1930 there would be 23 million.[66] The auto, the work schedules of a modern economy, and the affluence of early twentieth-century society meant that it was becoming possible for the park agency to fuse the two publics: the public at large, for which national parks were attractive symbols of nature at its most spectacular, and the public which the parks served directly by accommodating their annual vacations. Such a fused constituency might become powerful and perhaps unbeatable.

Mather and Albright's policies, especially those of park development and management, aimed at acquiring and satisfying as many of these potential clients as possible. Initially, they worked at improving park access, thereby putting the national parks in a position to take advantage of increased western travel. They encouraged auto travel within the parks by extending and upgrading roads. Mather himself led a highly publicized auto tour to mark the opening of Yellowstone's Grand Loop Road, a circuit of over a hundred miles designed to take the motorized tourist past many of the park's prominent features. He also enthusiastically supported the plan advanced by the western tourist industry to build a highway system which would connect all the western parks into one grand touring circuit. The Service's leaders recognized the importance of the national parks as secondary as well as primary destinations, and Mather sold the railroads on the idea of a through-pass plan, one by which a person traveling across the continent could stop at Yellowstone for a while and then continue on at no extra cost. Tioga Road, which traversed Yosemite, was also intended to take advantage of cross-country travel. With its completion, a traveller from the east heading to San Francisco or heading east from that city could visit Yosemite without having to make a large detour. A road through Glacier National Park served the same purpose; it made the park attractive as a secondary destination to those crossing the continent.

Interior Secretary Lane's 1918 letter of instruction to Mather on how to manage the parks, commonly believed to have been written by Mather himself, was a concise expression of Mather's management philosophy, whatever its authorship. It read in part that, "Every opportunity should be afforded to the public, wherever possible, to enjoy the national parks in the manner that best satisfies individual tastes."[67] Mather certainly made the effort to accommodate every taste. To accommodate the highbrow, he arranged for summer lectures on natural subjects to be given in Yosemite by University of California faculty members. In Yellowstone, he instituted field trips, nature plays for children, and educational campfire talks by rangers. According to Albright, "National Park visitors wanted a good time as well as education," so accommodation did not stop with uplift and education.[68] To provide the good time, the Service perpetuated the Yosemite "firefall," a nightly show, long antedating the establishment of the agency, which involved building a bonfire on the edge of Yosemite Falls at dusk, then pushing it over the edge to create a spectacular cascade of flame and embers. Albright had the streams of Yellowstone heavily stocked for the pleasure of anglers, and there was serious thought of running a cable car across the Grand Canyon. (Albright favored the project, but Mather's objection killed it.) In lodging facilities as well as park activities, the National Park Service tried to accommodate all tastes. On this point, Lane's letter to Mather read that rustic campsites were to be maintained, but so were "comfortable and even luxurious hotels wherever the volume of traffic warrants the establishment of these classes of accommodation." [69] Albright prided himself on the degree to which the Park Service honored its charge by providing a wide, even extreme, range of accommodations.

Mather's biographer states succinctly that his subject was "no primitivist who wanted to curb mass use." [70] But it is hard to tell exactly what Mather was, or more accurately, it is hard to determine exactly where Mather put the limits of propriety in efforts to acquire a mass constituency for the National Park System. His tolerance of slovenly behavior would make a permissive parent blanch. "The Parks belong to everyone," he said when confronted with the trash of a particularly careless group of park visitors. And he made it clear that "everyone" included the litterers. "We can pick up the cans, it's a cheap way to make better citizens." [71] Mather's evaluation of national parks was also conditioned by their volume of use and popularity among the public at large. Ise was perplexed at the way Mather insisted on upgrading Hot Springs, a site with little in the way of scenery or much of anything else, to the status of national park. [72] The fact that Hot Springs enjoyed twice the annual visits of any other unit under Park Service management undoubtedly weighed heavily in Mather's thinking.

There were definite limits to Mather's efforts to court popularity, however. He did oppose the Grand Canyon cable car. He also said, "Our job is to keep the parks close to what God made them." And on visiting Coney Island, Mather exclaimed, "This is exactly what we don't want in the national parks." [73] One might suspect that Mather's high-principled comments were a bit disingenuous since he also suggested attracting more people to the parks by installing swimming pools, golf courses, and tennis courts in them. But whatever one concludes, Mather and Albright's efforts to attract masses to the national parks cannot be dismissed as acts of pure opportunism, undertaken to ensure the survival of the Park System (something which later students of the System would suggest). Nor is it likely that these policies were devoid of any moral commitment. After all, in the early 1970s, long after any need to promote national park use for their survival had passed, Albright was still advocating maximum park access. [74]

No, as a general policy, popular and little-restricted access to the national parks fit in with the progressive tenets which permeated the atmosphere of government in the early twentieth century, tenets which clearly guided Mather and Albright.[75]

Formula for Success

To recapitulate briefly, then, Mather and Albright's search for a sustaining coalition to support their new agency was wide-ranging. It involved sales-manship and public relations, and more than anything else it involved finding wise management policies for the national parks themselves. In retrospect, we must judge Mather and Albright successful. They managed to enlist the support of the nation's financial, political, professional, and social elites. They found friends in the business community, and they kept the support of the public interest organizations which figured so large in progressive conservation. They spun a favorable image among the general public, which they also courted and won as a client group. Nevertheless, they had a supporting coalition of diverse parts, and keeping it together produced some problems for the agency.

Perhaps the most fundamental problem was reconciling park preserva-tion with park development, that is, reconciling the agency's role as the preserver of the great scenic places with its role as a provider of recreation opportunities to much of the American public. During the early years of the twentieth century, most of the upper-class citizens involved with parks and public open space leaned toward minimal development and preserva-tion-oriented management.[76] The NPA soon adopted this attitude and as early as 1923 it was voicing its concern in its publications over the increas-ing numbers of autos in the national parks. The NPA feared that what it saw as the true worth of the national parks—their value as places for communion with nature and appreciation of its grandeur—would be dimin-ished by the flood of auto campers, and that the Park Service would lose sight of its proper, i.e., its preservationist, management goals in its efforts to accommodate the flood.

On the other hand, there were pressures to increase visitor accommoda-tions. First, individual park users, tourist organizations, and auto clubs sometimes objected to what they saw as a lack of adequate tourist facilities and an overemphasis on preservation in park management. Second, con-gressmen and the secretary of interior were often more impressed by reports showing great use than by assurances that wilderness values had been maintained.[77]

Mather and Albright steered a middle course in their park planning and tried to please both sides. Usually they would allow nuclei of intensive visitor services in the parks, make those nuclei and some of the most spectacular sites in the park accessible by high-grade roads, and leave the rest of the park land—most of it—as wilderness. Defending his approach to park development, Mather said:[78]

> It is not the plan to have the parks gridironed by roads, but in each it is desired to make a good sensible road system so that the visitors may have a good chance to enjoy them. At the same time large sections of each park will be kept in a natural wilderness state without piercing feeder roads and will be accessible only by trails to the horseback rider and the hiker.

Fortunately, most of the great western parks were large enough to accommodate such an approach, and as a strategy aimed at holding together the Service's supporting coalition, it was successful. Tourist organizations, Congress, and the visiting public were usually kept happy by the increased park use it accommodated, while the more preservation-minded groups, although by no means completely satisfied, were at least placated. However, agency survival and well-being required more than skillfully drawn park management plans and an alliance of support for the park system as the agency inherited it. It also required astute decision making with regard to System expansion and the assumption of new responsibilities.

EXPANDING THE SYSTEM

Competition with the Forest Service

As successful as early Service leadership was in establishing a base of support for the agency, it would have been unwise for Mather and Albright to have sunk into self-congratulatory inactivity. The agency was still a small one, small enough to serve as a pawn of little worth in interdepartmental rivalry. Moreover, it still had functional rivals; the army managed the Civil War battlefield parks, the Forest Service treated areas of the national forests almost as if they were national parks, and scenic land approaching national park quality remained under the General Land Office. Expansion of the agency's size and its range of responsibilities could help solve these problems.

Students of public administration have found virtue in size. Some suggest that pure size is important for agency survival and that there is a survival threshold for new agencies.[79] If an agency is smaller than a critical size, it cannot render enough useful services to its clients to ensure its own survival. Others suggest that large organizations have a better chance of survival than small ones because they will have greater resources at their disposal. This in turn enables them to gain economies of scale in their operations and to take advantage of greater specialization.[80] Looming above these general advantages, one specific consideration probably prompted agency leadership to adopt a policy of aggressive, opportunistic Park System expansion during the agency's first years: fear of the Forest Service.

As mentioned above, a mission which did not extend much beyond minding the great national parks was intrinsically a modest one in terms of the size of the agency required to carry it out. As long as the much larger Forest Service was still identified with utilitarian management, this limited mission posed no problem since it made the Forest Service appear to be, by dint of its basic philosophy, an inappropriate agency for the management of the national parks. Those who favored strict preservation of the national parks might have disapproved of Mather's schemes to promote what seemed to them like indiscriminate tourism in the parks, but at least he was not cutting down all the trees.

Through the 1920s, however, the Forest Service responded to the increased mobility and recreation demands of the American public by making recreation opportunities and aesthetic qualities increasingly important forest management considerations. By 1928, Chief Forester Stuart could write in a letter to his regional foresters that recreation's "rank in national forest activities will in large degree be a major one and in a limited degree a superior one to all other management considerations."[81] The Forest Service took a page out of Mather's book on how to build a recreation client base. For example, it encouraged the founding of the Trail Riders of America, an organization of riding enthusiasts which sponsored horseback trips led by forest rangers through the national forests. Also during the Mather years, the idea of maintaining the forests to preserve ecological integrity was finding acceptance in the Forest Service. Aldo Leopold, a forest ranger, was writing of the need to maintain areas of the forest system as wildlife habitats and for the other benefits unaltered forest land conferred. He was finding disciples within the Forest Service and influencing its policy.[82]

For the Park Service, the dangers in these Forest Service policy changes were many. First, they might prompt elements of the Park Service's supporting alliance to defect. From a preservationist's point of view, perhaps the Forest Service would be a better manager of the national parks if it really had shed its utilitarian philosophy. One could make the argument that if the Forest Service were the manager of the entire public domain, it could satisfy the timber production and resource extraction interests, consign part of the public domain to satisfying the recreating public, and then manage the national parks themselves along strict preservation lines. This was something the Park Service, with its need to court the public by promoting access, could not do. Second, the recreation interests might be prompted to abandon the Park Service for the opposite reason. Under the

Forest Service, a wider range of activities just might be possible in the national parks. The Forest Service allowed hunting in its wilderness areas; it might allow it in the national parks. Camping regulations were generally more liberal in the national forests; they might become more liberal in the national parks if they were under Forest Service management.[83] Third, the Forest Service was offering unified management of the public domain. The notion of unified resource management has a certain intrinsic appeal to the reorganizers of bureaucracy, even if the benefits actually to be derived from the unification are never made clear.

Mather and Albright were certainly aware of the danger from the Forest Service. How could they not be? The Forest Service was never coy about its ambitions with regard to the national parks, nor its conviction that it could manage the parks more efficiently than the Park Service. Park Service leadership dealt with the danger in several ways. As a defensive tactic, Mather and Albright cast the competition between the two agencies as one between antithetical philosophies and functions, as one, in short, between materialist utilitarianism on the one hand and preservation for mass appreciation on the other. In fact, Park Service leadership even tried to retard the Forest Service's abandonment of the utilitarian imprint Pinchot had given it. Mather denounced any recreational uses of the national forests, insisting that recreation was exclusively the responsibility of his agency and that the mission of the Forest Service was just the one advocated by Pinchot, i.e., growing timber, providing livestock grazing and, in some limited cases, providing watershed protection.[84] In one instance, Mather even turned the arguments of efficiency through unified responsibility against the involvement of the Forest Service in recreation. In 1922, Mather blocked congressional appropriations for campground development in the national forests by convincing the chairman of the House Appropriations Committee that two agencies should not both be doing the same job, i.e., providing recreation on federal land, since it would put the government in competition with itself. (Incidentally, Mather did his convincing on one of his personally led V.I.P. tours of the national parks.) As one of the Forest Service's recreation specialists remembered the episode, Mather "imbued the committee very strongly with his convictions on this point. We attempted to get an appropriation for . . . recreation. It was turned down flat, we didn't get a nickel."[85]

Mather and Albright must have known that their tactic of trying to keep the Forest Service hemmed up in utilitarianism could never lead to complete security in spite of the early successes it brought them.[86] As long as the Park Service was small and its entire charge could logically be fit into another agency's range of responsibilities, their agency was a tasty morsel. Only increased size, broadened scope of operation, and diversified responsibility would make it less appetizing. There were several directions of expansion possible. An obvious one was through adding more spectacular pieces of western landscape to the National Park System.

Adding Natural Areas

Such an expansion had two major virtues from the agency's viewpoint. First, it increased holdings in a noncontroversial manner, i.e., noncontroversial in the sense that there was no question over whether areas of great western scenery were appropriate additions to the System. Second, many such sites outside the National Park System were under Forest Service management, and each acquisition of one of these by the Park Service undercut the Forest Service as a rival. Because of these advantages, Mather and Albright were unequivocally committed to such expansion, especially when it was at Forest Service expense.[87] During the Mather–Albright period, several of the greatest national parks, including Mt. McKinley, Zion, Bryce Canyon, and Grand Canyon were added to the System. Frequently, as was the case with Grand Canyon, the new parks were carved out of national forest land. The foresters in turn defended themselves from these raids by classifying potential national parks under their jurisdiction as "primitive areas" and by administering them in a manner similar to the national parks.

Questions of Standards

There were only so many truly sublime sites, however, and administering all of them would not take a very large agency. But there were innumerable other sites on the public domain, those which were less unique or smaller or perhaps slightly spoiled by the inroads of human activity. The addition of such sites to the System would expand it, but, unlike the unquestionably great places, such sites raised the question of standards; in one form or another this question has persisted to the present. Lane's 1918 letter of instruction to Mather was explicit on the question of standards: "The National Park System as now constituted should not be lowered in standard, dignity, and prestige by the inclusion of areas which express in less than the highest terms the particular class or kind of exhibit which they represent."[88]

Such an exclusionary policy had many advantages. From at least as far back as the midnineteenth century, the spectacular western sites which were to become the national parks were wrapped in American patriotic mythology.[89] They were seen as places of spiritual refreshment, wells from which draughts of patriotic inspiration could be drawn. When, in the 1930s, the National Resources Planning Board referred to the "Sacred Areas" of the National Park System, it was referring to the great western parks, and one suspects that its meaning was literal.[90] In managing these areas, the National Park Service was itself infused with a spiritual mystique, something with advantages for both internal cohesion and external support. The accumulation of many lesser areas could contaminate this image, reduce its power, and thereby lessen support for the entire Park System.

Second, an exclusionary strategy would please many of the agency's supporters, including preservationist organizations and some congressmen. For example, the National Parks Association linked its demands for preservation-oriented park management with calls for highly selective policy on new parks. The 1920s saw frequent demands in the *National Parks Bulletin* for the Service to eliminate the lesser parks and to resist the temptation to add new ones that did not measure up.[91] Although Congress had members who were anxious to push their own park proposals, it also had strong supporters of a policy that admitted only the most sublime sites to the National Park System. When a bill to establish Rocky Mountain National Park was proposed, objections from Wisconsin's Representative Stafford, who thought the site did not have the stuff of true national parks, killed it. (The bill passed the next year anyway.)

Thus, adhering to high standards in practice was a problematic matter in which the expansion of the System had to be weighed against some real costs, and Mather and Albright's behavior reflected this. In his history of the Park Service, Ise discussed the hard fight Mather and Albright sometimes put up against inferior parks. But Ise also shows how they tended to be selective in their opposition to legislatively initiated proposals, balancing the pros and cons, which included those of political expediency, before deciding which way to go.[92] The result was an inconsistent adherence by Park Service leadership to the standards it espoused. For example, under Mather's leadership, the agency was little interested in adding the Mineral King Valley to King's Canyon National Park because the valley had been extensively mined, and this ruined it as a national park in their eyes. Yet, in other cases they were willing to accept areas encumbered with mining claims and even ones where active mineral extraction was still in progress. There was the case of Wind Cave National Park in South Dakota, where fear of offending a state's congressional delegation outweighed any consideration of standards or quality. Ise feels that in the balance, "Mather was perhaps too ambitious for more parks and not quite discriminating enough as to their quality."[93] Departure from high standards, however, was the price of expanding the natural wing of the National Park System. In a sense, there was no place to go but down.

It should be noted that while actual adherence to strict standards of park selection conferred a mix of advantages and disadvantages, espousing such standards conferred mostly benefits. It cost little to take an exclusivist line when particulars were not involved; it was good for the image, and it pleased organizations such as the National Parks Association. Besides, espousing strict standards did not really cost much in flexibility; standards could be invoked to reject those proposals the agency opposed and quietly forgotten when it came to units which political expedience or ambition caused the agency to support. Accordingly, both Mather and Albright formally espoused an exclusionary national park policy. Albright spoke of "rounding out" the National Park System, which implied that there were only so many potential additions. Perhaps his use of the term "rounding out" also implied that most potential units were already in the System. Mather spoke of a sorting out process which would bring the great places still outside the System into it while removing those parks which did not measure up. "Every suitable site in, every unsuitable one out" was the phrase he used.[94]

Other New Responsibilities

Although the addition of natural sites in the West might have raised questions of standards, it did not involve the Park Service in responsibilities qualitatively different from those included in its original charge. There were avenues of possible growth for the agency, however, which did involve acquiring responsibilities which were clearly extensions of the agency's charge. Students of bureaucracy are of differing opinions on the virtue of diversifying agency responsibilities. Some see a positive value in it since the more diverse an organization's responsibilities and clients, the less it has to depend on any one of them.[95] In short, diversity means flexibility. Others have reservations about the advantages of diversity per se. For example, Rourke writes that in the quest for power, an agency's strategy has to be one of optimizing its jurisdiction.[96] This means that activities without political support, which consume an inordinate amount of energy, or which divert an agency from its important functions, are liabilities rather than assets.

Drawbacks notwithstanding, from their first days in office, Mather and Albright sought to expand the park system so as to accumulate new responsibilities in all parts of the country. Although Lane's 1918 letter of instruction to Mather was concerned only with the great national parks, by Albright's retirement fifteen years later, the Service was managing a diverse range of places nationwide and overseeing programs that had little direct connection with its original management charge. First, there was geographical expansion.

Eastern Parks

It is probably natural for an agency to be interested in increasing the spatial scope of its operation and with it the geographical spread of its clientele and supporters, since in the federal system formal power over an agency is widely distributed through congressional representation. In spite of the fact that the Park Service was a western agency by dint of its holdings and the regional orientation of its parent department,[97] the mass use and appreciation of the parks gave it a geographically broad-based constituency. Nevertheless, Mather and Albright wanted a System whose holdings as well as constituency were nationwide. Specifically, they wanted footholds in the East. If they could get national parks in eastern states, the strength of the System as a whole would be increased. First, congressional appropriations would be easier to come by, and second, it would undercut what they saw as a dangerous alliance forming between westerners who wanted the parks opened to resource development and easterners who felt that the National Park System, located in the West, was of primary benefit to westerners.[98] Moreover, beginning with the Weeks Act of 1911, and the federal policy of buying abandoned eastern farmland to add to the National Forest System which the act permitted, the Forest Service had been expanding its holdings and shifting its scope of operation to the East. If the Park System remained heavily concentrated in the West, federal responsibility for preservation and recreation land management in the populous East might fall to the Forest Service by default.

It should not be thought that Mather originated the idea of eastern national parks; there were moves to establish national parks in the East even before the founding of the Park Service. For example, in 1899, an association was founded in Asheville, North Carolina, to promote the idea of a national park in the southern Appalachians, an idea which eventually gathered support from much of the East. However, it took Mather's prompting to get the first eastern park established. Acadia National Park, on the coast of Maine, was authorized three years after the Park Service was created. From his first days in office, Mather had made this park one of his top priorities. In its first year of operation, Acadia had 64,000 visitors, making it the third most visited national park in the nation and illustrating the value of eastern parks.

Mather and Albright then threw their efforts into establishing the Great Smoky Mountains National Park in the southern Appalachians of North Carolina and Tennessee and the establishment of Shenandoah National Park in Virginia's Blue Ridge Mountains. Their efforts paid off when the two parks were authorized in the late 1920s. Two years after its opening in 1931, Great Smoky Mountains became the most visited park in the System, and the annual number of visitors climbed rapidly through the 1930s. What the popularity of Acadia illustrated about the value of eastern parks in constituency building, the Great Smoky Mountains confirmed with finality. In the late 1930s it was predicted that the few eastern national parks would soon outdraw all of the western ones combined.[99] As tables 2-2 and 2-3 show, such predictions were close to the mark.

Table 2-2. The Ten Most Visited National Parks in 1941

Park[a]	State	Visits[b]
Great Smoky Mountains	Tennessee-Kentucky	1,310
Shenandoah	Virginia	1,071
Rocky Mountain	Colorado	681
Yosemite	California	598
Yellowstone	Wyoming-Montana-Idaho	580
Mount Rainier	Washington	477
Grand Canyon	Arizona	437
Acadia	Maine	409
Platt	Oklahoma	317
Sequoia	California	303

Source: National Park Service, *Public Use of the National Parks: A Statistical Report (1941-1953)*.
[a]The five eastern national parks accounted for 35 percent of all visits to the 26 national parks in 1941.
[b]Visits in thousands, rounded off to nearest thousand.

Table 2-3. The Ten Most Visited Units of the National Park System in 1941 (Excluding National Parks)

Unit[a]	State	Visits[b]
Lincoln/Washington National Memorials	Washington, D.C.	2,735
Blue Ridge National Parkway	Virginia-North Carolina	895
Lake Mead National Recreation Area	Arizona-Nevada	845
Fort McHenry National Monument	Maryland	686
Gettysburg National Military Park	Pennsylvania	670
Statue of Liberty National Monument	New York	452
Mount Rushmore National Memorial	South Dakota	393
Chickamauga-Chattanooga National Military Park	Georgia-Tennessee	367
Lee Mansion National Memorial	Virginia	357
Castillo De San Marcos National Monument	Florida	307

Source: National Park Service, *Public Use of the National Parks: A Statistical Report (1941-1953)*.
[a]Of the 114 units of the National Park System that were not national parks, 51 were east of the Mississippi. However, the 51 eastern units accounted for 80 percent of the visits to the entire set of 114.
[b]In thousands, rounded off to the nearest thousand.

Historic Sites

While agency leadership put its efforts into the creation of parks in eastern natural areas, it also worked to add the nation's premier historical sites to the National Park System. Since most of these sites were in the East, one of the advantages to be gained by putting these sites into the System was the same as that gained from bringing in the eastern natural areas, i.e., they would bring more easterners into direct contact with the Park System. There was an additional and very compelling reason to move the agency into historic site management, however. As Albright acknowledged, a role in historic preservation would give the Park System a type of park which was beyond any possible claim by the Forest Service.[100] With historic preservation added to goals of the National Park System, neither the System nor its managing agency could be swallowed whole.

Albright was an amateur history buff, and he did more of the work involved in getting the great historic sites into the Service than did Mather.[101] Albright based the Park System's claim to historical sites on the idea of parallel responsibility. If the System was the repository of the nation's great natural sites, it was the logical place for the great historical sites as well. Above all, Albright was interested in the major battlefields of the Civil War—Gettysburg, Chattanooga, Vicksburg, and Antietam— which he offered to take over from the army. In the decades after the Civil War, the army had used the battlefields for instruction and training, but the rapid evolution of military strategy and technology in the early twentieth century had greatly lessened their instructional value. This left the War Department indecisive about Park Service offers to take over their management.[102] Some army officers, thinking mostly of efficiency, were in favor of the transfer; the battlefields were of no use to the modern army. Others saw maintaining the battlefields as the army's sacred trust and so opposed the move. Mostly because of this ambivalence, Albright's efforts at effecting a transfer throughout the 1920s and early 1930s were unsuccessful.

Albright was successful in acquiring historic commemoration responsibilities for the Park Service elsewhere, however. Early in 1930, President Hoover established the Washington Birthplace National Monument and assigned it to the National Park System. Later that same year he signed legislation authorizing the public acquisition of the Yorktown battlefield, and its establishment as a national monument in the Park System. According to Albright, with the authorization of the Yorktown Battlefield Monument on July 3, 1930, the Park System's historic preservation wing was "in business."[103]

State Parks

In the 1920s, while Mather and Albright were expanding the Park System in many directions, the states were establishing park systems of their own, something Mather strongly and actively encouraged. (Mather considered the National Park System a model which should be emulated by the states.) He convened the first National Conference of State Parks under the sponsorship of the National Park Service in 1921, and under him the Service offered advice and guidance to state park commissions on a variety of matters, as well as serving as a national clearing house for information on state parks. This assumption of an active role in relation to the state park systems had clear strategic advantages for the agency. It created a new class of clients among state and local park officials, who benefited from agency aid and advice. It supported the agency's claims to a leading or even exclusive role in federal recreation policy. Furthermore, by encouraging state park systems, the National Park Service might enable the states to accept as parks many of the inferior sites which state congressional delegations were proposing as national parks. In fact, Mather's biographer gives this last reason as the principal one for Mather's vigorous support of the state park movement.[104] Without doubt, these practical reasons were important to Mather; he was a very pragmatic administrator. But he was also very much imbued with the progressive vision, and for him a system of state parks had an important place on the ideal landscape of a motorized, affluent America: "I believe we should have comfortable camps all over the country, so that the motorist could camp each night in a good scenic spot, preferably a state park. . . . I hope some day the motorist will be able to round up his family each night on some kind of public land."[105]

The Capital

These extensions of the Park System through the addition of less-than-spectacular natural areas and historic areas, and the Park Service's assumption of a role as federal advisor to lower levels of government on open space matters could all be seen as, in some way, logical extensions of the Service's original charge. However, under Mather and Albright the agency also sought responsibilities which were not logical extensions of its charge but rather jobs which simply needed to be done. For example, shortly after the Park Service was founded, Mather and Albright began campaigning to bring the public parks and monuments of Washington, D.C. into the National Park System and under agency control. And as Congress began authorizing parkways for the D.C. area, the Park Service leadership sought to bring these into the System as well. Albright's argument here stressed simple practicality; since the Park Service was the repository of the necessary skills, in his words, "a professional bureau with administrative, engineering, and interpretation experts," it was well suited for the job. Naturally, Albright was aware of the political advantages of acquiring such responsibility. He wrote that "National Park Service management directly under the eyes of Congress and the President, would broaden the interest of members of Congress in the whole National Park System," and he predicted that "this would bring stronger support in basic legislation and appropriations."[106]

The culmination of more than fifteen years of expansionist policy came in 1933 when President Roosevelt signed two executive orders which, together, enormously expanded the Park System.[107] They transferred all national monuments under the control of the Forest Service to the Park System, and in doing so brought twelve large new areas under Park Service management. The orders also added over fifty historical areas, including the great Civil War battlefields Albright had long sought. Most of the historical areas were in the eastern states and their inclusion in the System meant that it became truly national, with a presence in most states of the union. It meant that the Park System's role in historic preservation was given an undisputed legitimacy. In a stroke, these additions more than doubled the size of the System and the number of visits the System received annually. The executive orders even gave the Service its long-desired control over the parks, monuments, and parkways of Washington, D.C. The agency and the System it managed now had a measure of security they had never before known. As Albright summed it up, the orders "effectively made the National Park Service a very strong agency with such a distinctive and independent field of service as to end its possible eligibility for merger or consolidation with another bureau."[108]

Although official recognition of the Park Service as the lead agency for federal recreation policy was still a few years away, with the executive orders of 1933 Albright accomplished almost all he had set out to do. He resigned from the directorship on August 9, 1933, the day before the orders were to go into effect. Few eras have such definitive endings.

The Accomplishments of Mather and Albright

Although many organizations have founder myths in which their first directors are viewed as nearly superhuman, the Service's high regard for its early leadership seems justified by the actual accomplishments of Mather and Albright. The two men had the array of problems facing any new agency plus some which were unique to the Park Service. They solved them, and the solutions they worked out were to be a source of strength and guidance to the agency long after they left it.

At the root of this success was their ability to fashion a set of policies for the Park System and a range of responsibilities for the Park Service which at the same time established a favorable public attitude toward the two, gave them appreciative clients and strong backers, and assured their survival. Downs suggests that any agency's set of responsibilities, its "turf," to use his term, can be mapped in a figurative sense.[109] In the center is a heartland of undisputed responsibility and beyond that are areas of progressively more disputed control or progressively less commitment. The National Park System, the "territory" of the Park Service, at the end of Albright's directorship can be described within this construct. At the core were great natural parks, the "crown jewels," as they are commonly referred to within the Park Service. Only agency extinction would be likely to shift this responsibility to another agency. Although visits to the great national parks were only a fraction of the total visits to the National Park System by the mid-1930s (table 2-4), they were what the public thought of

Table 2-4. Visits to the National Park System, 1917–36

Year	Total visits to units under National Park Service management[a]	Percent of total visits accounted for by National Park visits	Percent of total visits accounted for by National Monument visits	Percent of total visits accounted for by National Historical Area visits
1917	—	99	1	—
1920	1,048	86	14	—
1925	2,053	85	15	—
1930	3,247	85	15	—
1932	3,755	78	11	11
1934	6,337	55	23	22
1936	11,990	48	14	38

Sources: National Park Service, *Public Use of the National Parks: A Statistical Report (1904–1940).*
[a]Visits in thousands, rounded off to nearest thousand.

when it thought of the Park System, and they were the stuff of which the Park Service formed its self-image. How much the agency continued to identify almost exclusively with the national parks is illustrated by an incident which occurred in connection with the executive orders of 1933. As part of the orders, the name of the National Park Service was changed to the Office of National Parks, Buildings and Reservations, a name which, in fact, much more accurately reflected the agency's true range of responsibilities than did the old one. The change was not well received in the agency, however, and it immediately set out on a campaign to change the name back. It succeeded the following year, and the name has remained simply the National Park Service ever since.

For the great natural areas of the system, Mather and Albright worked out policies which were successful on several levels. By opening up the parks to the public at large, they made the Park Service an agent of modern society; they provided America with the Great Destination as the annual vacation became part of the rhythm of its life. At the same time they gained the support of those who made their living off these larger and larger annual migrations. By resisting resource exploitation in the parks, they became heirs to John Muir and the preservationist wing of the conservation movement. The great sites under the control of the Service would be preserved for their aesthetic and spiritual value. Thus in its policies toward the great parks, the agency was at the same time a booster, a professional land manager, and a priest of American nature worship. Although it was not quite the problem it would later become, it was no mean feat to steer a safe course between preservation and mass use, between boosterism and professionalism, and between the lowest of park taste and the most refined.

While securing this core of responsibility through its management policies, Mather and Albright sought to advance the System and Service by moving into and staking out recognized claims to the marches. With the executive orders of 1933, we find arranged around the core all the new responsibilities which the Service claimed as logical extensions of its original charge. Historical preservation was justified as the parallel to, and therefore a logical extension of, the Service's mission to preserve nature. The agency's role as advisor to the state parks was also interpreted as a necessary complement to national park management and as a logical outgrowth of the agency's own accumulation of management skills. The argument for the assumption of national monuments was similar, i.e., the Service was already managing large natural parks; managing monuments was but a natural extension of this responsibility. Farthest out on the periphery of agency responsibility was the maintenance of the parks, buildings, and parkways of the city of Washington. This was connected to the main less by logic of mission structure then expedience—the National Park Service had the resources and the skills to do the job, so it should do it.

This, then, was the territory Mather and Albright bequeathed to their successors; it was topographically rich terrain and it formed a strong base of support for the agency. The National Park System as they fashioned it was made up of distinct types of places, each of which provided some strategic service to the whole. However, it brought with it supporters and clients who were sometimes in conflict and responsibilities which were occasionally contradictory. The task of keeping everything in adjustment would not be a simple one, and two questions in particular would require the constant attention of those who would run the Service after 1933. The first concerned the management of the crown jewels. The middle ground between preservation and mass use would constantly shift as conditions in the parks or the relative strength of the preservation and use advocates changed and opinions within the service rank and file (which were never undivided) changed with them.[110] The second concerned the peripheral responsibilities; more specifically, how the agency should treat them. How much of the Service's energy should be put into them and what, if any, new responsibilities should be acquired? Although this kind of problem is a common one for bureaucracies,[111] it was nevertheless a matter requiring continual attention and adjustment to changing demands and opportunities.

Mather and Albright also bequeathed a legacy of style and power. Under the two of them, National Park Service leadership was aggressive and entrepreneurial, and under Mather it was even flamboyant. Through charisma, cajolery, hard work, and the impression of confidence they exuded, they stayed on top of the Service's relationships with its public supporters and with Congress. To be sure, the Park System had to accept some units they did not want, but by and large these were exceptions.[112] Under Mather and Albright, the agency was the chief arbiter of its own course; it had become what its leadership wanted it to be.

FROM ROOSEVELT TO KENNEDY

The Park Service in the New Deal Era

Arno B. Cammerer succeeded Albright to the directorship of the National Park Service in 1933. Cammerer had entered the National Park Service in 1919 and as assistant director under Mather and as associate director under Albright he had long been close to top agency leadership.[113] So, in a sense Cammerer represented continuity. However, the agency which Cammerer led would find itself operating in a very different milieu than that of his predecessors. Although the Depression had begun with the crash of 1929, it was not until 1931 or 1932 that its true depths and the need for major remedial action became apparent. When it did, coping with the Depression became a major demand, perhaps the major one, on elements of the federal bureaucracy.[114] During the 1930s, the agency found an important place in New Deal efforts to cope with the wounded economy and its social consequences.

The Civilian Conservation Corps

Fortunately for the Park Service, Roosevelt's relationship to his secretary of interior, Harold Ickes, was a close one, and Ickes convinced the president that unemployed labor could be put to good use in carrying out conservation and recreation projects for Interior's agencies. Conservation and the provision of recreation opportunities were areas in which the government could spend much money and employ much labor with relatively little controversy, since they were widely recognized as legitimate concerns of government. Moreover, it could be argued that the public provision of recreation facilities complemented and in some cases stimulated private business. Accordingly, the National Park Service was assigned a large role in managing the Civilian Conservation Corps (CCC) program and soon that role came to eclipse all other agency responsibilities in terms of men and resources employed. By 1935, the agency was running 118 CCC camps in the National Park System. The enrollees worked on hundreds of park development projects, including a large backlog of projects which had been planned under Mather and Albright but which had not been carried out for lack of money and manpower.

Stimulated by the availability of federal emergency funds, the states expanded their interest in conservation and recreation. During the first three years of Roosevelt's presidency, seven state park systems were established while nationwide a total of 350 new state parks were created. The National Park Service drew up development plans for the new state parks, and carried out many of the plans with the CCC labor at its disposal. The Service was soon running almost 500 CCC camps in the state parks (figure 2-4). Under the Emergency Relief Administration, submarginal farming

Figure 2-4.
A Civilian Conservation Corps crew at work making shingles for park buildings.

and grazing land was acquired by the federal government, and much of the land deemed of value for recreation was turned over to the National Park Service. Some of this land was developed into parks with the idea of eventually turning them over to the states in which they were located and for whom they would serve as model state parks. These "demonstration parks" were developed with CCC labor as well. In fact, the agency's use of CCC labor in its many projects was so lavish that at the program's peak size in 1935, the Park Service was employing 6,000 supervisors and 120,000 enrollees in connection with it.[115]

Dams and Seashores

Thus, for a mix of reasons—the new public land, the labor-absorbing qualities of park projects, and the Roosevelt administration's view of recreation as an important component of the good life—the development of recreation facilities outside the traditional national parks became increasingly important to the National Park Service during the 1930s.[116] In keeping with these trends of physical and functional expansion, the decade saw the Park Service begin to take over the management of recreation facilities at dam sites through cooperative agreements with the Bureau of Reclamation. In 1936, the first unit based on a reclamation project was added to the National Park System when Lake Mead National Recreation Area at Boulder Dam was established. In a strategic sense, these recreation areas were practically pure benefits for the Park Service. They usually brought it a large number of clients, while at the same time they involved little expense and none of the use versus preservation questions often raised by the management of the great national parks.

Also during the mid-1930s, the Park Service conducted surveys of the Atlantic and Gulf coasts for their recreation potential. The survey reports recommended that several large coastal areas be acquired by the federal government and managed by the Park Service as national seashores. The agency first tendered a claim to responsibility for urban recreation with these reports, since one of the principal arguments the agency advanced for the national seashores was that most were only an hour or two by auto from a metropolitan area. (Ironically, the only one established in the 1930s, Cape Hatteras National Seashore, was nearly as inaccessible to major population clusters as were most national parks.)

Statutory Recognition

The increased emphasis on recreation outside the national parks and monuments was both confirmed and encouraged by Congress with the Parks, Parkway and Recreation Act of 1936.[117] The act clearly established the Park Service as the preeminent federal recreation agency, and it singled out the agency as the proper conduit of federal aid to lower levels of government for recreation projects. In fact, it made the agency responsible for determining the recreation potential of all federal lands except for those already under the jurisdiction of the Department of Agriculture. In order to carry out this charge, the secretary of interior was authorized, through the National Park Service, to "seek and accept the cooperation and assistance of federal departments or agencies having jurisdictions of lands belonging to the United States." Thus the 1936 act confirmed the claim of the Park Service to a recreation coordinating role among federal land management agencies, as well as among states.

This statutory recognition of the Park Service as the premier federal recreation agency brought clear advantages to the agency. First, it provided a new means of reaching the old goal of a geographically broad base of constituent support. The additional charge of ensuring sufficient recreation facilities for the nation gave the Service much more geographical flexibility than did just the management of great natural areas or the historic sites. Historic sites and great (or even somewhat less than great) scenic areas are where you find them, but parkways and pleasant parks with camping facilities, picnic tables, and baseball diamonds are where you put them. Second, in the 1930s, recreation was a rapidly expanding federal activity. By 1938, promoting recreation in one form or another was the overt object of over sixty federal programs, some of which sheltered important latent functions such as regional development and economic stimulation.[118] Thus recreation's potential as a bureaucratic power base seemed vast, something confirmed by Robert Moses's rise to power.

Finally, the Act of 1936 strengthened the independence of the National Park System in the face of threats from the Forest Service, which, far from ending, had taken a new twist during the New Deal era. Under Interior Secretary Ickes, the rivalry between Interior and Agriculture reached new heights. Agriculture maneuvered to get control of the western grazing lands from Interior, while Ickes tried to combine the Forest Service with Interior's agencies into a new Department of Conservation—which he would head.[119] For a while, it looked like Ickes would be successful in his plans, but for the Park System and Service, success could have had disastrous consequences. With the Park Service and the Forest Service in the same department, a merger of the two would have become subject to the whims of a single secretary and could be brought about by a simple act of intradepartmental reorganization. Such a merger might have resulted in institutional oblivion for the Park Service. Although Roosevelt supposedly promised Ickes control over the Forest Service, such a transfer never took place. When it became clear that his gambit had failed, Ickes, perhaps to console himself, engineered the transfer of several large and spectacular tracts of national forest land to the National Park System. Two of the premier parks of the National Park System, Olympic in Washington and Kings Canyon in the California Sierras, were established on these transferred lands. The Forest Service in turn defended itself against further losses by increasing the protection and recreation management of its primitive areas.

Opposition to Agency Policy

The National Park Service's move into recreation was not without its costs. In the 1920s, the agency's expansion of its holdings had made it suspect in the eyes of the other departments, and in the 1930s, distrust grew with each Park Service success. The Service's sister agencies bitterly resented its claims that management of the National Park System automatically conferred jurisdiction over recreation planning on all federal lands.[120] The Park Service's quest for new responsibilities in the 1930s came to be viewed as a classic case of bureaucratic imperialism.[121] In Congress the agency's motives were questioned. For example, Senator Adams of Colorado declared of National Park Service leadership that there was "no limit to their ambition."[122] The agency's movement into recreation outside the national parks had further costs in the disaffection of its preservationist supporters. According to Ise, by the late 1930s it was "under heavy fire from the purists of some of the conservation organizations," particularly the National Parks Association and the Wilderness Society.[123] The purists charged that the Park Service had violated true national park standards and accused it of giving too much attention to recreation areas, state and local parks, and recreation in general while ignoring the great national parks and monuments.

So disenchanted were some of the Park Service's long-time supporters with its policies in the 1930s that the National Parks Association bulletin called for splitting up the National Park System. On one side would be "national primeval parks," i.e., the great natural areas. On the other side would be the rest of the National Park System and the other responsibilities that the Park Service had acquired. According to the proposal, a separate agency should manage each of the two charges.[124] Although nothing came of the idea (perhaps it was intended as more of a rebuke to the Park Service than a serious proposal), it did reflect a straining of the close ties Mather established between agency leadership and the National Parks Association and kindred organizations. This strain became more apparent when, in 1938, the Save-the-Redwoods League rejected a proposal to bring together California's several state parks in the redwood country into one unified national park. The League did not fully trust the National Park Service, and it was afraid of what the parks would be transformed into under its management.[125] The agency's preservation-minded allies would soon have their day, however.

In 1940, Cammerer resigned and Newton Drury accepted the directorship of the Park Service. Drury had been the director of the Save-the-Redwoods League.

Policies Under Drury

Drury was cut of very different cloth than were his predecessors (figure 2-5). When Mather was not ignoring the conventions of proper administrative behavior, he was flouting them. While Albright and Cammerer were less flamboyant than Mather, they were nevertheless aggressive, opportun-

Figure 2-5.
Newton Drury (center), director of the Park Service through the 1940s.

istic administrators. Drury's behavior, on the other hand, was constantly restrained by his stringent sense of bureaucratic propriety and his dislike of the rough and tumble world of Washington politics. He was never comfortable before appropriations committees, whose chambers Mather had viewed as practically his second home. As his years at the Save-the-Redwoods League showed, he preferred to accommodate opponents with quiet compromise rather than confront them in public contests. He always seemed more aware of the weakness of his position than its strength.[126]

He was also strongly opposed to the way the Park Service and the National Park System had evolved over the previous decade, and he was determined to right things. Drury, like most of the Service's preservationist allies, had been repelled by what he termed the Service's "cheap showmanship" in its efforts to increase park use.[127] He also opposed what he saw as the dilution of the System's high standards through the indiscriminate creation of new parks. His conservative, Republican sense of the proper distribution of powers among the levels of government was offended by the Service's aggressive moves into state park matters. In essence, Drury believed in a passive, caretaker role for the National Park Service. He would sum up his view thus: "I firmly believe . . . that the intent of the Basic Act of 1916 which created the National Park Service was to establish it primarily as the custodian of the masterpieces of nature and sites that are of great historical worth because they are significant examples of the greatness of America." [128] A corollary of this view was his rejection of the "tendency to dissipate its resources by calling on it to . . . undertake custody of areas or performance of functions that are not signifi- cant from a national viewpoint and are not logically a responsibility of the service."

Drury wanted only minimal development of visitor facilities in the na- tional parks; public accommodation would be provided in the parks only when nearby outside enterprises failed to adequately meet public needs. He summed up his guiding principle on park development with a single word, "restraint." [129] Although Drury's attitude did not represent a com- plete rejection of the niche Mather and Albright had established for the Service and which Cammerer had extended, it did involve a clear shift of emphasis. Albright, now a private businessman but with a continuing interest in the National Park System, knew Drury and had strong reserva- tions about him as agency director.

Deterioration of the System

The new director had little chance to turn his philosophy into manage- ment policies; World War II began the year after Drury took over the Park Service. During the war, Drury's major problem was not one of righting agency philosophy but rather one of protecting the parks from those who saw them as a source of raw materials for the war effort. [130] Whereas World War I produced Secretary Lane's suggestion that Yellowstone's elk herd be turned into tinned food for the boys at the front, World War II brought demands that the parks be used as sites for mountain warfare training, as grazing areas for cattle and, in the case of the parks of the Pacific North- west, as sources of sitka spruce, a light wood important in airplane con- struction. The position of the Park Service was a weak one. Much of its staff had been drafted into the armed services, its concerns were of low priority to the administration, its appropriations were minuscule, and, if that were not enough, its headquarters was moved to Chicago to make room in Washington for the war effort. Under the circumstances, Drury defended the parks as best he could. He successfully resisted encroach- ments when possible and quietly accommodated them and tried to mini- mize the damage when he could not.

After the war as government, economy, and society felt their way back into a peacetime world, Drury's dislike of politics and his disavowal of bureaucratic entrepreneurship hurt the System. He failed to gain sufficient appropriations to accommodate, or protect the Park System from, what was virtually an explosion of National Park use in the years after World War II (table 2-5). Facilities throughout the System, designed to accommodate prewar levels of demand, were degraded and destroyed by a combination of overuse and neglect. With the Korean War, the economy was partially returned to a wartime footing, and the opportunities of the late 1940s were lost. The System continued to erode. In 1953 Bernard De Voto, long a supporter of the national parks, wrote a widely quoted article in *Harper's* entitled "Let's Close the National Parks." In the article, he discussed and deplored the state of neglect into which the national parks had been allowed to fall and although he put most of the blame on Congress, he did not excuse Park Service leadership.

Drury reminds one of Marcus Aurelius, a man in the wrong times. Perhaps Drury would have been the right man to have led the Park Service into the decade of the 1940s had it been a tranquil one. He could have tempered the agency's aggressiveness, quietly shifted the balance of management toward presentation, and restored good relations with the preservationists and with its sister bureaus. But the decade was not tranquil, and the problems it presented the National Park System were not the ones which Drury seemed best equipped to handle. His handling of the Echo Park controversy was probably his greatest failing as a director, and that

Table 2-5. Visits to the National Park System and Selected System Units, 1941–55 (in thousands)

Year	Bryce Canyon	Gettysburg	Mount Rushmore	Total system[a]	Total system visits as a percent of 1941 visits
1941	125	670	393	21,237	—
1942	30	118	140	9,237	44
1943	8	76	31	6,878	32
1944	10	121	30	8,340	39
1945	33	252	85	11,714	55
1946	126	509	325	21,752	103
1947	163	644	427	25,534	120
1948	176	631	571	29,859	141
1949	193	669	657	31,735	150
1950	213	656	740	33,253	157
1951	225	684	741	37,106	175
1952	225	755	835	42,300	199
1953	243	681	914	46,225	218
1954	238	706	910	47,834	225
1955	254	720	895	50,008	236

Sources: National Park Service, *Public Use of the National Parks: A Statistical Report (1941–1953); Public Use of the National Parks: A Statistical Report (1954–1964).*
[a]Excluding National Capital Parks.

failing provided a lesson in System management which would shape agency policy over the next decade.

Echo Park Controversy

Briefly stated, the Echo Park controversy involved a proposal advanced by the Bureau of Reclamation to build a dam and create a backup lake in the Echo Park area of Dinosaur National Monument. (It was but one part of the ambitious Colorado River Storage Project.) Although as far back as 1920, it was the declared policy of the Park Service to oppose any surveying for dam sites within the National Park System, Drury, presumably in the interest of interagency cooperation, allowed the Bureau of Reclamation to survey Dinosaur National Monument. While plans for the dam were taking shape, the Service under Drury pursued an uncertain course. It did not support the project nor did it appear to oppose it. Drury must have been concerned about the damage to the monument that a dam would cause but he did not voice his concern.[131]

It was not until 1949, when the Bureau of Reclamation had almost finished its surveying in the monument area that the Service's misgivings about the dam moved it to prevent further surveys. By then, however, the Bureau had lined up a powerful supporting coalition behind the project. It was well defined and there was political momentum behind it. For its part, the Park Service failed to keep its allies, the preservation groups, informed of the issue. Although Drury was on a first-name basis with preservationist leaders, most of them were not even aware of the dam project before 1949. When, in the following year, Interior Secretary Chapman decided in favor of the dam, the Service's room to manuever in opposition to it narrowed; it was too late for Drury to bring the Service into the coalition forming against the dam as a full partner. Outsiders, citizen groups, and congressional friends of the Park System would have to carry the fight.

Although the dam was eventually stopped, the Park Service hardly came out of the controversy a winner. It had lost mastery of its own house. The fate of a unit of the National Park System was decided by the interplay of public interest groups and their congressional allies on one side and the Bureau of Reclamation and its allies on the other. The agency was a bit player in the drama. An agency always pays a price for the support of its allies. The greater the relative strength of the allies, the greater the restrictions they will be able to impose on an agency and the greater will be the consideration of their interests in the formation of common goals.[132] The Park Service was diminished relative to its allies by its behavior in the Echo Park controversy and the diminution would have its costs. But Drury did not remain to see the consequences or even the final resolution of the controversy. Apparently for reasons unrelated to the dam, Secretary Chapman asked Drury to step down, and in January 1951 he did.[133]

Mission 66

Preservationists regretted the removal of Drury, who had been one of them, and were apprehensive about his successor, Conrad Wirth, a career Park Service man who had established a reputation as a talented administrator while running the agency's CCC program in the 1930s.[134] They feared that Wirth, being a career civil servant rather than a private citizen turned public administrator out of noblesse oblige, would show less independence of mind than did Drury. An editorial in *The New York Times* expressed the fear of the preservation minded that change in leadership would lead to an emphasis on the promotion of organized recreation in the Park System, and to a lessened interest in the preservation of "untouched areas of wild and natural beauty for themselves alone."[135] But Drury's style and policies, be they ever so principled, had involved real costs to the Service and the System. Connections with Congress were so frayed and relations with the appropriations committees had deteriorated so much that, shortly after Wirth took office, De Voto wrote that "the lack of money has now brought our National Park System to the verge of crisis," and that "so much of the priceless heritage which the service must protect [was] beginning to go to hell" as a result.[136]

In some parks, lack of adequate facilities meant hopelessly overcrowded campgrounds complete with broken equipment and irritated visitors. Perhaps worse, in others poor, unrepaired roads discouraged people from visiting them at all. The inaccessibility of Dinosaur National Monument and its lack of visitor facilities might have been compatible with managing the park for its wilderness values, but it also gave ammunition to those who wanted to turn the park into a big lake. Proponents of the dam had made a point of how inaccessible the monument was by circulating a photo which showed several men pushing and hauling a car over a road in the monument that looked very much like a boulder field. How few visitors the park received was apparently an important factor in Interior's initial decision to go ahead with the dam.[137] Ise, looking back on the controversy, concluded that Dinosaur National Monument would have been safer if there had been good roads into its canyons.[138]

Without public support and the good will of congressmen and other high officials who were affected by public support, the Park Service would have to rely heavily on the strength of the conservationist public interest groups as it had done in the Echo Park controversy.[139] When Wirth took office, the conservationists appeared to be a slender reed on which to lean. In the early 1950s, the Echo Canyon Dam fight seemed as good as lost to all but the most optimistic. The conservation movement was viewed as a relic of Teddy Roosevelt's era. Grant McConnell wrote in 1954 that "the fires of the earlier movement have subsided and still burn only in scattered groups." He concluded that the movement was small, divided and frequently uncertain.[140]

Wirth had seen the National Park Service weakened and the System threatened under Drury. He also had an example close to home of what could happen to an agency weakened by failure and suffering from an inability to adjust to circumstances. Even as the Service was fighting the Bureau of Reclamation at Echo Park, the latter was in its "slide into poverty."[141] As the number of good unused dam sites in the West became increasingly scarce, the Bureau failed to find something to do other than building dams in the West. As a result, its appropriations dropped drastically, more than 50 percent between 1950 and 1955.

Wirth's strategy for reestablishing agency independence and shoring up the Park System's defenses against external assaults was embodied in Mission 66, a massive, decade-long program of improvement and construction of facilities throughout the National Park System. The goal of the plan, named Mission 66 because the program would finish in 1966, the fiftieth anniversary of the National Park Service, was to bring all units of the Park System, regardless of origin, "up to a consistently high standard of preservation, staffing and physical development, and to consolidate them fully into one national park system."[142] Such an ambitious plan would require congressional and White House approval and a level of annual appropriations many times higher than that of the late 1940s and early 1950s. Wirth managed to get that approval by personally convincing President Eisenhower of the merits of the plan and by carefully cultivating the good will of the powerful men in Congress, many of whom, including Virginia's powerful Senator Harry Byrd, Sr., he came to consider close personal friends.[143] In fact, the program had so much support that in 1955 the administration and the House Appropriations Committee got into a little contest over who would authorize the most money to get Mission 66 started. Once under way, appropriations for park development increased steadily; 49 million in fiscal year 1956, 68 million, 76 million, and 80 million in the three succeeding years.

Mission 66 was to increase public support for the System and Service by reviving Mather's policy of accommodating as wide a range of public tastes as possible. The program was especially aimed at increasing the visitor use of parks threatened by resource exploitation schemes and thereby strengthening the Service's hold on them.[144] Washington's Olympic National Park had been carved out of national forest lands, and the northwest timber industry was never reconciled to its loss. To counter pressure for timber harvesting with public use, the Service built the high-grade Hurricane Ridge Road through the park. The construction of scenic

Steven's Canyon Road in nearby Mount Rainier National Park was for the same end. Ironically, Mission 66's commitment to upgrading access to the parks was a strong factor in killing the Echo Park project once and for all. In 1955 when Interior Secretary McKay announced that the department was dropping plans for the dam, he also announced that as part of Mission 66, the Park Service would build an improved road into the monument "which will permit the public to get into the heart of the area and enjoy the beauty of this wonderful canyon country." [145] In all, the National Park Service built or upgraded more than 2,000 miles of roads in the National Park System as part of Mission 66.

One of the most important innovations of Mission 66 was the "visitor center." Whereas previously, education facilities ("interpretation facili-ties," to use the Service's euphemism) had been housed in small, museum-type buildings, the new visitor centers were ambitious modern structures that housed information and interpretative facilities, exhibits, rest areas, and administrative offices. In the larger parks they might also include souvenir stands, cafeterias, large audio displays, and even auditoriums. At Gettysburg a large modern visitor center was built within a few hundred yards of the Bloody Angle, completely dominating the site where Pickett's men reached the Union lines, wavered, and fell back—the highwater mark of the Confederacy. Soon large visitor centers perched just as ostenta-tiously at other historical sites and in many of the system's natural areas. During Mission 66, 114 visitor centers were built in the National Park System, centers which closely identified the Park Service in the public eye with the parks under its charge. As part of Mission 66, the agency also embarked on a program of building the facilities it thought it needed to properly run the System. It built two large training centers for agency personnel, one at the Grand Canyon and the other at Harpers Ferry, West Virginia. It greatly upgraded administrative and maintenance facilities, and employee housing was improved throughout the System.

When it planned its improved visitor and service facilities, the agency aimed not only at accommodating present use but also at accommodating and even encouraging increased future use. All the indicators of increased potential demand for what the Park Service offered were there—increasing per capita income, increasing numbers of cars on the nation's roads, increasing average length of the annual vacation—and Wirth intended to meet this demand. [146]

With Wirth, as was the case with Mather and Albright, it is difficult to fully understand his notion of what was appropriate in accommodating increased park use. On airports he seemed undecided. Official agency policy under Wirth was to bar the construction of airports in the parks (while encouraging them close enough to the parks to make the parks easily accessible by air). Yet the Service did not make efforts to remove the airports from Mount McKinley, Grand Canyon, or Death Valley. Under Wirth, the Service generally increased visitor facilities in the System, but it also planned on removing them in some places. Wirth hoped to phase out overnight accommodations in the main valley of Yosemite National Park, and, in spite of local protest, he insisted that overnight visitor accommodations in the popular eastern end of Rocky Mountain National Park be reduced. In response to the demands of preservationists and others, Wirth scuttled a plan to build a parkway northwest of Washington, D.C. on the bed of the old Chesapeake and Ohio Canal, but this did not stop him from aggressively promoting parkway construction elsewhere. He talked about the sublime characteristics of the crown jewels, but he sought to bring into the System reservoir-based recreation areas with few if any traces of the sublime. Moreover, once they were in the System, he often treated these areas like national parks, building ambitious visitor centers and interpretative facilities in them. By doing so he obscured the difference between the grand and the banal in the System. Wirth allowed an interdenominational "Shrine of the Ages" to be built close to the rim of Grand Canyon, something Ise, who was usually reserved in his criticisms of agency leadership, called "a rather cheap and unseemly business." [147] But Wirth vigorously discouraged efforts to establish such shrines in other parks.

As impossible as it may be to fully understand Wirth's sense of what constituted proper Park System development and management, one thing was clear: it was not in accord with that of the preservation-minded groups. To them, Wirth's Mission 66 was the essence of those things they detested; it was the sacrifice of preservation to mass use; it was catering to the lowest common denominator of park taste; it was the submergence of the unique qualities of the individual parks under the weight of seemingly interchangeable, undistinguished development plans. If Mission 66 was a cure for that which ailed the System, it was worse than the disease. For his part, Wirth maintained proper but stiff relations with groups like the Sierra Club and the Wilderness Society. While he was agency director, he held his criticisms of the preservation groups in check, and he usually praised them in public for the support they gave the National Park Service. Beneath the surface, however, the relationship was not all sweetness and light. [148]

Wirth and the First Directors Compared

Shankland described Wirth as a "true believer in the policies of the Mather–Albright era" [149] and in looking at his park management policies one must agree. Although it seems unlikely that Mather's protective sense of the System's image would have allowed him to support such projects as the Shrine of the Ages, in the main Mission 66 did smack of the agency's early efforts to secure a mass constituency.

The period of Wirth's directorship, the Eisenhower era, was similar in many respects to the era of Mather and Albright. Both periods were ones of a great optimism about the future. Both had a faith bordering on the smug in the soundness of American institutions, and both trusted science and professionalism. Both had a homogenized rather than a pluralist view of the public and its needs. (In reality, their view tended to obscure needs and values which were not those of the middle class.) In many respects, the later era was the embodiment of the progressive vision that guided Mather and Albright. In the 1920s, the first directors planned for the vanguard of the modern life; in short, for those in nuclear families, with dependable incomes, a family car, and a paid annual vacation. Although such Americans did not constitute an absolute majority of the population then, faith in progress led Mather and Albright to believe that they were the clientele of the future. By Wirth's time the vanguard had come closer to being the norm.

Mission 66 succeeded for the same reasons the policies of Mather and Albright did; they were consonant with the dominant social values of the era and they catered to the articulated demands of the times. And succeed they did. In a strategic sense, the late 1950s were halcyon years for the National Park System. System-wide, annual visits rose rapidly. The parks seemed secure, and their managing agency seemed to have a comfortable niche in the federal bureaucracy. If imitation is a form of flattery, the Forest Service paid the Park Service a rare compliment when it launched its Operation Outdoors in the late 1950s. The program was patently modeled on Mission 66.

For all the similarities between the Wirth directorship and those of Mather and Albright, there were critical differences which would bring down Wirth and lead to a string of sharp reverses for the Park Service. First, there was the agency's relationship to the preservation groups. While Mather and Albright could usually convince the preservationists that the Park Service's policies of mass access were necessary to strengthen the political defenses of the National Park System, Wirth could not. If the Echo Park controversy had led many preservationists to believe the agency was incapable of defending the parks, then Wirth's policies led them to believe it was not even inclined to do so.

In the early 1950s, when groups which spoke for preservationist policies were relatively weak, cooperation with agency leadership was probably their best strategy in spite of their misgivings. They grew in strength as the 1950s progressed, however. The membership of groups like the Sierra Club started to increase rapidly and their politics became more aggressive under new and younger leadership. Second, the National Park Service under Wirth put almost all of its energies into Mission 66, to the conspicuous neglect of its advisory responsibilities to lower levels of government. This neglect, although a continuation of Drury's policy, was at odds with Mather's. It was a source of continuing frustration to those who wanted to improve state and local recreation and who looked to the National Park Service for aid in doing so.

Not only were the directorships different in some very important ways, but so were the eras in which they were set. During Mather's era, modernism was a vision attached to a set of profound socioeconomic transformations which were still in their incipient stages and therefore little ramified. The automobile was a means of liberation; it would make both the city and the country better places to live in and perhaps even bring about a near-utopian synthesis of the two in what came to be called the suburbs. The ascendancy of the middle class would wipe out class differences and with them the class-based antagonisms which haunted the late nineteenth century social landscape. Increasing levels of production would free society from material want.

During Wirth's era, many aspects of the progressive vision became reality, and when they did, their failings and unfortunate side effects became manifest. Far from creating a classless society, the ascendancy of the middle class simply pushed the needs of the rest of society out of view. The problems of social maldistribution seemed to have been little affected by increasing levels of material wealth, and the production of that wealth exacted its toll in environmental degradation. The automobile, on which so much hope had been pinned, was at best a mixed blessing; while it permitted suburbs to grow around the cities, this growth left social segregation and economic dislocation in its wake. Expressways tore the life out of the neighborhoods through which they passed and created backwaters in the areas distant from them. They showed no respect for either nature or history which happened to be in their path. Nature, which increased mobility was supposed to make accessible to everyone, was somehow lost or was made counterfeit in the process. In short, what had been a utopian vision during the Mather era had come to look more like the temptation of Tantalus several decades later. Although these failings did not seem like very acute ones to the Eisenhower administration, they would shortly explode into public consciousness, and onto the nation's political agenda. The National Park Service would be caught up in that explosion and it would never quite recover from its effects.

3 The Modern Era

THE CRISIS OF THE EARLY 1960s

Loss of Faith in Progress

The 1960s and 1970s saw the Park Service attempting to adjust to changed circumstances. During these two decades the agency found its traditional goals devalued, the premises on which it had built past successes and managed the Park System rejected, and its relationships with its traditional sources of support either eroded or reordered. To understand the agency's policies during this period, we must first look at these altered circumstances.

On the most profound level, there was accumulated discontent with the progressive vision which had formed the base of so much of the agency's sense of purpose and so many of its policies. In 1962, Clawson presented a schematic landscape on which recreation facilities were ideally located. The landscape, shown here in figure 3-1, was an ordered and balanced one of complementary land use. A contained, discrete city gave way to compact fields, and they in turn gave way to undisturbed nature. The works of man and those of nature were in balance on this landscape, and technology was comfortably accommodated. Recreation opportunities were abundant and well integrated with other uses. Conflict and tension seemed absent. The mix of activities found on it is a good one, with each activity in its proper place for the greatest social good. The landscape portrayed can be read as a philosophical statement about the relationship of nature to civilization, and, beyond that, as a moral assertion. Mather and Albright would have recognized this landscape and would have been comfortable on it.

Figure 3-1.
The ideal recreation landscape. (From Marion Clawson, *Land and Water for Recreation*, Chicago, Ill., Rand McNally, 1963.)

Indeed, it was the landscape of the progressive ideal, and one they and their successors through Wirth planned for.

By the time this schema was published, however, the assumptions about nature and civilization implicit in it were already being rejected by the preservation activists, who were winning more and more converts to their cause. The preservation-oriented wing of the conservation movement had always had a faction who held a fundamentally negative view of civilization. Perhaps Bob Marshall, a founder of the Wilderness Society in the 1930s, gave this view its most extreme expression with passages such as the following, in which he complained that:[1]

Highways wound up valleys which had known only the footsteps of the wild animals; neatly planted gardens and orchards replaced the tangled confusion of the primeval forest; factories belched up great clouds of smoke where for centuries trees had transpired toward the sky, and the ground-cover of fresh sorrel and twinflower was transformed to asphalt spotted with chewing gum, coal dust and gasoline.

The factories and asphalt were part of a tyrannical civilization which hoped to "conquer every niche on the whole earth." But Marshall was in the minority then. Most preservationists of Marshall's era were optimistic about civilization, and they were primarily concerned about development in inappropriate places, not development itself.[2] The way the Sierra Club dealt with what it considered inappropriate ski development proposals in the wilderness areas of California's national forests reflected this dominant attitude. It simply proposed alternative sites in other, more appropriate areas of the national forests.[3] It did not reject ski areas, nor hydroelectric dams for that matter, as bad things, just inappropriate things in some places. During the 1950s and 1960s, however, a change in Sierra Club thinking took place, and this benign view of civilization lost adherents. By the early 1960s, faith in progress no longer guided the actions of the Sierra Club nor the more active of its companion groups.[4] As a sign of the times, the Wilderness Society, founded thirty years earlier on an assumption of extreme incompatibility of civilization and nature, had grown into one of the most important preservation organizations.

This loss of an earlier optimism about progress had a corollary in the loss of belief that, somewhere, there was a limit to civilization's tendencies to make over nature. The dominant view now seemed to be that, left to its own devices progress would, as Marshall had warned, fill every nook and cranny on the landscape. Writing in the early 1960s in favor of the Wilderness Act, Wallace Stegner (who was to become a foremost voice in preservation) wrote of "a headstrong drive into our technological termite life, the brave new world of a completely man-controlled environment."[5] As he saw it, the consequences of letting this trend play itself out would be awful. This negative view of progress, which increasingly dominated preservationist thinking, was not one of stasis or complementarity. It allowed little chance for harmony between nature and civilization, nor was an equilibrium, short of the complete destruction of nature, likely to be arrived at. Given this perception, development itself, not just isolated instances of inappropriate development, became a threat. One of the ramifications of this view was that the facilities which made the national parks accessible to people appeared to be just as destructive as those which aimed at physically exploiting their natural resources. In fact, by the 1960s the Sierra Club, which had earlier encouraged roads to the parks, came to see new and upgraded park roads as the greatest of threats to the parks.[6]

The task for environmentalists became not so much one of guiding development into its proper locations but rather one of opposing development wherever it would take place at nature's expense. An often-repeated story among preservationists illustrates this point. When David Brower, then executive director of the Sierra Club, was asked which of two proposed dams the Sierra Club would find more acceptable, he refused to choose; neither was acceptable. Given his view of development, each represented an intolerable diminution of the wild. Any possible differences between the environmental impacts of the two dams paled before that fact.[7]

Loss of faith in progress also meant time became an enemy rather than a friend. Instead of each year bringing the landscape closer to a beneficial modern equilibrium, it brought more ruinous development. One critic of the preservation activists wrote that they behaved as if the present was "critically important for all time to come."[8] Indeed, there was much truth in this observation. In 1960, Brower wrote an article in the Sierra Club magazine entitled "A New Decade and a Last Chance: How Bold Should We Be?" The gist of the article was that since "what we save in the next few years is all that will ever be saved . . . much boldness [is] called for."[9]

This new world outlook was at variance with that of the Park Service leadership under Wirth, which still took a more traditional view of progress. From these opposite world views the preservationists and agency leadership would develop different frames of reference from which to assess just about everything the Park Service did. Thus, while the decade of the 1950s began with Park Service leadership and the leading preservationists differing occasionally on tactics but in basic agreement philosophically, by the early years of the next decade the gulf between them was vast.[10]

Outdoor Recreation "Crisis"

In the late 1950s and early 1960s, a second but related phenomenon had a strong impact on thinking about the National Park System—a "crisis in outdoor recreation."[11] True crisis or not, there was undeniably a crest of concern for the nation's recreation facilities during this period, and it was prompted by several factors. First, although the Park Service had embarked on Mission 66, increases in park visitor use were so great that even the System's expanded facilities were being overrun. Second, state and local park systems were also experiencing rapid increases in visitor use.[12] Finally, in the nation's metropolitan areas, lack of open space became a public issue. Linked with it were concerns over unregulated city expansion and the absence of adequate urban land use planning.[13] To study this recreational "crisis," Congress, with Department of Interior backing, set up the Outdoor Recreation Resources Review Commission (ORRRC) in 1958.

Park leadership was not enthusiastic about the establishment of the study commission. A Park Service administrator later recalled the agency's attitude toward ORRRC: "The National Park Service objected to the study itself, which it saw as a threat to its position. It cooperated [with ORRRC] only when it had to."[14] In the agency's view, the commission was unnecessary, and if any such study were needed, the Park Service was the one to do it. After all, the Parks, Parkway and Recreation Act of 1936 explicitly had given it federal responsibility for recreation planning. Since the idea of the commission had the support of the Department of Interior, however, the Park Service had little choice but to go along with it.

Agency leadership had a good reason to greet the establishment of ORRRC without enthusiasm: it represented a real threat to a part of the agency's domain which it had poorly defended and to which its claims were eroding. In 1950, the Park Service led federal land-managing agencies in the number of recreation visits to its lands, but by 1960 it had fallen to third place, behind the Corps of Engineers and the Forest Service (table 3-1). This loss of preeminence undercut the Park Service's claims to responsibility for overall recreation planning on federal lands. Further-

Table 3-1. Annual Recreation Visits to Federal Lands, 1950 and 1960

			Percent of total[a]	
Agency	1950	1960	1950	1960
National Park Service	32,780,000	72,288,000	33	21
Forest Service	27,368,000	92,595,000	28	27
Tennessee Valley Authority	16,645,000	42,349,000	17	13
Bureau of Reclamation	6,594,000	24,300,000	7	7
Corps of Engineers	16,000,000	106,000,000	16	31
Total	99,387,000	337,532,000		

Source: ORRRC, Report 13, p. 1, Table K.
[a]Rounded off to the nearest whole percent.

more, the Service had paid little attention to its role as federal recreation advisor to the other levels of government over the previous fifteen years.

Perhaps this neglect was a natural outgrowth of appropriations problems in the postwar period. As mentioned above, after the war the National Park Service could not even command sufficient funds to keep its parks and monuments in decent repair.[15] It is not surprising, then, that the agency did not press Congress to provide the money for recreation planning on non-federal lands. Nor could it have been expected to divert its hard-pressed personnel to what the agency considered peripheral responsibilities. By the late 1950s, the agency's financial picture had brightened. By then, however, it was involved in Mission 66, and this left little energy to devote to recreation planning outside the parks or to interagency policy coordination. In sum, this neglect might have been logical given the circumstances, but it left the agency very exposed when ORRRC undertook its investigations.

ORRRC Findings

If the Park Service disliked the idea of the study commission, ORRRC's findings were even less to its liking. ORRRC found policy chaos and wrote that "lack of anything resembling a national recreation policy is . . . at the root of most of the recreation problems of the federal government."[16] Given the Parks, Parkway and Recreation Act of 1936, the Park Service would have been the logical place to put the blame, but ORRRC did not. What it did do was worse; it ignored the agency's claims to this area of responsibility. In fact, ORRRC's summary report mentioned the Park Service only in passing when it said that "the National Park Service was not formed to provide recreation in the usual sense but to preserve unique and exceptional areas."[17] Thus the report relegated the Service to a rather small niche in the whole field of outdoor recreation and did not even acknowledge, much less deny, the agency's claims that management of the National Park System gave it a natural hegemony in this area.

Moreover, the report asserted that such preservation as was the Park Service's proper charge was not one of the nation's leading recreation problems. The problem with recreation was not so much the total number of acres available for recreation nationwide but rather the lack of what it called "effective acres," i.e., those close to where most Americans lived, and thus available to them on an everyday basis.[18] To bring order to federal policy and to fill what it saw as the pressing need for "putting recreation opportunities into the environment of people's daily lives," the report had two major recommendations.[19] The first was the establishment of a fund which the federal government could use to purchase land and out of which it could make grants to state and local governments for the purchase and development of open spaces. The second was the creation of a federal bureau of outdoor recreation. The new bureau would be the conduit of federal recreation aid to the states. It would coordinate the recreation activities of the federal agencies, and it would even undertake studies of the national significance of areas being considered for inclusion in the National Park System.

In 1962, when the ORRRC report was released, Eisenhower was out of office, Kennedy was president, and Stewart Udall was his secretary of interior. Udall had followed the study's progress, and when its final report was released he was ready to act on its recommendations for a matching fund and an outdoor recreation bureau. In the same year the ORRRC reports were released, the Bureau of Outdoor Recreation (BOR) was established by an act of Congress and placed under the secretary of interior.[20] To ensure against undue influence by the Park Service, a Forest Service administrator was selected to head the new bureau. He in turn was careful not to go to the Park Service when filling the new bureau's top positions.[21]

Bureau leadership, still fearing the Park Service, moved to secure a base of power for itself among the states and other federal agencies. It funnelled increased amounts of money for recreation to agencies such as the Bureau of Land Management and the Bureau of Sport Fisheries and Wildlife. It was very careful not to give other agencies the impression that it intended to impinge on their responsibilities or their freedom of action. In the Bureau's first policy guidelines, it disavowed any intention of becoming a land-managing agency. It described its coordination responsibilities in a carefully circumscribed manner: "no power has been conferred on the Bureau by statute or by Executive fiat to impose its will on any other government entity." [22] The guidelines also declared that "there will be no empire building in this Bureau," and indicated that BOR had no intention of placing "the clammy hand of restraining bureaucracy on the initiative of other federal bureaus, states or the private sector."

To build support among its clients at the state and local level, BOR strained to make compliance with federal grant requirements as easy as possible and funded a large number of projects for which minimal justification could be shown. [23] Although this would later seriously hurt the Bureau's reputation for competence, it did shore up its support among state and local recreation administrators. Thus the survival of the Park Service's "none-too-friendly sibling" was assured, at least for the time being. [24]

The establishment of the Bureau of Outdoor Recreation effectively repealed the charter that the Parks, Parkway and Recreation Act of 1936 had given the Park Service and by doing so, deprived it of a large, if neglected, responsibility. It was the realization of the threat first articulated by Chief Forester Graves in 1920 when he called for a center of federal recreation planning outside the Park Service. This, however, was not the end of the agency's misfortunes under Secretary Udall.

The New Conservation

Stewart Udall, one of Kennedy's bright young men, is often considered one of the great secretaries of Interior. [25] Be this true or not, Udall certainly was the most active and visible secretary of interior between Harold Ickes and James Watt. The Department of Interior can be, and frequently is, a quiet corner of the federal administration where routines are followed with a hint of drowsiness. Under Udall this was definitely not the case. Udall was a strong department head and a proselytizer who saw to it that his ideas about conservation (which taken together he called the New Conservation) were incorporated as major elements of both the New Frontier and the Great Society. As a consequence, the Department of Interior under his direction came to play an important role in both the Kennedy and Johnson administrations.

Udall's New Conservation, as elaborated in his 1962 best seller, *The Quiet Crisis*, involved at least a partial rejection of the vision which informed the conservation movement of the early century. Udall's use of the term progress was often ironic; for example, he wrote that "the vision of a Frederick Law Olmsted was too advanced for the apostles of 'progress.'" [26] He, like many preservation-minded conservationists of the 1960s, saw technological innovation and the economic growth it engendered as mixed blessings. His view on this was clear: [27]

In a great surge toward 'progress,' our congestion increasingly has befouled water and air and growth has created new problems on every hand. Schools, housing, and roads are inadequate and ill-planned; noise and confusion have mounted with the rising tempo [of] technology; and as our cities have sprawled outward, new forms of abundance and new forms of blight have oftentimes marched hand in hand. Once-inviting countryside has been obliterated in a frenzy of development that has too often ignored essential human needs.

Also, like preservation activists outside government, he felt that the passage of time could not be counted on to bring only benefits and the solution to problems. He wrote that "with the passing of each year neglect has piled new problems on the nation's doorstep."[28] Our collective carelessness and inattention allowed these problems to accumulate while we were concerned with other things.

An important part of Udall's New Conservation was its conceptual link between the problems of traditional conservation and those of the cities and minorities, which were finding high places on the federal government's agenda in the early 1960s. In The Quiet Crisis, the squandering of human potential and natural resources were set up as parallel problems, and the imagery of urban decay and natural resource abuse was transposed for striking effect.[29] For example, Udall wrote that "the prime business of those who would conserve city values is to affirm that such human erosion is unnecessary and wasteful." Then, reversing the imagery, he wrote that unwise use of the land resulted in "slum valleys" and "regional slums."[30]

By linking social and urban problems with the concerns of traditional conservation, Udall was making a politically appealing statement about the former. His view of the causes of the nation's conservation problems was a traditional, politically innocuous one, i.e., the problems were ultimately the result of indifference and short-sightedness. Clearcutting for short-term profit damaged watersheds. Lazy ways of getting rid of industrial by-products caused water and air pollution. By establishing a close parallel with conservation problems, Udall seemed to suggest that the causes of social problems also had their roots in wastefulness and negligence, what he called "our national sloth."[31] Such a view was more appealing (and less revolutionary in its implications) than one which saw these social problems as resulting from the fundamental dynamics of the American socioeconomic system.

The link also conveyed optimism about solutions. Conservation was an area of American public policy in which professional and technical solutions to problems were seen to have been effective in the past. If social and conservation problems were similar, there was reason to believe that professional and technical solutions would work on social problems as well. Moreover, this view of the nation's problems as largely technical was compatible with much thinking within the federal government during the

1960s. By making a conceptual link between the two areas, Udall was also staking a bold claim for an expanded role for Interior, one which would move the department far beyond its traditional concern for natural resources and into an active role in achieving social equality and, in general, improving the quality of American urban life.

President Johnson accepted most of the tenets of the New Conservation, including Udall's contention that conservation, properly understood, dealt with city problems as well as rural ones.[32] Having accepted this, Johnson directed Udall to put his department and its agencies in the service of the cities. The establishment of BOR allowed Udall to increase Interior's emphasis on urban problems and increased the flexibility of the department in dealing with them.

Udall's view of recreation was, like his view of conservation, expansive and integrated. Recreation was not an incidental pastime; it was something to be woven into daily life. Likewise, parks were more than discrete and incidental refuges on the modern cityscape. It is difficult not to read Udall's own ideas and ambitions into his discussion of Frederick Law Olmsted's vision. "As his vision broadened, Olmsted became more than a mere planner of parks, he saw that urban design should include the whole city."[33]

Out of Step

From Udall's perspective, the Park Service under Wirth could not have been less suited to carry out policies in accord with these views. If anything, under Wirth the Park Service had increased the distance between recreation and the rest of life with its concentration on its own national parks, to the neglect of its advisory responsibilities. Indeed, in its very world view, as expressed in Mission 66, the Park Service was out of step with the New Conservation. Mission 66 reflected a great faith in progress rather than a healthy distrust of it. Its building program reflected assumptions about the harmony of development and wilderness which were no longer in fashion. The Park Service under Wirth was not a useful instrument toward Udall's ends, and the establishment of BOR was a way around this obstruction. Wirth's resistance to the ORRRC study and to the formation of BOR further undermined Udall's confidence in him.[34]

Relations between Udall and Park Service leadership went from bad to worse during the early 1960s. Wirth gave an interview to the press and neglected to mention that the Service was part of Interior. Udall promptly sent a reprimand in which he questioned Wirth's loyalty to the administration and to the new policies it was developing.[35] During the early 1960s there was a general feeling among the new men Kennedy had brought to Washington that the federal bureaucracy was intrinsically conservative and in need of a thorough shake-up if it was to be innovative, and perhaps this too contributed to Udall's decision to remove Wirth.[36]

In any case, Udall made his decision: Wirth would have to go. Assistant Secretary Carver was given the hatchet to wield.[37] At a large gathering of Park Service personnel in Yellowstone, with Wirth in attendance, Carver took the podium and lashed out at the Service for its egotism and its "public-be-damned" attitude, for its unwillingness to accept the Bureau of Outdoor Recreation and, in general, for failing to stay in step with the administration or the times. Publicly humiliated, Wirth felt he had no choice but to resign, which he did four days later. In a letter explaining the affair to Albright, Udall summed up his feelings toward the Service under Wirth when he wrote that "we have sometimes been critical, even strongly critical, of the stiff-necked attitude of the National Park Service. When it stands like Horatius at the bridge, patience runs low."[38]

The stiff-necked attitude referred to policies which, with little variation, had ensured agency success and System security for over fifty years. But the removal of Wirth indicated how inappropriate these policies had become and how their continued pursuit had brought the Service low.

Udall replaced Wirth with George Hartzog, a lawyer and former Park Service administrator who had left the agency to enter private law practice only a few months before Wirth's resignation. Under Hartzog and his successors Walker, Everhardt, Whalen, and Dickenson, the Park Service would struggle to bring itself and the Park System it managed back into step with the times, to restore the agency's spirits, and to bring it the full measure of authority and security it had known earlier. Unfortunately, this was not to be an easy task. The world in which these things had to be accomplished, that of the 1960s and 1970s, proved to be less hospitable to the agency than earlier ones had been. Policy guidelines for the Park System were unclear and difficult to follow. The old relationships on which the agency had depended for strength and freedom of action and which served as foundations for park management were altered or no more. We now turn to these relationships.

THE EROSION OF OUTSIDE SUPPORT

Environmental Groups

Mather and Albright depended heavily on extragovernment support in achieving their ends, especially on preservation groups and the nation's business, scientific, and social elites. Recent directors, however, could not count on this support. From the mid-1960s onward, the relationship of the

Park Service to the national preservation organizations was very different from that of any preceding era, and much of the difference was due to the increased strength of these organizations. Whereas McConnell could characterize the conservation movement of the early 1950s as weak and divided, a decade later it was anything but weak. The activist generation of the 1960s joined the long-established conservation movement, and the combination proved to be a formidable national movement.[39] Old agency allies like the Sierra Club, Wilderness Society, and the Audubon Society found themselves not only poles apart from agency leadership on philosophical issues but also riding a crest of increased environmental interest on the part of a large segment of the American public. At the end of World War II, the Sierra Club was largely a California organization with 3,500 members. By 1965, it had 30,000 members nationwide, and the following decade would see membership more than double again.[40]

The increasing strength of the environmental movement was reflected both in new legislation and the behavior of federal agencies. In 1964 the passage of the Wilderness Act, long the primary goal of the Wilderness Society, established a system of congressionally designated wilderness areas on federal land. Just a few years earlier, the wilderness system, opposed by both the Park Service and the Forest Service, seemed like a pipe dream. In 1969, the National Environmental Policy Act became law, and the Environmental Protection Agency was established shortly afterward. Even the powerful and independent Army Corps of Engineers was shaken by the vigor of environmentalist assaults on its traditional mainstay, large waterway "improvements," so much so that it decided that "increased environmental sensitivity [in its projects] was the price of continued agency well-being."[41]

In reviewing the balance of power between public agencies and interest groups, students of bureaucratic politics conclude that the advantages usually lie with the interest groups.[42] Although this was probably not the case with the Park Service in the 1950s, when the agency could afford to ignore the wishes of its preservationist supporters in carrying out Mission 66, the rise of environmentalism in the 1960s, linked with the Service's fall from grace, certainly strengthened the relative position of environmental groups. While this was happening, several things led them to view their relationship with the Park Service, not as the easy one of allies, but more as one of conflict. First, the environmental movement, which was transforming and invigorating the preservation groups, was in turn linked to a broader public interest movement which included among its demands more direct access to decision-making arenas.

Environmentalism shared with this broader movement a deep bias against both private enterprise and government organizations.[43] It was

suspicious of both the businessman and the bureaucrat, and assumed that
the two would take advantage of any chance to join forces against the
public weal. The Park Service did not escape this suspicion. There was the
feeling among environmentalists that the Park Service was a typical public
agency, interested primarily in its own aggrandizement, and that it could
be counted on to honorably discharge its responsibilities only if it served
ends of convenience, expansion, or security. An environmental activist
explained Park Service opposition to the designation of a wilderness area
in one of the National Parks thus: "Wilderness means less management
and that means they can't expand their bureaucracy, something they, like
all bureaucrats, love to do." Another related view was that bureaucratic
opportunism had so imbued the Service that learning how to abandon
principles and bow to pressures were big, if covert, elements in Park
Service career training, so much so that in the words of a prominent
environmentalist, "when you get to superintendent level, you've had the
[moral] stuffing knocked out of you."

As the old conservation movement was infused by new leadership and
expanded its membership, it became more militant in its tactics. In part
this was because the new generation of leaders was more inclined toward
confrontation and in part because the groups felt compelled to keep up
membership interest with a steady diet of conflicts, skirmishes, and victories. The language of their newsletters and bulletins seemed more like
communiques from the Western Front than messages about litigation and
lobbying efforts. Symptomatic of this changed attitude was the way the
National Parks Association, because it tended to be nonconfrontational in
its relationship with the Park Service, came to be viewed by its fellow
public interest groups. It was considered both old fashioned and ineffectual, an Uncle Tom among environmental organizations. The judiciary
became the public interest movement's favorite branch of government,[44]
which was not surprising, given its view of bureaucracy. The environmental movement shared this liking for the courts, and the federal land
management agencies, including the Park Service, were frequently bought
before a judge by the environmental groups.

Although a general inclination to distrust public bureaucracies was a
characteristic of the entire public interest movement and neither the Forest
Service nor the other conservation agencies escaped the aggressive tactics
it prompted, those concerned with the National Park System saw special
cause for distrusting the Park Service. In their eyes, the Service had used
Mission 66 to ensure its own political security at the expense of both the
Park System entrusted to its care and the values it was founded to uphold.
With its ambitious development plans, the Park Service had pandered to
the lowest common denominator of the public taste. In an effort to court
popular favor, it had closed its eyes to accumulating environmental prob-

lems. There was the feeling that long after Mission 66 ended the program still reflected the deepest and truest inclinations of the Park Service, and, if given a chance, the agency would revert to Mission 66-style policies. One environmental activist told the author that "the Park Service never abandoned its big facility, [Robert] Moses approach. When the pressure is on, they'll come out in favor of wilderness, but they'll back off later when the pressure subsides."

These two inheritances—one a set of attitudes taken from the public interest movement, and the other a set of lessons learned from recent Park Service behavior, led these groups to see themselves, rather than the Park Service, as the best judges of what was right for the National Park System. An informal coalition was formed by several groups, including the Sierra Club, Wilderness Society, National Parks and Conservation Association, Izaak Walton League, and, later, Friends of the Earth. Representatives met periodically in Washington to hammer out common positions on national park matters and to act, as much as possible, as a policy directorate for the National Park System. Acting together, the environmental groups have frequently side-stepped the Service on issues, going to the courts or sometimes even directly to Congress to get what they want for the National Park System.[45]

In sum, the environmental groups have come a long way from the days of amicable alliance with the Park Service. Since the early 1960s, the agency has faced powerful, self-assured organizations which, backed by the weight of public opinion, have chosen as a modus operandi conflict and litigation as frequently as quiet persuasion.

Elite Support

The 1960s and, to a greater extent, the 1970s, saw the agency lose much of the active support of the nation's business and professional elite, support present at the founding of the Service and carefully nurtured by its directors from Mather through Hartzog.

During the first twenty years of the Park Service, the relationships between the agency and those prominent citizens interested in its welfare were based largely on the relationships between Mather and his friends and peers: eminent financiers, scientists, journalists, publishers, etc. In the 1930s, a more formal element was introduced into the agency's relationship with the nation's elite when the Advisory Board on National Parks, Historic Sites, Buildings and Monuments was established. The board was to advise the secretary of interior, who, under advisement from the director of the Park Service, appointed its members for indefinite terms. The first board appointees were a prominent lot, many of whom had been Mather's friends and most of whom had been previously active in national park affairs. Through Wirth's directorship, the board remained stocked with powerful men actively interested in the national parks.

The advisory board tended to be conservative on standards of national significance, and therefore it occasionally objected to new parks which it felt did not measure up.[46] It also differed occasionally with the more politically aggressive directors like Wirth on park development plans. In reality, however, it was more often a source of support than advice for the agency. Because both the board and Park Service leadership had a very slow turnover before the 1960s, the two maintained a close working relationship. Since the board was a departmental board, technically attached to the office of the secretary of interior rather than the Park Service, it had the appearance of detachment when it advised on national park matters. This made the board very useful to the agency leadership in its dealings with Congress. If the Service did something which aroused Congress's objection, it could protest that it was simply following the advice of the advisory board. The board, working closely with agency leadership, would go along with the story.[47] It was a prestigious "bucket of fog," to use the words of a former agency administrator, convenient when needed to obscure agency leadership's true motives and intentions.

Erosion of Power

The board's formal attachment to the secretary of interior, while a source of its strength, was also one of the causes of its diminution as a useful instrument of Park Service leadership. Although secretaries of interior had always appointed the board's members, they had relied on the advice of the Park Service in doing so. Perhaps Udall saw ending the close relationship between the board and the Service as a way of undercutting agency independence, or perhaps the administration saw seats on the board as plums to be given to political allies. In any case, he broke this tradition by appointing a prominent union official (with no prior connections with the Park Service) to the board, and he did so without consulting the agency. This appointment in and of itself did not damage the board or reduce its usefulness since, according to a top Park Service administrator who worked closely with the board, its members gave the labor leader "a crash education in being a national park statesman." But the precedent was damaging and, in the words of the same administrator, "under Nixon there were just too many political appointees who weren't very educable."

Since Nixon, the tradition of political appointments to the board has continued. As a result, in the words of another administrator who worked closely with the board, "today's members just don't carry much weight,"[48] and the board has lost the respect of Congress. Although Hartzog managed to use the board as effectively as altered circumstances would allow, agency directors who followed him were often clumsy in their relationship with the board, which both alienated it and further undercut its usefulness. Walker, Nixon's politically appointed director, was never comfortable with the board, and vice versa. The board members did not like him very much, and he ignored them. Walker's successor Everhardt appeared unsure of himself in the world of Washington politics, and his inability to make the most of what remained of the board's power was but a small piece of a larger picture. According to one top administrator, Everhardt did not understand the importance of leaving it attached to the secretary, so he arranged to have it attached directly to the Park Service. By removing the appearance of impartiality it had when it was a departmental rather than an agency board, he destroyed what little remained of its credibility and power.

While the board was withering as a source of support, agency leadership failed to go beyond it and reestablish the powerful but informal ties to the elite that characterized the Mather era. There were undoubtedly several reasons for this failure. For one, after Drury, none of the directors were of the elite themselves. One might argue that by virtue of their high positions within the federal bureaucracy, Park Service directors were ex officio members of the nation's narrow pinnacle of power. However, they were not viewed as true peers by the agency's elite supporters if they were career Park Service men, and even less so if they were political appointees. Nor is it likely that the later directors felt an easy relationship born of social equality with those men who traditionally were so important to the Service.[49] This social gap must have been a strong inhibiting factor in establishing close informal relationships to compensate for the weakening of the advisory board. But whatever were the reasons, there was a sense within the agency of an opportunity missed. An agency administrator who had worked with Laurance Rockefeller on several national park projects asserted that "the Service had a golden opportunity to use the tremendous resources of the third generation Rockefellers the way Mather and Albright used the second, but it didn't. [These Rockefellers] were willing to help [the Service], but they were never asked."

Even if the recent directors had asked for the kind of elite support that characterized the Mather era, it is not certain they would have received it. Things were different in the 1960s and 1970s. The pervasiveness of environmentalism meant that many of the nation's social and professional elite were probably more in sympathy with the views of the environmental movement on national park policy than with those of agency leadership. Therefore if they became involved with the national parks, it most likely would have been as supporters of at least the more moderate of the environmental groups. Also, parks had increasing competition from museums, orchestras, and other claimants for funds as objects of elite support. Whether parks were intrinsically poor competitors, or Park Service leadership competed poorly, the fact remains that other institutions like the National Endowment for the Arts and the National Trust for Historic Preservation were more successful in obtaining elite support in the 1960s and 1970s.[50]

Thus, we see that the Park Service lost two important elements of its extragovernment support, and when combined with other events which were taking place, this profoundly affected Congress's relationship with the agency and its charge, the National Park System.

CONGRESS AND THE NATIONAL PARKS

The Shifting Balance of Power

Congress is potentially the most important organization in the life of a federal agency because it has the greatest power to advance or reverse agency fortunes. This power has two distinct sources. First, agencies derive their basic powers from laws enacted by Congress. Second, Congress controls an agency's purse strings through the appropriations process, and no matter how broad its formal authority, an agency's real power rests ultimately upon its fiscal resources.[51] This constitutionally granted power is reinforced by established practice, which accords Congress a role in determining political values as well as in passing laws. In the American system of government, the legislature is generally seen as the principal organization for setting the values by which the nation is governed.[52] Congress's claims to power were also reinforced by political theories which encouraged it to enter into the details of administrative policy making.[53]

In practice, however, legislators seldom wielded such power over public administrators. There were practical limits to their capacity to exercise such control. Above all, Congress usually lacked the expertise of the agencies and had only a limited amount of time and energy to devote to agency oversight.[54] The result was that into the middle decades of this century there was a de facto balance of power, with Congress expressing its will only in the most general terms, and leaving the specifics of interpretation up to agency administrators. Out of this evolved a system of interdependence and reciprocity in which legislative authority was balanced by administrative experience and professional competence.

In the early days of the Park Service's relations with Congress, the agency's hand was a strong one for several reasons. The Antiquities Act of 1906 gave presidents the power to set aside sections of the public domain as national monuments through a simple proclamation. Before the Second World War, presidents were not adverse to using this power, so agency leadership had at their disposal a means of expansion which circumvented Congress entirely. Second, before the era of large congressional staffs, the agency had a near monopoly of expertise on national park management. Congress might question the agency on professional matters, but it could seldom do so from a position of equal knowledge. Moreover, Congress was inclined to respect the Park Service's judgment. As discussed above, early leaders of the agency worked hard to cultivate congressional friendship and trust, and they were successful. The combined result of these factors was a great degree of agency initiative in its relationship with Congress, so much so that occasionally legislators felt their prerogatives were being infringed upon. For example, Representative Rich of Pennsylvania, speaking against the establishment of Everglades National Park in 1934, objected to what he saw as excessive administrative initiative: "It is time we woke up and used our own gray matter, and cease to follow some professor who is telling us to enact certain bills into law because some secretary wants it or some department head may have a fancy for the law." [55]

In 1954, Long suggested that the extreme view of legislative supremacy, which held agencies accountable to Congress for the minutest detail of administration, had been abandoned except as folklore and "political metaphysics" because it was so out of step with what was possible. [56] Nevertheless, when the means of establishing such supremacy and extending congressional control deep into agency operations became available, the folklore and political metaphysics remained to justify it. The growth of the congressional staff, which between 1950 and 1980 had tripled to more than 18,000 persons, provided some of the means. Staff members had the time to become experts in the operation of the agencies their committees or subcommittees oversaw, time which a busy congressman did not have. Congress now also had an expanded and upgraded Congressional Research Service and the increasingly professional General Accounting Office at its disposal. These new means of exercising oversight heightened expectations, and Congress came to expect more control over the federal bureaucracy. [57]

Two factors heightened this trend of congressional ascendancy with regard to the Park System specifically. First, the reluctance of recent presidents to make use of the 1906 Antiquities Act to declare new national monuments reduced its value to the Park Service. [58] Second, during the intragovernment conflicts of the early Nixon years, the Park Service was immobilized by White House orders that it not cooperate with Congress. This shifted much initiative on national park matters, especially on new park proposals, to the Interior committees of Congress and to their staffs. These factors combined to give Congress a greater sense of its right to closely oversee the Park System plus the experience to do so.

During the 1960s these trends were not yet fully developed, however. Hartzog was friendly with important congressmen, and his performances before committees were models of finesse.[59] He was trusted and liked in Congress, and his friends on the Hill stood up for him when he needed them. For example, McPhee writes that, "in 1969 when rumor spread that Hartzog was about to be replaced, congressmen and senators heated up in sufficient numbers to evaporate the rumor."[60] But the Service's relationship to Congress under Hartzog was profoundly different than it had been thirty years earlier. The balance of power had swung against the agency. Hartzog used his deferential humor and his sly wit to get whatever latitude he could get in what was basically a subordinate role. The semblance of parity which Hartzog's virtuoso appearances before committees produced was illusory. When Hartzog was removed by Nixon in 1972, the degree to which the fundamentals of the agency's relationship to Congress had changed became apparent as the latter came to exercise its perogatives to the fullest. The atmosphere of the handshake and verbal assurance which had characterized dealings with Congress under Hartzog was replaced by a stiff formality and frequently by ill-tempered grillings of the Service administrators by legislators. Furthermore, although those congressmen with whom Hartzog dealt frequently were his friends and fishing chums, Sidney Yates, a frosty ex-public prosecutor, headed the House Appropriations Subcommittee for Interior that Hartzog's successors had to face. Several things would determine how Congress would use its newfound power over the System and Service.

New Congressional Interest in Parks

As Congress increased its sway over the Park System, the environmental groups in turn increased their influence over Congress. With their large memberships and penchant for aggressive politics, the wishes of these groups were ignored by congressmen at their peril. A congressional staff member with long experience in environmental legislation put it succinctly: "Wilderness issues can make or break congressmen in elections." The unexpected 1972 primary defeat of the chairman of the house committee with national park jurisdiction, Wayne Aspinall, was largely at the hands of environmentalists who objected to how he handled park and wilderness matters.[61] It brought this lesson home to any who might have been slow learners.

After 1976, changes in lobbying laws lifted many of the restrictions on the political activities of public interest groups, further increasing the ability of environmental organizations to put pressure on congressmen.[62] Soon environmentalists were wielding both the carrots and the sticks of political influence. During election campaigns, some organizations would send workers into the districts of friendly representatives to bring out the constituency, a favor especially appreciated by those in tight races. Because environmentalism had become a national movement whose moral import affected congressmen along with everyone else, the environmentalists could sometimes effectively use appeals to idealism when appeals to political reason failed. For those impervious to entreaty, there was the well-publicized "dirty dozen" list, a rogues' gallery of the twelve representatives with the worst voting records on environmental issues.

In its dealings with the Park Service on legislative matters, the environmentalists held two strong cards. First, they had greater power to shape the issues. Second, although Congress usually views public administrators as more legitimate spokesmen for the public interest, when it came to the national parks, the Park Service was undercut by the widespread view, lingering from the Wirth era and the events surrounding the founding of BOR, that it was an overly aggressive, empire-building agency. As a result, the environmental organizations were often on an equal or even firmer moral footing than the Park Service when they went to Congress with their wishes for the National Park System.[63]

The Pork Barrel Park

Beyond any questions of environmentalism, however, Congress had good reasons to concern itself with national park matters and to disallow the opinions of the Park Service when doing so. National Park politics can be lofty and farsighted or it can smack of every man for himself.[64] The politics surrounding the establishment of many of the great national parks was certainly of the lofty variety. It derived its rhetoric from the broad movement of progressivism and fitted into a program of resource conservation which had many visionary—one might even say utopian—aspects. Yet, even while the great early parks like Yellowstone, Yosemite, and Sequoia were being established, Congress showed an appreciation of the pork barrel potential of national parks. As we have seen, Platt National Park was established through the efforts and log-rolling skills of Oklahoma's congressional delegation, which was intent on bringing the bacon home to local constituents. Mather and Albright tried to keep the pork barrel politics in connection with the Park System to a minimum by emphasizing high (and therefore exclusionary) park standards. By and large they were successful. Although they could not prevent an occasional pork barrel park from coming into the System, their usual control over their agency's relationship with Congress prevented it from happening frequently. Until recently, their successors had similar success in holding the line.

Through the 1970s, however, Congress, for several reasons, turned increasingly to the Park System as a source of local benefits for their constituents. First, there were changes in the way Congress itself worked. Reforms to democratize the House weakened the seniority system by eliminating many of the sanctions used by House leadership and committee chairmen to keep junior members in line. With party discipline weakened, other means had to be found to maintain party unity and to gather support for what congressional leadership viewed as key bills.[65] Thus, the value of all constituent benefit bills increased. They became a necessary lubricant for the machinery of Congress. Also, as identification with a particular party declined among the public at large and more voters viewed themselves as independents, being in good standing with the party lost importance as a factor in reelection. As a result, incumbents generally had to survive by their own wits.[66] They were on their own to persuade their constituents that they should be returned to office, and bringing home the bacon in the form of local benefits could do a lot of persuading.

While these changes were taking place within Congress, the great interest in outdoor recreation made recreation facilities an especially attractive distributive good. Outdoor recreation activities are so varied and broad that they can furnish something for almost everyone. As a result, "the promotion of outdoor recreation through governmental action [became] one of the more interesting of God's gifts to politicians."[67]

In the meantime, other important types of pork barrel benefits were losing their appeal. The large public works projects of the Army Corps of Engineers started running into serious grass-roots opposition in the late 1960s. At the same time, environmentalists stepped up their opposition to many of the Bureau of Reclamation's western irrigation schemes. When they began to encounter strong opposition, these projects became divisive issues rather than untarnished benefits, and many congressmen lost their enthusiasm for them. A Park Service administrator recalls that during this period, many congressmen switched their interest from river and harbor "improvements" in their districts to federally sponsored parks. Thus, as Congress increased its control over the Service and the Park System it managed, the national park became a highly prized distributive good.[68] By the early 1970s, both trends—increasing respect for the arguments and muscle of environmental groups and increased inclination to use parks as distributive goods—led Congress to promote the expansion of the Park System.

As the Park Service lost its exclusive moral authority over the System in the eyes of Congress, it also lost its ability to say "no" on what it considered unsuitable parks and to have its opinion accepted. The agency never shifted from its formal position that only the highest quality parks should be in the System. However, every time a park with less than sublime natural or historical features was added and the gap between the espoused standards and the actual system broadened, the appeal to quality became a little more hollow. When, during a year of fiscal retrenchment, $10 million was spent to build 10 miles of the very incomplete and marginally useful Natchez Trace Parkway in the home district of the chairman of the House Appropriations Committee, the System must have indeed seemed up for grabs.[69]

The agency itself contained mixed feelings about this pressure for expansion. In some quarters it met with resentment. There were those who saw rapid System growth as a Faustian bargain which involved too great a compromise of standards. "We'll become the federal government's surplus property dump" complained one agency official who bitterly resented what he saw as the abandonment of standards of quality for admission to the System. One person expressed his criticism this way: "The Park Service has become a servant of Congress in the worst sense. It has become Congress' flunky in carrying out its pork barrel chores while it is supposed to be the guardian of the national interest. Unfortunately the Service doesn't have the power to uphold that interest." An administrator volunteered that "if things don't change, we'll wind up with a National Park System by congressional donation." A law which required the director of the Park Service to submit annually to Congress a list of twelve candidates for admission to the system was viewed by some as unjustified and destructive to the System. It was disparagingly referred to within the agency as the "park-of-the-month plan."

Along with this resentment, however, was an attitude of accommodation. A Park Service administrator with long experience in congressional liaison said, "I've always been an expansionist because, among other things, it means going with the flow of political reality." One manifestation of this accommodation was an attitude which avoided questioning congressional wisdom in establishing parks but instead concentrated on giving them expert management once they were established. Reflecting this attitude, an agency administrator said, "Mt. Rushmore is a travesty but Congress put it in the System. [So] we did the best job possible with it. That's our approach to new units. It's one of the strengths of the Service."

This "can be" attitude was buttressed by a healthy respect for congressional capacity for retribution. Arousing legislative hostility is a sure way for an agency to get itself in trouble, and the Service administrators seemed well aware of the dangers.[70] When the author asked a high Park Service official why the agency didn't take a more aggressive stand in relation to Congress, he replied that "as hard as we hit they can always hit back harder." When the author suggested to another agency official that perhaps the Service could discourage the addition of sites of little merit by giving less than its complete attention to some of the unworthies already in the System, she replied that it would be a dangerous strategy: "No one would have the nerve to play that game, the risks of arousing congressional wrath are just too great." In fact, there were many paths to congressional wrath. When Congress dealt with national parks, its internal dynamics prompted distinctly different viewpoints and generated important conflicts over the future of the System.

Congressional Park Politics

The center of congressional interest in the national parks has traditionally been in committees responsible for overseeing the Department of Interior. Since their inception, these committees have been dominated by westerners who saw Interior as a "western" department. In the Senate, because representation is not dependent on population, the West is more heavily represented, and historically this made the Senate the more active chamber in matters concerning Interior.[71] In the 1970s, however, the initiative on national park matters shifted to the House for two reasons. First, the increasing appeal of parks as distributive goods was felt more by representatives who, with two-year terms, were under more pressure to keep up a flow of benefits to their constituents than were senators with their six-year terms. Second, the Senate committee responsible for national parks was reorganized in the 1970s and was given responsibility for overseeing energy policy. Once this happened, most of its time was taken up with high-visibility energy issues, leaving only enough time for an incidental interest in park matters.

While this shift of interest was taking place, changes in committee rules and procedures were greatly enhancing the power of the House subcommittees at the expense of the committees as a whole. The discretionary power of the committee chairman over subcommittees was reduced while subcommittees were given their own staffs. As for the effect of these changes on the House Interior committee, an observer of Congress in the mid-1970s pointed out that the Interior committee had become balkanized, with subcommittees taking on a life of their own.[72] The parks subcommittee, rather then the Interior committee as a whole, became the center of congressional action on the national parks.

With the seating of the 95th Congress, Representative Phillip Burton of California became the subcommittee chairman and in that position he became the key congressman on national park matters. Burton was as much in sympathy with the environmentalists as anyone in Congress. "When the environmental slobberers have seventy votes, I'm one of them," is the way he described his attitude toward environmental issues.[73] He subscribed to the environmentalist view that time was not on the side of the environment.[74] Burton was also an astute politician who realized that parks, because of their distributive value, were good bargaining chips and that therefore his subcommittee had control of a powerful political currency.

Whereas in earlier years the Interior committee had seen itself largely as an exclusionary gatekeeper to the National Park System, under Burton that role diminished and the subcommittee placed emphasis on getting new units into the System.[75] In 1976 Congress passed a bill at the subcommittee's prompting that established the park-of-the-month club.[76] This ensured a steady flow of park proposals for the subcommittee to act on, a flow amply supplemented by park proposals introduced on the House floor by individual congressmen.

The Question of Standards

While Burton's eye for political opportunities encouraged him toward an expansionist view of the System, it also encouraged him not to be especially mindful of standards of park quality. His environmentalist view of the future also inclined him toward a policy of rapid and relatively unhibited expansion of the system. If time were the enemy and development a malevolent juggernaut, then arguing about whether proposed parks measured up to the highest standards would be the height of foolishness. Better to get them safely into the System now, and worry about intrinsic merits later.[77]

Burton's chief method of adding new parks was to bring many park proposals together in large omnibus bills, thus ensuring sufficient support for the passage of the entire batch.[78] In principle, this log rolling in connection with national parks was nothing new. Ise writes of such deals in the 1930s, for example, when the supporters of a park in the Great Smoky Mountains and the supporters of one in the Shenandoahs joined forces and made a united fight for both the parks. Two things were new with Burton's log rolling, however. One was the scale at which he operated. His omnibus bill of 1978 established a dozen parks, expanded twice that number, and raised spending ceilings on almost three dozen others.[79] A second was his inclination to disregard the opinions of the Park Service when putting together his omnibus bills. An environmental activist who worked with Burton said of him, "He really doesn't give much of a damn about what the Park Service thinks. I guess if they screamed about something he might back off, but they'd really have to scream."

Among the reasons for this disregard was undoubtedly the fact that Burton relied strongly on the environmental groups for advice on national park matters. But also Burton and agency leadership were not always in agreement on the basic wisdom of such ambitious legislative constructs as the omnibus bills. Burton viewed the inclusion of clearly inferior units in the omnibus bills as a necessary evil to ensure their passage and thus get the meritorious units. Park Service leadership feared that the precedent established by the wholesale entry of units of little value might outweigh the benefits of the bills. There was little the agency could do, however;

Burton had much power in the House and the support of the environmental groups, with whom he worked closely. He was, in the words of one administrator, "the tiger that the Service rides."

The Politics of Accommodation

Although the House subcommittee on parks could mark up bills that authorized the establishment of new parks (and usually see that they were passed on the House floor), authorization in and of itself could not ensure the actual establishment of parks. Money was needed for purchase of the land and the development of facilities, and the appropriations committees of the Senate and, more important, of the House had control over the money. Congress operates under two conflicting demands.[80] One is general demand of the American public for efficient, even frugal, government. The other is a composite demand, made up of thousands of specific requests for programs and benefits—all of which cost money. The principal means by which the legislature handles this conflict is a balanced advocacy system: legislative committees (and their subcommittees) tend to support increased spending in their fields of oversight while appropriations subcommittees (whose recommendations are seldom challenged by the full appropriations committee) become advocates of spending restraint and will support programs only to the degree that they conform to their notions of fiscal responsibility and efficient government operations. Chairmen of appropriations subcommittees are usually senior House members and they usually hold safe seats. Because their reelection prospects are favorable, these members can afford to resist pressure from interest groups.[81]

As a result of these differences, conflict between the two committees, with an agency caught in the middle, is an ever-present possibility. Fortunately, there are congressional norms which bank the fires of conflict, and perhaps the most important is the willingness of the legislative and appropriations committees to concede a legitimacy to each other's goals and to respect each other's domains. In the House there normally exists a mutual recognition that the appropriations committee should not define legislative programs in an appropriation bill and that the legislative committee will accept the level of spending set by the appropriations committee.[82] Another thing which acts to dampen conflict is the fact that both appropriations and legislative committees are sensitive to the will of the House as a whole. Usually the House is of two minds on the matter of program expenditures versus frugality; that is to say, it wants both, and it exists within a balance of tension between its two desires. Occasionally, however, the demand for programs in a certain area will overwhelm its impulses toward fiscal restraint. Conversely, at times demand for general spending cuts will create a climate unfavorable to program expansion. When the former is the case, appropriations committees will tend to relax their restraining role, as they did during the days of the Great Society legislation.[83] When budget cutting becomes very important, as it did for Reagan's 97th Congress, the legislative committees trim their sails. As a result of this interplay of forces and expectations, accommodation between the legislative and appropriation committees is usually reached.

While the House Interior subcommittee responsible for park authorization played the role of conservative gatekeeper to the Park System, its dominant values were largely compatible with those of the Interior subcommittee of the House Appropriations Committee and there was little conflict between the two.[84] When the parks subcommittee abandoned its conservativism in the 1970s, however, and made adding new parks its primary business, the goals of the two subcommittees became much less compatible. Unlike the Great Society programs of the previous decade, these new authorizations were not accompanied by a collective sense of the House that their importance outweighed all considerations of fiscal restraint no matter how much individual legislators might have wanted them. If anything, the contrary was true; the dominant rhetoric of the Carter administration and the 95th and 96th Congresses was that of economy. Moreover, many of the units being authorized were clearly not of top caliber and were blatantly the result of pork barrel politics. Thus, the increased appropriations demands of the expanding system went against the norms of the whole appropriations process. They also went against the personal inclinations of Congressmen Yates, chairman of the House Interior Appropriations subcommittee. According to a former Park Service director, "Yates had a real statesman's sense of priorities, he tried to fend off pork in his subcommittee." Furthermore, the split between the philosophies of the two committees was heightened by what an observer called "a chemistry problem" between their two chairmen, Burton and Yates.

Burton's subcommittee assumed that by adding more parks, a broad enough congressional constituency for the Park System would be created to force the appropriation subcommittee into more generosity. This assumption was not to be borne out by events. Congressmen who successfully pushed for park authorization found the appropriation process less responsive to their demands. For one, Yates was well insulated from any importuning. In the words of an agency administrator, "He's in a safe district, he's 70, and he's his own man. [As a result] all sorts of lobbying pressure just rolls off his back." Second, park appropriations were usually marked up in large heterogeneous appropriation bills in which funds for many programs in many different budget areas were voted on together. This meant that legislators seldom got the same clear vote on park appropriations that they got on park authorizations. As a result of this conflict, by the early 1980s a backlog of well over a billion dollars in unfunded acquisitions had accumulated.

We see, then, that the agency's loss of control over its relationship with Congress had serious consequences. It meant the loss of much of its control over the gate to the Park System and the new congressional gate-keepers chose to exercise their power to expand the System with less of an eye toward standards and more of an inclination to use the System as a repository for threatened areas and as a source of distributive goods. This, in turn, provoked the hostility (and parsimony) of the Interior appropriations subcommittee, which became reluctant to fund the expanding System. Thus the System found itself in a cross-fire between two antagonistic elements of Congress, neither of which much respected the opinions or interest of its manager, the National Park Service.

THE PARK SERVICE WITHIN THE EXECUTIVE BRANCH

Prior to Udall's term as secretary of interior, the Park Service maintained a considerable degree of independence within the executive branch, although this did not guarantee that its wishes would always be respected.[85] Its good relations with Congress helped it maintain that independence, as did the long tenure of its directors, who were considered professional administrators rather than political appointees. Wirth served for twelve years, Drury eleven, and Cammerer for seven before that. On the other hand, except for Ickes, the terms of the secretaries and assistant secretaries they served were short; Wirth and Drury each saw a succession of bosses come and go.[86] Agency independence was further strengthened by its sources of support in the conservation movement and by the way it maintained an image of professional competence in an era when the respect for professionalism built by progressivism had not yet been completely eroded by time and politicians sniping at "the bureaucracy."

Shifts in Control

In the years which followed Wirth's retirement, several changes encouraged Interior leadership, and sometimes the executive office itself, to exert greater influence over park policies. First, views of the proper exercise of power had changed. In previous eras, respect for professionalism would have made a politically appointed cabinet head hesitant to become too blatantly involved in the charge of a department with a strong reputation for professionalism. Within the past few decades, however, there has been a quest for broader perspectives, and what was once viewed as specialized professional competence often came to be seen as narrowness of vision.[87] Udall certainly held this view of the Park Service, and he forcefully expressed it through his proxy, Undersecretary Carver, at the Yellowstone meeting of 1964. Udall saw the Park Service responding not to broad public needs but rather to its own narrowly defined self-interest.

An obvious solution to the problem of agency self-interest and narrowness of vision was to remove power from the parochial agencies, and move it upward so that it was centered in the hands of departmental executives, where the interests of wider segments of the public could be taken into account in making policy choices.[88] This was exactly what Udall did when he forced Wirth's retirement. As Udall viewed it, real public service would require that the Park System be under close scrutiny of departmental leadership, and he remained in office for two full presidential terms, until 1968; thus he had the chance to see that his wishes were put into practice.

In the 1970s, Interior's close supervision continued. In 1972, Nixon sacked Udall's appointee, George Hartzog, and replaced him with Ronald Walker, a White House staff aide who knew nothing about either agency administration or national parks.[89] When this happened, Interior was forced to take an even more direct hand in national park matters, and since the secretary was occupied by more pressing business, the burden fell to the assistant secretary in charge of national parks, Nathaniel Reed. Reed soon found himself performing by default many of the functions normally left to the agency head. In supervising the agency, Reed frequently ignored the inexperienced director and went directly to agency offices for information and advice. He also made his wishes known to agency administrators directly, thereby bypassing agency leadership. Whereas either the agency director or assistant director customarily appeared before Congress to defend agency policies,[90] Reed often took this task upon himself. By background and inclination, Reed was an environmentalist and he took a personal hand in ensuring that the agency managed the Park System in accord with the Wilderness Act and the National Environmental Policy Act (NEPA).[91]

Yet another factor came to encourage the upward drift of agency leadership during the Nixon era—the disposition of Alaskan public lands. How the permanent status of the vast public domain in Alaska was settled is a long and complicated story going back to the forty-ninth state's admission to the union. We need not go into it here. Suffice to say that in the 1970s plans were being drawn up to carve vast new national forests, parks, monuments, and game refuges out of the domain. Since two different departments, Agriculture and Interior, and many separate agencies were involved in overall planning, decision making took place at a high level within the departments. Thus, the Park Service found itself a bit player in a process which was of major importance to the future of the System in its charge.

When Carter succeeded Ford, Robert Herbst replaced Reed as the assistant secretary responsible for the National Park System. Like Reed, Herbst was an environmentalist. He had been director of the Izaak Walton League, and as head of the Minnesota Department of Natural Resources he had won the praise of the state's environmental groups. His background inclined him to view his role as that of the spokesman for environmental values within Interior and, if necessary, as the enforcer of those values in the agencies under his charge.[92] Like Reed, Herbst served under a secretary (Cecil Andrus) who was generally sympathetic to environmental causes, but whose broad range of responsibilities, including increasingly insistent questions about mineral and energy resources on the public domain, left him with little time to be concerned about the National Park System. Finally Herbst, like Reed, took responsibility for a Park Service which was under a weak director. Walker had been replaced by Everhardt who, although he had long experience in the Service, appeared to many to be out of his depth as its director. Thus, many of the factors that drew Reed into an active role in national park matters were present when Herbst replaced him. But there was now one more factor, the demonstration effect. According to an agency administrator who worked with both assistant secretaries, "Reed ran the Park Service with Walker as director and when Herbst came in he saw this and decided to do the same."

Partly to ensure a strong hand in park affairs, Herbst installed as Park Service director a young service administrator with ideas very similar to his own but who lacked an independent base of support within the agency. The two, Assistant Secretary Herbst and Director William Whalen, formed a captain and executive officer relationship, with Herbst handling overall policy and external relations, including many committee appearances, while Whalen tended to day-to-day System management. Herbst himself saw his relationship to agency leadership as one of policymaker to policy executor.[93]

True to his background, Herbst worked closely with his friends in the environmental movement.[94] Part of this close working relationship included giving the environmentalists informal veto power over certain agency decisions. For example, the environmentalists objected to the Park Service's first choice for its representative to the interagency Alaska planning group, and by doing so they successfully blocked the nomination.

The rapid tempo of national park authorization by Congress under Representative Burton, like the ongoing Alaska lands question, seemed to be too important a matter to be left to the Park Service leadership.[95] Herbst dealt directly with Burton on park legislation and, more important, he worked with him rather than against him. Like Burton and most of the leaders of the environmental movement, his views of the Park System were expansionist; he took a broad view of the types of units which should be included in it and he was little concerned with conventional national park standards. Putting areas in the System was a good way to save them from development. In his opinion, the widening rift between the park authorization and the appropriations subcommittees was neither a serious nor a long-term problem. He subscribed to the view that as new park authorizations built a large enough demand for appropriations in the House, the problem would take care of itself.

Finally, there is the executive office itself to consider. Its Office of Management and Budget (OMB) also became an important player in national park matters in the 1970s. Since the 1920s, there has always been a unit within the executive office responsible for reviewing agency budgets with an eye toward keeping them lean and compatible with administration priorities. Until 1970, the Bureau of the Budget had this role. Its strength relative to that of the departments, with their independent constituencies, was weak though, and the budget bureau could not bring much coordination to the executive branch. In 1970, however, Nixon reorganized the executive office. He created the powerful OMB and filled its ranks with trusted supporters and administration inside men. The OMB, with its unqualified loyalty to the president, was to be a useful instrument toward two goals important to Nixon—control of the budget and control of the agencies themselves.[96] Its director, Roy Ash, became part of Nixon's inner cabinet of trusted officials. Meanwhile, politically endorsed appointees recruited by the deputy director of OMB and operating under his general supervision were placed in the budget sections of the cabinet level departments in order to ensure their cooperation with OMB.[97]

As a result, during the Nixon era, the Park Service, like many agencies, lost much control over its budget to the newly powerful OMB, whose examiners were in a position to closely supervise each step in the annual budget cycle and thereby enforce compliance with administration wishes. With Nixon's OMB so closely involved in its budget process, the Service had little chance of asserting independence on policy matters which had budgetary impacts—which was to say almost everything. After Nixon's resignation, OMB's size, intimate involvement in the budget process, and organizational proximity to the president guaranteed it considerable residual powers over the agencies, and the OMB budget examiner for the Park Service remained an important actor in national park affairs.

Thus, the past two decades have seen considerable doubt about the Service's traditional goals and considerable erosion of the strength and independence of agency leadership when it came to dealing with outsiders in its policy environment. Both of these trends had an impact on the internal dynamics of the agency.

SEARCHING FOR NEW ROLES

Internal Tensions

Scholars have noted that human organizations exist in a state of internal tension arising from differences over organizational goals to be pursued and the means of pursuing them, over clients to be served and environmental signals to be heeded.[98] But just as centrifugal forces give rise to unique perspectives, encourage the pursuit of parochial interests, and engender conflict, organizations as a whole have the means of forcing their subunits into at least a working coalition, if not a team. Most obviously, there is the hierarchical chain of command in formal organizations. It ensures that responsibility for the pursuit of broad goals is linked with the authority to direct the activities of those pursuing narrower subgoals. Conformity to broad organizational goals is also encouraged, however, by a knowledge that one's own occupational well-being is usually linked to the general success of one's organization.[99]

Any organization is likely to possess both centrifugal and centripetal forces, and exist in a state of internal tension as a result. Because the strengths of these pulls depend on circumstances unique to each organization, the balance differs from organization to organization. Since the environments of organizations are dynamic, over time the balance also changes within an organization. When centrality is ascendent, the ability of incumbents to shape their jobs according to their own inclinations and perspectives will be reduced, the activities of different subunits will be kept in accord with the wishes of leadership, and administrators will be encouraged to assign their primary loyalties to their organization and not to their external contacts. One would expect exactly the reverse where centrifugal forces are ascendant.

The events of the 1960s and 1970s appear to have heightened the centrifugal tendencies within the Park Service and to have reduced the agency's capacity to act in a unified manner. First, the goals pursued by the Service came to be viewed by many of its personnel as the wrong ones and therefore in need of changing, while those advocated by the agency's external critics in the environmental movement or those committed to the cities seemed more correct. Second, as agency leadership lost dominance in its relationship with Congress and came under increasingly restrictive supervision by Interior and OMB, it lost its exclusive control over internal rewards and sanctions and could no longer use them to enforce conformity within the ranks.[100]

The Park Service had always been part of the networks formed around environmental issues.[101] Agency employees were frequently active members of environmental organizations, and the flow of personnel from the leadership ranks of the environmental groups into administrative positions within the Park Service had always been a strong one. Newton Drury's move from the directorship of the Save-the-Redwoods League to the directorship of the Park Service is but the most prominent example of what was a fairly common move. When the goals of the environmental groups and agency leadership conformed, being linked to the environmental groups in issue networks was a source of strength for the agency, as it clearly was during the Echo Park controversy. However, when the values of the environmental community and those of agency leadership diverged, the agency employee who was also an environmental activist had conflicts of loyalty. Divergence took place under Wirth when the environmental groups adopted increasingly strict preservation norms while the Park Service embarked on Mission 66. The divergence continued under Hartzog. These two directors were very much in control of their organization, however, and both had the ability and inclination to enforce conformity with leadership goals. One service administrator recalled, "if Hartzog found out someone here in D.C. was cutting a deal behind his back, that someone would wind up cleaning out garbage pails in a small park in the middle of nowhere."

The values of the environmentalists became more socially dominant, however, and there was a widespread acceptance of environmentalism by the agency's rank and file. According to an agency professional with many years of service, "environmentalism in the late 1960s and 1970s buried the Mission 66 spirit, it happened to me personally and throughout the agency." Once this happened, enforcing conformity with leadership goals when they differed from environmental values became a more difficult matter. The cost of inducing conformity against prevailing social values is high since pure coercion must be used and even the most powerful authorities have only a finite stock of this kind of power available to them.[102] As a result, Service directors after Hartzog were not able to ensure that the primary loyalty (in thought or action) of those employees with close contact with the environmental groups was to the Service's policies.[103]

Shift in Authority

Another cause of centrifugality stemmed from the fact that during the period of weak directors beginning with Walker, more autonomy went to agency subunits. Once routines of autonomous operation and interaction were established below leadership level, especially in the regional offices and in units responsible for congressional liaison, it was very difficult for directors to reassert their authority. The legislative units could and did use their contacts in Congress to defend their autonomy within the agency's Washington headquarters, while regional directors and the superintendents of major parks established independent power bases through their dealings with local congressmen. Because of increased congressional dominance of agency affairs, it became a relatively easy strategy to pursue and a favored one of ambitious policy entrepreneurs within the Park Service.[104] The temptation for an administrator to strike out on his own was encouraged both by the tradition of bureaucratic entrepreneurship within the federal government and by the recent infatuation with "organizational democracy," which encouraged each member of an organization to let his voice be heard on even the most basic of policy questions.[105]

There was considerable willingness on the part of congressmen to act as independent sources of support for policy entrepreneurs within the Service, especially for those advocating a stricter adherence to environmental principles or a redirecting of the Park System toward serving urban populations. Occasionally congressmen were even willing to back these advocates against agency leadership when it tried to suppress their activities. One agency administrator who was vocal in his advocacy of an increased urban role for the System and who enjoyed the support of a powerful congressman operated independently of the wishes of agency leadership for years. Of his relationship to agency leadership he told the author, "I'm both useful and dangerous to them. I'm dangerous because of my political connections but I've used them to get more beneficial bills passed than anyone else." The administrator was one of the important actors in the recent congressional authorization of a major new urban park, a park that was in fact authorized in spite of the misgivings of agency leadership about its value.

According to agency sources, similar entrepreneurship occurred in the regional offices, where congressmen and regional Park Service administrators, or even officials within the parks themselves, collaborated on new park authorizations. Together, the legislator and agency administrator would hammer out the details of the plan, quietly gather local support, and devise a legislative strategy—all without informing Service leadership. According to an agency administrator in Washington who was responsible for new park legislation, "our job was to anticipate that kind of legislation, which could come at [us] from any direction."

Modern Demands

We have seen, then, that in the past two decades the Park Service experienced a loss of steering capacity. The organization was no longer wholly capable of controlling its own behavior or fortunes because those who formerly sustained the agency in the pursuit of its goals increasingly imposed their own wills on it. This loss can be partly traced to the widespread sense that the Service had been pursuing the wrong goals; ones which were irrelevant to the agency's supporters or which were outdated from the broad perspective of societal wants and needs. Udall made this point when he established the Bureau of Outdoor Recreation and forced Wirth's resignation. Environmental groups implied it when they criticized the agency for what they saw as its lack of commitment to environmental principles in its management policies. The message was between the lines but obvious when the National Parks for the Future study group said "The nation has changed radically since the Park Service was founded. We are faced with different needs, and the Service can and should respond to these needs." [106] In 1972, the year Hartzog left, the National Parks for the Future Study Group told the agency that "bold new policy directions" were needed if the agency was to be of service to modern American society. [107] Udall had said the same thing when he appointed Hartzog in 1964.

Hartzog's successors, Walker and Everhardt, were, like Hartzog, concerned with restructuring the agency's mission in conformity with modern demands. Under Whalen, the efforts to adjust became intense. He felt that earlier agency leadership missed out on the major changes in values that swept American society in the 1960s. As a result, the Park Service was "one of the last vestiges of the old order, totally out of step with the times." [108] Whalen subscribed to the view, widely held outside the agency, that the Service had fallen prey to bureaucratic hardening of the arteries and that all its problems stemmed from this key fact. [109] Accordingly, his biggest concern was to bring the agency in line with modern times, socially, economically, and environmentally.

This then was the central task which occupied, and sometimes even obsessed, Park Service leadership in the modern era. If it could understand exactly what the modern era demanded from the Park System and Service, its problems of internal control and external dependence would be mitigated and both the parks and their keepers would be restored to their former high places in American life. The following four chapters examine how the agency has grappled with this problem in the various areas of its responsibility and what the results have been.

4 Nature Policy

Great Natural Areas as the Base of the System

If there is one charge wholly unquestioned by Park Service personnel or the agency's critics, it is the agency's responsibility for the management of America's great natural areas, the several dozen national parks commonly referred to as the "crown jewels" of the National Park System (figure 4-1). Russell Dickenson, agency director during the last year of the Carter administration and then under Reagan, instituted a review of the units of the National Park System for their national significance as one of his first official acts. He dismissed any need to examine the crown jewels, however; they were beyond question.[1] Robert Cahn, a journalist who has covered the National Park System extensively, recently wrote that the great natural areas "remain the keystone of the National Park System."[2] A Park Service administrator at agency headquarters in Washington expressed this sentiment forcefully: "The great national parks are what the Service is all about; the rest is just management chores."[3] This view is certainly not a new one. Ise wrote that ever since the 1930s when the Park Service greatly expanded its range of responsibilities, many "considered everything but large natural areas to be merely side issues for the National Park Service."[4] Even Secretary Herbst, who under Carter pushed the Service toward new urban commitments and encouraged the rapid addition of historical areas, could say that although many agency responsibilities were undertaken in response to "changes in society as it evolves, the core of the National Park Service's mission is the crown jewels, this is absolutely timeless."[5]

Figure 4-1.
Glacier National Park showing upper St. Mary's Lake and Mahtatopa, Little
Cheese Citadel, and Fusillade mountains.

Not surprisingly, those most involved with the urban national parks, the
System's historic sites, or the Service's forays into regional planning do
not always share this view. As we shall see, Service historians frequently
stress that preserving the historical is equally as important as preserving
the natural. Similarly, those involved with urban parks are likely to consider serving urban areas as important as preserving natural resources.
Even those outside the natural wing of the Service, however, concede at
least a parity to the great natural areas; that they are as important as
anything else in the System, and that managing them is as important as
anything else the Park Service does.

As Mather and Albright intended, the Park Service and the Park System
are most closely identified by the public with the great natural areas of the
West. Indeed, this view seems so pervasive that there is often a clearing-
the-record quality about statements which aim at educating outsiders about
the range of agency responsibilities or about the true composition of the
National Park System. For example, a ranger said in a published interview,
"We find that over half of our Park Service total employment is east of the
Mississippi. You'd never think so, but that's the fact."[6] Thus, unlike many
other things the agency is involved in, there are no questions of purpose
associated with its management of the nation's great natural areas. They
are at the center of the agency's image and everyone concedes that their
care is a prime agency responsibility.

While this issue is and always has been closed, questions about exactly how the agency should define and treat the great natural areas are very much open. Moreover, these definition and management questions go to the heart of the System's identity and the Service's sense of mission. However, there are really two separate but related issues here. The first is that of values to be emphasized in managing natural areas already within the System. The second concerns the selection of new units and, by implication, the qualities which make a natural area outstanding and worthy of inclusion in the System.

Management Questions

Questions about managing natural areas in the System and about adding new areas are as old as the National Parks themselves. To what degree should management stress the strict preservation of nature in the parks? To what degree should the agency accommodate and encourage expanded use of the parks through the construction of roads, accommodations, interpretative facilities, and so forth? Although Mather and Albright were extremely successful in working out a niche for their young agency which included a range of recognized responsibilities, a network of support, and a high level of public recognition, they never left their successors a clear policy guide on park use versus preservation.

As we have noted above, Mather and Albright, who were usually in close agreement, differed completely on a proposal to build a tramway across Grand Canyon. Albright thought it would improve the visitor's park experience; Mather thought it would demean a natural wonder. We see evidence of uncertainty and debate on similar questions throughout the history of the Service. For example, although the Park Service's founding act of 1916 authorized it to remove whatever trees were necessary to build accommodations and to keep vistas open, in 1934 the superintendent of Yosemite National Park was censured by headquarters for carrying out a vista-clearing operation which diminished the primeval quality of the park.[7] If proper behavior was unclear to field personnel facing immediate management decisions, it was equally perplexing to those in Washington responsible for drafting field guidelines. A former Department of Interior administrator recalls of Park Service leadership, "they were forever kicking around use versus preservation in the 1940s and it was nothing new then."[8] In fact, Stratton and Sirotkin suggest that during the 1940s, the Service's lack of resolution in the face of the Echo Park dam proposal was due in part to its failure to settle for itself the proper balance between wilderness and recreation values.[9] Thus it always seems to have been difficult for Park Service administrators faced with a use versus preservation question to know what was the politically proper decision for the

Service or, indeed, what was the right decision for the parks in a deeper sense.

Before the early 1960s, there was a common thread in the agency's handling of use versus preservation questions—a sense that all possible policies were firmly bounded by two agency responsibilities. On one side was the Service's responsibility to make the crown jewels and other natural areas of the System as accessible as possible to the bulk of the American public. As Albright made clear in his autobiography, this responsibility was to be defined liberally and meant accommodating a wide range of preferences in activities, lodgings, and entertainment. On the other side was a responsibility for preserving the parks from undue alteration and degradation. One could not push policy too far toward either use or preservation without engendering resistance from within agency ranks or from its supporters without, and this ensured that a rough mix of use and preservation values usually prevailed in decisions.[10]

Even Director Wirth's Mission 66 program, generally considered the highwater mark of user-oriented park policy, contained such preservation elements as phasing out the crowded overnight accommodations in Yosemite National Park's Yosemite Valley and transferring them to the less environmentally sensitive Big Meadow area. In the early and mid-1960s, however, preservation took on new meanings and the goals of public access were redefined. This produced marked changes in the pressures brought to bear on the agency, as well as in its own referents for preservation in System management.

A New View of the Parks

In 1964, shortly after the resignation of Director Wirth, Congress passed the Wilderness Act, establishing a system of wilderness areas on federal lands, including those managed by the Park Service, the Forest Service, and the Bureau of Land Management.[11] The passage of the act is considered a milestone in U.S. conservation history and rightly so. It put more than 10 million acres of land into wilderness status and established a procedure by which millions more could be added. By statute, these wilderness areas were to remain free of development and be managed exclusively to maintain their wilderness characteristics. The act was also a definitive legal expression of the pessimism which had come to dominate much of the thinking about the relationship of modern material culture to the natural world. The preamble of the act makes its assumption on this point clear: "In order to assure that an increasing population, accompanied by expanding settlement and growing mechanization, does not occupy and modify all areas within the United States and its possessions, leaving no land designated for preservation and protection in their natural condition. . . ."[12]

The Wilderness Act was the culmination of more than a decade of lobbying by environmentalists, so in that respect it can be seen as a conclusion. It was also a beginning—the start of a decade and a half in which demands for nature-oriented public land policies were increasingly articulated through the political process at the federal level. During this period, the Park Service, whose Mission 66 program had done much to prompt the Wilderness Act in the first place, had its management of the great natural areas subjected to much critical scrutiny, and natural parks came to acquire new meanings and sources of value which were very different from those derived from the progressive vision of the Mather and Albright era.

With increasing environmental awareness in the 1960s and 1970s, and a shift in ecological thought among environmentalists and the public at large,[13] the major national parks came to be valued both as important parts of the global ecosystem and as unique, distinct areas where nature-altering human activities must not be allowed to take place. Simmons wrote of the national parks as important regulators of global atmospheric systems, such as the carbon dioxide and oxygen cycles.[14] He also saw them as important biotic reserves, genetic storehouses, and natural laboratories. When the Sierra Club fought for an expanded Redwood National Park, their aims were to include in the park boundaries the greatest possible species diversity and complete biological communities.[15] Other environmentalists stressed the importance of national parks as bellwethers, i.e., as early indicators of any change in national or even global environmental conditions. For example, Assistant Secretary of Interior Herbst wrote that "the national parks are the crown jewels of our enviable U.S. environment, and any dimming of their luster signals similar slippage of the quality of the national environment as a whole."[16] Paul Pritchard, Executive Director of the National Parks and Conservation Association, expressed a similar idea when he told the author that "we can look at the parks as laboratories; they'll tell us how we are doing environmentally as a nation."[17]

This emphasis on the use of the parks as instruments to attain very broad environmental goals led naturally to placing a high value on scientific investigation into the biological and ecosystemic properties of the national parks, and to management based on the results of such investigation. The National Parks for the Future (NPFF) study group saw past levels of Park Service research funding as inadequate, especially "when one grasps the magnitude and complexity of the Park Service's resource-management task and the quality and quantity of information required to fulfill its assigned mission." To prove its point about inadequate research efforts, the group compared "ecological" research in the Park Service with that of the Forest Service and the Bureau of Sport Fisheries and Wildlife (table 4-1). From a policy analyst's point of view, what is most interesting about the table is not the figures it contains but its implication that a comparable

Table 4-1. Ecological Research Budget—Fiscal Year 1972

Agency	Total operating budget[a]	Research budget[a]	Research as percent of total operating budget
U.S. Forest Service	509.6	54.3	11.0
Bureau of Sport Fisheries and Wildlife	82.0	15.7	19.0
National Park Service	133.4	1.02[b]	0.7

Source: Conservation Foundation, National Parks for the Future, p. 93.

[a] In millions of dollars.

[b] Natural science research only. Does not include archaeological and historical research.

amount of research is necessary for the Park Service and the other agencies to do their jobs. Much of Forest Service ecological research is in fact forest product oriented, and much of that of the Bureau of Sport Fisheries and Wildlife was directed toward putting a target in front of a sportsman's gun or lure. The message here seems to be that the Park Service is responsible for delivering a similar "product" in ecological maintenance.

The NPFF study group went on to assert that "for the past fifty years the need for such information has been clear."[18] In fact, the need had not been clear for fifty years. The perceived need for such information reflected modern ideas about ecosystems and ecology and the agency's responsibilities toward them. Certainly it was not a responsibility of which Mather and Albright seemed aware; they frequently enlisted the aid of the Bureau of Biological Survey for help in predator control, i.e., the extermination of wolves and mountain lions in the parks, and they ordered vegetation clearing operations with little thought of their ecological implications. When Mather and Albright were working out their formula for agency success and the security of the System, the natural world was seen as more robust, and civilization less inclined to mischief toward nature. In the early days, it was usually sufficient to protect the parks from the gross degradations associated with mining, timber harvesting, and dam construction.

Mather's modern biographer, apologizing for his subject's failure to conform to contemporary notions of ecological responsibility, writes that "Mather was not in the ecological know, his love of birds and beasts was deep and genuine, but it was not erudite."[19] But even erudition about the natural world in the 1920s and 1930s would not have included being "in the ecological know" as we would define it today. The founders of our modern sensibility, philosophers like Aldo Leopold and scientists like F. E. Clements, were just beginning to find audiences[20] and most educated men were only dimly aware of the dynamic interconnections of living things to which we have become so sensitized. Thus the demand for increased scientific management of the national parks and the assumption that deft ecosystem management was necessary for their well-being was founded on a contemporary notion that ecosystems were, by and large, very fragile things, while civilization's power to damage the natural world, either through malevolence or carelessness, was very great.

Agency critics in the environmental community wanted a change in park selection criteria as well as in management criteria. For example, the NPFF's report read that: [21]

> New natural-resource parks should be established and land should be acquired to provide a system representative of all principal physiographic regions in the nation. There are a large number of representative vegetative-physiographic land types, characteristic of broader regions, not represented in the park system. Some of these areas are scientifically valuable for their undisturbed vegetation and as the habitat of threatened wildlife species.

The new environmental view was behind this recommendation. Given that nature was under assault, the Park System had to be pressed into duty as a giant ecological lifeboat. [22] Preservation of biotic assemblages or representatives of physiographic regions, rather than popular appeal or conformity to traditional landscape aesthetics, had to be the most important criterion for selection. The areas the NPFF study group identified as major gaps in the National Park System, the short-grass prairies and the arctic tundra, were gaps largely because they were examples of large American biomes unrepresented in the System and not because of any inherent visitor appeal. Gardon reflected similar assumptions about proper selection criteria when he argued that it was inconsistent to omit a great plains national park from the System because "the prairies are as unique and distinct as other major ecosystems such as the redwoods in California and the everglades in Florida." [23] Some environmental groups came to see the Park System as a convenient shelter for any lands which in their view ought to be spared from resource extraction or development because of their biotic characteristics.

Demands for Reduced Public Access

A logical concomitant of increased concern for the ecological properties and biological value of the National Park System was dissatisfaction among many of those close to the System—environmental activists, scientists, journalists, scholars, and some politicians—with how the Park Service handled park visitors. If one assumes that nature is robust and the natural parks resilient to the stresses that visitors put on them, then the agency's founding mandate to encourage use of the parks while preserving them for the future involved little more than threading policy between excesses of unrestricted use and preservationist zeal. This is exactly what Service leadership tried to do throughout the Wirth era. Sometimes policy shifted toward one pole, sometimes toward the other, and sometimes the Service was unsure of just where the proper course lay.

Use vs. Preservation

With increasing park use and an emerging view of nature which no longer assumed hardiness in the face of such use, what had been seen as concurrent mandates came to look like a logical contradiction. In fact, McCool, surveying contemporary writing on the National Park Service, notes how frequently the founding mandate of 1916 is referred to as inherently contradictory.[24] If use destroys, how can a management policy both accommodate use and preserve the natural area? A mandate which is inherently contradictory must, by logical extension, become a management dilemma—a problem for which there is no solution that does not violate a constraint.[25] The use and preservation question has widely come to be seen as just such a dilemma for the agency. For example, Irland insists a contradictory mandate gives the agency the task of "harmonizing the unharmonizable."[26] Rowntree, Heath, and Voiland see the use versus preservation question as a "fundamental dilemma" for the Park Service.[27]

In a dilemma, something has to give way, and what most of the Park Service's critics wanted to see give way was its traditional policies of accommodating visitors. As noted earlier, in the early twentieth century, environmental groups such as the Sierra Club, the Save-the-Redwoods League, and Manzamas encouraged the construction of roads into wilderness areas. Such roads, they argued, would increase the use of wilderness and therefore increase public support for wilderness protection. Furthermore, mass access to the wild places fit neatly into the progressive view of the proper landscape for modern life. As Wirth pointed out, this attitude had lost much of its original support among preservation activists by the 1960s, and access came to be seen as much more of a threat than a blessing to the great national parks.[28] Just how great this shift was can be seen in the retrospective judgments rendered on Mission 66. Writer after writer criticized the program for encouraging heavier use of the parks at the risk of the parks themselves.[29]

Joseph Sax, one of the most lucid of the environmental critics of the Park Service, seems alone in recognizing that the "popular threat" to the parks has roots that go deeper than Mission 66, and that popular use might ultimately be connected with the well-being of the National Park System. He writes that Mather "spent his life winning friends and popular support for the park system; the measure of his success is that the most serious problem of the parks is that they risk being loved to death."[30]

Others have simply overlooked or rewritten this crowd-pleasing aspect of early Park System management. Udall in *The Quiet Crisis* spoke fondly of the early days of the Park Service when "the National Parks were administered under wilderness protection principles."[31] Perhaps they were, but if so, the wilderness principles of Mather's day were similar to the Mission 66 thinking which encountered such strong criticism during Udall's era.[32]

Visitor Accommodations

No matter how they viewed the history of agency visitor policy, however, most critics of Park Service policy linked demands for an increased ecological orientation in park management with an insistence that the agency change its treatment of park visitors. The NPFF study group saw little need to accommodate a wide range of visitor tastes. It wrote that "necessary in-park accommodations should be designed and operated to fit into the natural scene. The plush resort hotel is simply inappropriate."[33] It added that the Park Service was not obliged to provide campsites equipped with electric outlets, running water, or toilet hookups. The group further asserted that camping vehicles were contrary to the park ethic and "those who wish to use them should be asked to leave them at the park boundary and visit the park on its terms rather than theirs." The assumption here seems to be that parks in and of themselves have terms, ones which do not include such things as running water and electric outlets.

To allow for management on the park's terms, the NPFF study group recommended creating a new park category for the large national parks and national monuments, the National Wilderness Park. The dominant purpose of each of these national wilderness parks would be "the preservation of an intact ecosystem."[34] This clear articulation of ecosystem preservation as the park's dominant purpose would then justify limiting visitor facilities. The NPFF report read in part that, "visitors to the national wilderness parks must be prepared to accept a wilderness lifestyle."[35] It added that those desiring more in the way of accommodations "have to accept the necessity of leaving the park at the end of the day to find such accommodations outside its boundaries."

Darling and Eichhorn argued that even when a small portion of a national park is devoted to visitor accommodations, it may be too much. They write that, "5% of Yellowstone National Park is taken up by development, a portion which seems to us inordinately high, for the traumatic influence of this 5% will be felt over a much larger area."[36] They further argued that "in a sense even the wilderness portions of the parks are developed since there are trails in even the most remote places." Although few would go as far as Darling and Eichhorn in defining development to include wilderness trails, many environmentalists agreed that the parks were overused and underprotected, and that the two formed one interrelated problem.

While a new awareness of ecological sensitivity led to a demand that access be circumscribed more severely, new views of the park-using public arose among environmentalists to legitimize such a demand. For example, Sax divided park visitors into two distinct categories, those roughing it in the parks and those visiting them in a nonprimitive manner. Sax went on to explain that the two styles were rooted in different sides of the American character and as he described it, the distinction clearly had its Manichean undertones:[37]

> on one side [is] a repugnance at the seemingly boundless materialism that infused American life, a spiritual attachment to untrammeled nature, and a self-congratulatory attitude toward preservation of nature's bounty; and on the other [is] a commitment to economic progress where it could be exacted, nationalistic pride, and the practical use of nature as a commodity supportive of tourism and commercial recreation.

The one style of park visit was based on such things as nationalistic pride and a commitment to economic progress whatever the cost. The other sprang from a spiritual attachment to untrammeled nature. According to Sax, a visit to the park which relied on developed facilities was, in essence, using the park as an interchangeable commodity since motels, restaurants, and the like are things which can be placed (and enjoyed) anywhere. The inference here was that those staying in less than primitive conditions were incapable of appreciating the natural beauty of the parks in which the accommodations are set. He argued that it was a waste of the unique, sublime properties of national parks to permit them to be used for what he defined as commodity-oriented visits, and all such visits should be discouraged.[38] The NPFF report had argued for similar restraints on park use.[39]

Advocacy of reduced access and of forcing the visitor to accept the parks on these circumscribed terms was sometimes justified by reference to democratic theory. For example, Sax, drawing on history, wrote that: "As Olmsted demonstrated, the question in a democratic society is not the acceptance or rejection of what people want. People get the recreation that imaginative leadership gives them."[40]

Limits on Use

Suggested solutions to perceived overuse, misuse, and all their associated problems usually involved either instituting visitor quotas or placing severe limits on facilities development in the parks, or both. The NPFF report told the Park Service that it should "take immediate steps to implement a policy of visitor limitation based on 'conservative best-judgment' criteria."[41] It argued that the problem of overuse was so great that limits had to be established quickly "without waiting for 'rational' scientific criteria to be devised." (The NPFF report suggested, however, that once these limits and quotas had been established, "the public should be informed that maximum visitor loads for any area are not set arbitrarily but are determined by professional resource managers with the advice of scientists.") Important among the criteria for determining visitor quotas was psychological carrying capacity, i.e., how many people an area could sustain before the quality of the visitor's experience was degraded by the presence of other people. According to the NPFF report, the concept of psychological carrying capacity was crucial in regard to wilderness because "wilderness is . . . a game, it cannot be played at any one time at any specific place by more than a few people." Therefore, "respect for the quality of a wilderness experience demands that we accept more regulated use and quotas." According to Sax, the very act of rationing visitor use of the national parks would improve the perceived quality of the visit since "the visitors' sense of anticipation is heightened and entry to the place is made more dramatic by 'rationing'."[42] Lest accusations arise that there is something undemocratic about this, Sax assured us that "in all these devices [which limit visitor use or freedom] there is equality in the right of access, but a reduction in the total quantum of access in order to exalt quality over quantity of experience."

The second widely advocated means of reducing visitor numbers and impact was reducing the number and range of facilities for visitors in the parks. Although accommodating a wide variety of visitor tastes was a key part of the agency's initial strategy for success, those who have recently advocated a reversal of this policy have argued that such a reversal would have little cost to the System in terms of popular political support. For example, the head of one environmental organization told the author that times had changed: "The middle class family used to go to the [national] park as a primary destination. [They] car camped in it a while, then came home raving about what a good time [they] had. Now it's camper vans whizzing through the parks on the way to the next K.O.A. They don't get much of an impression [of the parks] and they don't become constituents of them when they get home."[43]

One commonly advocated form of this strategy is to move visitor facilities beyond the park boundaries. According to the NPFF study group, the first management question the Park Service should ask itself is: "What is the *minimum* level of facilities necessary within the park?"[44] When this has been decided, all visitor services which can be provided outside of park boundaries should be built there rather than inside the parks. Not satisfied with precluding new facilities, it added that "present facilities, such as lodging, curio shops and parking lots, now located inside park boundaries, should be phased out wherever practicable." The 1971 NPCA study, *Preserving Wilderness in Our National Parks*, took a similar stance when it suggested that the Park Service concentrate on management of the parks for preservation, and that lodgings and recreation opportunities not directly based on wilderness be provided outside the parks, either by private enterprise or other government agencies.[45]

Many of the attitudes revealed by recent critics of an open visitor policy have long pedigrees. What former Secretary of Interior Ickes wrote in 1938 about the characteristics of the American public and the demands they made on the national parks would not sound out of place today. Compare the tone of Ickes's writing with that of the passages quoted above:[46]

> When a national park is established, the insistent demand is to build roads everywhere, to build broad easy trails, to build air fields, to make it possible for everybody to go everywhere—without effort. [Some people] feel that they are roughing it if they twist their necks in a sightseeing bus, or expose their adenoids to the crisp air while gazing through field glasses at some distant scene. And these are the vast majority.

There was an important difference between Ickes's attitudes and the agency's more contemporary critics, however. While the latter told the Service to discourage most of its potential visitors, Ickes, curmudgeon that he was, would not abandon tolerance or the idea of the sovereignty of popular choice, no matter how distasteful he personally found that choice:

> I am in favor of opening a liberal and representative section of every national park to those who, because of physical limitations, are confined to motor roads. I am even willing to make this same concession to those who cling to motor roads as a matter of choice. But let us preserve a still larger representative area in its primitive condition, for all time.

But it was now forty years since Ickes had written the above, and political, and one might even say moral, force was behind the preservationist views. Whereas wilderness advocates tended to be weak when compared with their opponents through the Eisenhower era, with the passage of the Wilderness Act in 1964 they showed that they had come of age politically. What had been suggestions became demands.

NEW PARK SERVICE ATTITUDES AND POLICY

Acceptance of Environmentalist Attitudes

The attitudes of Park Service personnel toward the natural areas of the System seem, by and large, confluent with the views of the environmentalists. A sampling of agency personnel showed that 84 percent of those polled agreed that "preservation is the major purpose of the National Park Service."[47] Sixty-eight percent felt that visitor quotas were a good thing and 76 percent felt that even increased numbers of park visitors would not justify the expansion of visitor facilities within the parks. On the reverse of this coin, only 9 percent of the surveyed sample felt that the Service's main responsibility was the provision of recreation opportunities.

Not only have agency personnel accepted environmental principles, but they also seem to have accepted as legitimate the criticisms that the environmental groups have leveled at the Service and even the right of these groups to guide agency policy. In a published interview, an agency administrator, referring to the Sierra Club, the Wilderness Society, and the National Wildlife Association, said: "In a sense these are pressure groups on us to keep us . . . in line, but I like to look at them as supporters, because really that's what they are. They give us a lot of flak and we get up in arms, you know, but they are our friends, too."[48] This willingness to accept advice and criticism on park policy extended to top agency leadership. For example, it was apparent in former director Whalen's attitude toward the environmental groups: "Our natural allies are in the conservation community. For example, just look at the help and good advice we've gotten from the Sierra Club . . . You don't run against your natural allies, you run with them."[49]

Willingness to accept the environmental groups as a legitimate source of advice on policy, however, has not always been matched by a willingness to accord the same legitimacy to the demands of the general public. For example, an agency administrator said that, "we don't think . . . that the public in general necessarily has realistic demands, and that we should bend to their demands any more than to political demands."[50] The superintendent of one of the great natural areas in the System expressed a similar opinion when he told the author that it was wrong for a park manager to assume that what visitors say they want was what the Service should provide. In fact, he was not even sure that what visitors said they wanted was what they truly wanted: "[If] you ask people what they want for lodgings they say a Holiday Inn or something like that. But if they get put up in a tent cabin, they are usually pretty pleased the next morning, they experience morning sights and smells they didn't know existed or had forgotten." In summing up his attitude toward accommodating the expressed preferences of park visitors, he told the author "We're behavior modifiers anyway. We put up road signs to tell people not to feed the bears, so why not be positive about it."

This attitude, which placed emphasis on ecosystem preservation or visitor behavior modification over relatively unimpeded visitor access to the great natural areas of the System, did not go entirely unopposed within the Service, or without for that matter. For example, an administrator in

the Executive Office with close professional ties to the Park Service felt
that such emphasis was not in the agency's long-term interest: "The Service likes to manage wilderness and pushes wilderness in the System. Why?
It keeps up the cozy relationship with the interest groups, even though it
reduces [the agency's] administrative discretion. This will hurt them down
the line."

But perhaps the most serious objection, given its source, came from
Horace Albright. Although Albright resigned as agency director in 1933,
he was still a young man when he did so and he remained active on the
Park Service's citizen advisory board well into the 1970s. From this position, Albright objected to applying the Wilderness Act of 1964 to the
National Park System, and both before and after the passage of the act he
spoke out against increased stress on wilderness management in the
parks.[51] Albright insisted that all the great natural areas of the System
were, in fact, wilderness parks because most of the lands then encompassed had no facilities or roads in them, and were accessible only by trail.
Furthermore, visitor accommodations occupied only a small fraction of
any park. Albright feared tipping the balance of management too far
against the ordinary tourist, who he insisted was out for "fun and excite-
ment as well as inspiration."[52] In 1976, on the occasion of the sixtieth
anniversary of the founding of the Park Service, Albright wrote of what he
saw as disturbing demands to limit visitor access.[53]

> Will [the Park Service] be able to resist the efforts of individuals or groups who
> seek changes in the administration of many of our great [parks]? There is an
> insistent demand that concessions be located outside park boundaries, and the
> park areas themselves turned back as completely as possible to the condition
> they were in when discovered, thus making them usable only by hikers, back-
> packers, and horse or mule riders, and preferably few, if any, of the latter.

Albright also questioned the judgment of the Park Service scientists who
wanted to limit visitor use to prevent environmental damage. "[The] ecologist who would greatly limit use of the national parks usually claims that
the summer crowds 'damage' or 'injure' the parks. This is simply not so."
Albright saw the environmental damage argument as a red herring, and as
an excuse for, rather than reason to, limit visitor access. He called for
more, not fewer, campgrounds in the parks, saying that, "while more
travel should not be promoted, neither should it be restricted beyond
possible institutional capacities." Finally, Albright objected to the way the
agency itself justified its current attitude toward the visitor through selective omissions when quoting from its founding legislation.

> Rarely is reference made to the provisions of the laws creating national parks
> which authorize the Secretary of the Interior to grant leases for the installation
> of visitor facilities for the enjoyment of the parks by the people. Instead, the

provisions quoted are those that direct the Department and the Service to keep the reserved lands in their natural conditions for the benefit of future generations, which has been done for at least 80 years.

Albright's criticisms were largely ignored by the agency and the environmental groups. His was the voice of a ghost, or perhaps that of a grandfather who clearly did not understand the modern world or the imperatives of System management in the 1970s.

Management Shifts

The documents and actions of the National Park Service from the mid-1960s onward reflected the increasing influence of the environmental movement and its world view of the agency. Perhaps the first change in policy caused by the ascendancy of environmental values came in 1964, the year of the Wilderness Act, when George Hartzog, newly installed as Park Service director, established a three-part classification system for the national parks. Henceforth, all units in the System would be categorized as either recreational, historical, or natural areas, "so that resources may be appropriately identified and managed in terms of their inherent values and appropriate uses." [54] Fifteen years later Nienaber would judge this step a milestone in park management policy because it prompted the Park Service to adopt a preservationist orientation in a large number of parks. [55] According to Nienaber, development policy, which in the past had simply been to expand anywhere within park boundaries to meet public demand, "now became linked to physical impact, or, as the agency put it, 'ecological health or repose'." With this new tripartite management guide, agency leadership also acceded officially to one of the principal criticisms of Mission 66, i.e., that it caused the Park Service to lose sight of the inherent qualities of the parks in its efforts to accommodate and even encourage more park visits. "The single abiding purpose of national parks is to bring man and his environment into closer harmony" an agency policy document of the period asserted. [56] "Therefore," it continued, "it is the quality of the park experience and not the statistics of travel which must be the primary concern."

By the same time, that is to say, by the mid-1960s, there had come to be a considerable feeling, both within the Service and within the environmental community, that parks and highways mixed poorly. The vehemence of public protests in the early 1960s over proposals to put a freeway through San Francisco's Golden Gate Park served as an object lesson. Whether responding to pressure or acting from personal conviction, or a combination of the two, Director Hartzog took strong public positions against increasing road access to the parks and against parkways managed by the Park Service. Hartzog was quoted as saying, "The automobile as a recreational experience is obsolete, we cannot accommodate automobiles in such numbers and still provide a quality environment for a recreational experience."[57] He also said, "No more roads will be built or widened until alternatives are explored. We want to give a park experience not a parkway experience."[58] Under Hartzog this attitude became policy, and the following two management guidelines regarding vehicle access to the parks were promulgated:[59]

> The National Park Service must not be obligated to construct roads, or to manage traffic, in order that new kinds of mobile camping vehicles be accommodated. The development of parking areas for trailers at park entrances, and the exclusion of vehicles from park roads not capable of handling them, are appropriate solutions.
>
> Faced with a choice of creating a severe road scar in order to bring visitors close to a point of interest, or requiring visitors to walk a considerable distance, or considering an alternate transportation system, the decision should be against the road.

Park Service policy has reflected Hartzog's disenchantment with the automobile ever since. The current agency policy manual not only takes a disapproving attitude toward new roads but directs the agency to take a critical view of roads already in the parks. The manual implies that past agency policy with regard to roads was overzealous when it reads that, "for most existing parks, a road system has usually already been constructed in accordance with previous policies."[60] It also seems to imply that amends are in order when, in the next sentence, it states that "in updating plans for these parks, the Service will question the continued validity of the existing road system."

Visitor Facilities

The current policy manual and other recent agency publications also show a shift of position on issues of preservation and quality of visitor experience. For example, the manual shows that the Service has adopted the minimum facilities principle as official policy: "Overnight facilities will be restricted to the kinds and minimum levels necessary to achieve each park's purpose consistent with the protection of park resources, and will be provided only when the private sector or other public agencies cannot adequately provide for them in the park vicinity."[61] The agency has clearly abandoned the old Mather and Albright principle of accommodating each according to his taste. Those whose tastes do not incline them toward facilities "consistent with protection of park resources" will be left to seek accommodations elsewhere.[62]

The Service has also adopted the view that the parks are important as environmental indicators and biological refuges, and that these are important criteria on which to judge a park's worth. As an illustration, a recent

Park Service report to Congress on the state of the national parks opened by asserting that "America's National Park System may be considered a barometer of the world's environmental conditions." [63] The report continued to argue that the parks, because of the diversity of plants and animals they contain, "perpetuate the genetic diversity to insure their futures and to act as harbingers for human futures as well." It concluded that one of the Park System's top priorities had to be the establishment of detailed monitoring programs in order to measure changes in these park resources and in the wider environments within which the parks were found.

In order to accomplish this goal and its related recommendations, the report suggested that the Service must "significantly expand its research and resource management capabilities." [64] In the late 1970s, the Park Service established an Office of Science and Technology whose primary functions were to ensure that park management was in accord with ecological principles and to expand the agency's detailed, scientific knowledge of the System's natural resources. The office was placed directly under an associate agency director who had considerable power to protect his office's functions and to see that its opinions and research conclusions were incorporated into management decisions throughout the System. Since then, many national park master plans have called for removing tourist accommodations to beyond park borders, either to adjoining public lands or onto private property. The list of premier parks where the agency has or intends to remove major facilities is a long one and includes Yosemite, Sequoia, and Grand Canyon national parks.

Removing lodging facilities was not the only step the agency took to reduce visitor access to, or general levels of development within, the parks. The Park Service recently developed a management plan for the Colorado River in Grand Canyon which aimed at phasing out motor boats and reducing the number of people running the river. The agency recommended to the secretary of the interior that a proposed second road through the Great Smoky Mountains National Park not be built. It opposed extending the runway of Jackson Hole Airport in Grand Teton National Park. It relocated a miscellany of nonvisitor-related facilities, such as Job Corps centers, outside of the parks. It expanded the use of visitor quotas in the parks and it experimented with various types of reservation systems to allocate camping spots and pushoff times at the trailheads which had become, in some parks, scarce resources.

Other specific, high-level planning decisions reflected the agency's reluctance to accommodate increased visitor numbers in the great natural parks and monuments. In 1968, Ross Lake and Lake Chelan national recreational areas were added to the National Park System. The two units were next to the North Cascades National Park and were established, in large part, to provide accommodations for visitors to the North Cascades park. Thus they relieved the Park Service of the need to provide tourist accommodations in the park itself. This was the first time such an arrangement had been made, and it set a precedent. Perhaps more than anything, however, the Service's policies with regard to the use versus preservation conundrum was reflected in a list drawn up by Assistant Secretary Herbst of what he considered the thirty major accomplishments of the Park Service during his incumbency (1977–81). They reflected a wide range of social and environmental achievements, but not one of the accomplishments involved increased general access to the parks.

Visitor Numbers

The preservation impulses which expressed themselves as new planning policies, altered parks, and reorganized agency structure also translated into unspoken guides by which managers ran their parks. For example, during the 1970s, superintendents of the large national parks were little inclined to encourage increased visits to their parks, and they felt under little pressure from Service leadership to do so. The superintendent of Grand Canyon told the author, "I'd be more concerned with a clear increase in the number of visitors than I would be with a decline." Superintendents of such widely differing parks as Channel Islands, Sequoia, Everglades, and Fire Island all said the same thing in one form or another. Not only did they feel no pressure from higher authorities to increase visits, but to a man they felt that a substantial increase in visitors would only create problems for them. The superintendents of the Channel Islands National Park and the Everglades both valued the International Biosphere Reserve status of their parks as a measure of protection against any future pressure to shift the management emphasis toward visitor accommodation.[65] The superintendent of Channel Islands also valued his park's low visibility with the general public for a similar reason. It meant he did not have to worry about large numbers of visitors and the environmental degradation they were sure to cause.

Perhaps the clearest indication of changed attitudes and policies was not found in specific policy directives but rather in what the agency told itself were its responsibilities toward the parks.[66] The first chapter of the agency's current policy manual states that:

> The purpose of the parks, as defined by the Act [Organic Act of 1916] is to . . .
> 'conserve scenery and the natural and historic objects and the wildlife therein
> and to provide for the enjoyment of the same in such manner and by such means
> as will leave them unimpaired for the enjoyment of future generations.'

What is most revealing here was what the policy manual did not quote from the act. It omitted the passage which read that: "the director of the National Park Service may also provide in his discretion for the destruction of such animals and plant life as may be detrimental to any said parks, monuments and reservations."[67] This was exactly the kind of omission Albright was talking about. It represented, if not a distortion of the agency's own past, at least a willful obscuring of it. Times and attitudes had changed so much, however, that the true past simply was not a respectable antecedent of current management policy.

New Natural Areas: Selection Criteria

When the Park System was established, many areas whose natural features made them the equal of established parks remained outside of the Park System; there were the great mountain areas of the Northwest and spectacular desert landforms of the Colorado plateau, for example. Because this was the case, expansion of the System could be brought about through the addition of these sublime places during the agency's first decades. Early on, Mather had recognized that System growth which resulted from the addition of new crown jewels was the best, most solid kind of growth. Accordingly, he devoted much effort to turning these spectacular natural areas into the national parks. During his term as director, he more than doubled the number of large natural areas in the System.

It was partly by turning east for prime natural areas that Mather's successors continued Park System expansion in the years before World War II. The Great Smoky Mountains National Park was added in 1934. Shenandoah National Park was added the following year. Isle Royale entered the System in 1940. Moreover, considerable acreage was added to existing national parks during this prewar period. Death Valley was expanded by the addition of nearly a third of a million acres in 1937. Shortly thereafter, Dinosaur National Monument was expanded with the addition of the canyons of the Green and Yampa rivers, and in Alaska the size of Glacier Bay National Monument was increased by nearly a million acres. Although the directorships of Drury, Wirth, and Hartzog saw some spectacular areas, including North Cascades National Park, added to the System, the rate at which parks of crown jewel quality were added was not nearly that of the prewar period. Of the 23,840,000 acres of land in natural areas of the System on April 1, 1971, 96 percent were in national parks authorized or established before World War II.[68]

This slowdown resulted from the increasing difficulty the Service encountered in establishing large new natural parks during the postwar period. The reason for this increased difficulty was a simple one: most of the appropriate areas had already been included in the System by then, and if a few prime natural areas did remain outside, political circumstances worked strongly against their inclusion. The long and taxing battle in the 1960s to transfer the North Cascades from Forest Service jurisdiction to the National Park System illustrated just how difficult the addition of areas of unquestionably high scenic quality had become.[69]

This forced agency leadership into a quandary that neither Mather, Albright, nor their immediate successors had faced; it either had to accept that the National Park System was largely rounded out when it came to great natural areas or it had to change the criteria of appropriateness to allow the addition of new natural areas. Agency leadership, with its traditions of aggressive expansion, was loath to accept the former, but how was it to make basic changes in significance criteria without blatantly eroding park standards, and with it, its own credibility when it spoke about park quality? The Park Service, guided by new environmental values, found a way: it changed standards of appropriateness by putting a new emphasis on representing the physiographic regions and the varied ecosystems of the United States. In essence, the new standards aimed at turning the natural wing of the System into a natural history sampler and an ecological lifeboat, something the environmental community very much wanted.

New Criteria

The 1972 National Park System Plan, the agency's seminal planning document of the modern era, showed the degree to which the agency adopted physiographic and ecological representativeness as the primary criterion in evaluating natural areas. It also showed just how many opportunities for System expansion could be derived from such criteria. What had been an essentially rounded-out set of natural areas judged by the criteria of scenic value and visitor appeal suddenly became a very incomplete set of representatives of the nation's biomes and physiographic regions. Figure 4-2, taken from the 1972 plan, is a physiographic map of the United States which shows the degree to which each province was adequately represented in the Park System. It shows a System strongly concentrated in the western, mountainous physiographic regions. It also shows that most of the country's major regions outside the West had relatively few of their natural areas represented in the System.[70] The Cascade Range and the Chihuahuan Desert had adequacy ratings of over 90 percent. The Sierra Nevada, the northern and middle Rocky Mountains, and the Mohave Desert were rated above 80 percent. Most of the eastern provinces scored below 50 percent. New England was at 29 percent; the Gulf Coastal Plain at 17 percent; the Appalachian Plateaus and the Piedmont were given zero ratings. The ratings system also gave low ratings to some western regions. For example, the Columbia Plateau and the Great Basin were at 12 and 10 percent, respectively. The Wyoming Basin got a zero rating. In sum, the list of physiographic provinces, with each assigned an adequacy rating, showed that the System was far closer to empty than full.

The agency then extended the representativeness criterion from physiography to what it termed natural history themes, i.e., landforms, ecosystem

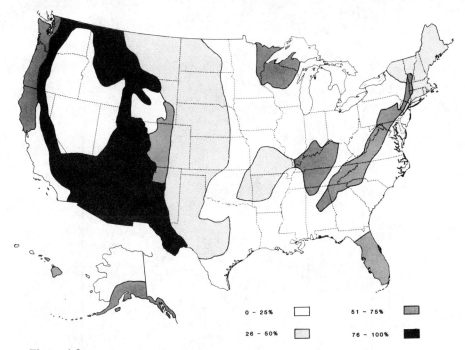

Figure 4-2.
Representation of natural regions in the National Park System. Shading indicates the degree to which a region's natural features are represented in the System. (From the National Park System Plan, Part II).

types, and examples of geologic history. On this criterion as well, the System showed many gaps.[71] For example, under landforms, mountain systems were given an adequacy rating of only 44 percent, which seems surprising since the National Park System is so intimately tied up with mountains in the public's mind. Seashores, lakeshores, and islands were even less adequately represented as a group, with a 42 percent rating. Under geological history, while the Precambrian era was well represented, the Cambrian, Oligocene, and the late Silurian were not. By combining physiographic and thematic criteria of representativeness, the Service identified what it considered the major gaps in the System. They were numerous and varied, as shown in table 4-2.

The degree of importance now accorded representativeness and ecosystemic values in the new unit selection was further illustrated by the formal guidelines for evaluating natural areas which the agency promulgated. Under these guidelines, the following characteristics were to be sought:[72]

Exceptionally high ecological or geological diversity, e.g., species, biotic communities, habitats, landforms, observable geological processes.

Biotic species or communities whose natural distribution at that location makes them of unusual biogeographic significance, e.g., high numbers

Table 4-2. Major Natural Area Gaps in the National Park System—1972

Terrestrial ecosystems	Aquatic environments	Landforms	Geological history
Boreal forest Interior and Western Alaska *Eastern deciduous forest* Appalachian Plateau Atlantic Coastal Plain Gulf Coastal Plain *Grassland* Central Lowlands *Desert* Great Basin *Tropical ecosystems* Oahu Kauai Puerto Rico Guam Trust Territories Samoa *Tundra* Interior and Western Alaska Brooks Range Artic Lowland *Coral islands, reefs, atolls* Leeward Islands Puerto Rico Guam Trust Territories Samoa	*Marine environments* New England—Adirondacks Leeward Islands Interior and Western Alaska Arctic Lowland Puerto Rico Guam Trust Territories Samoa *Estuaries* New England—Adirondacks Interior and Western Alaska *Lakes and ponds* Central Lowlands *River systems and lakes* Central Lowlands Gulf Coastal Plain Interior and Western Alaska	*Plains, plateaus, and mesas* Great Plains Central Lowlands Interior and Western Alaska Arctic Lowland *Mountain systems* Great Basin New England—Adirondacks Brooks Range *Sculpture of the land* Columbia Plateau Kauai *Works of glacier* Brooks Range	*Paleocene-Ecocene* Wyoming Basin *Oligocene-Recent* South Pacific Border Interior and Western Alaska

Source: National Park Service, *National Park Plan*, Part II, p. 7.

at range limits or of diverse geographic affinities, relicts, endemics, extreme disjuncts.

A concentrated population of rare plant or animal species, particularly those officially recognized as threatened or endangered.

A critical refuge necessary for the continued survival of either common or uncommon wildlife species.

Rare or unusually abundant fossil deposits.

Outstanding scenic area.

An invaluable ecological or geological benchmark due to an extensive and long-term record of research and scientific discovery.

An outstanding example of a geologic landform or biotic area that is still common or of broad distribution.

A rare extant remnant geologic landform or biotic area of a type that is now vanishing due to human despoilment, although once widespread.

An extant geologic landform or biotic area that was extremely unique in the region or nation during presettlement times.

These criteria were very different in substance and spirit from those contained in Lane's letter of instruction to Mather; they were much more scientific and biocentric. They emphasized the refuge quality of the System and only one of the ten could unquestionably be considered visitor-oriented in the sense of a traditional national park.

There were still those in the Service, however, who felt that great scenery was the primary prerequisite of a great national park. For example, when the Park Service was deciding which of the Alaskan lands it wanted for national parks, there was unanimous support among agency personnel for such mountain parks as Gates of the Arctic National Park in the Brooks Range but far less enthusiasm for the flat tundra parks. Many Service employees felt the latter would be unworthy additions to the System because they lacked scenic appeal. Some of those within the agency who still held scenery in high regard felt that although there might be justification for preserving natural areas because they had unique ecosystem qualities or represented a biome, the National Park Service was not the agency to manage such areas if they did not have scenic appeal as well. Such areas, the argument went, more properly belonged in the National Wildlife Refuge System, since an area managed for ecosystem reservation was more like a national wildlife refuge than a national park. It must be stressed, however, that these views did not represent official agency policy, but rather dissent from it, be it ever so loyal.

Making Things Fit

The shift of emphasis to comprehensive representation led to some curious assignments of units already in the System. For example, Platt National Park, a collection of hot springs and trailer hookups within the corporate limits of a small city in Oklahoma, was made to serve as a representative of the entire Central Lowland physiographic province. Mammoth Cave and the Natchez Trace Parkway were pressed into duty as representatives of the Interior Low Plateaus province.[73] The attempt to apply, post facto, the comprehensive logic of thematic categories to the System also created some strange peers among the natural themes. For example, hot water phenomena, i.e., hot springs (the most adequately represented "landform" in the System, thanks to the early twentieth century's penchant for making the sites of hot springs into national parks) was accorded the same taxonomic rank as mountain systems. And underground ecosystems (i.e., caves, another product of an earlier era's sense of values) was given status equal to that of marine environments. Representativeness as a criterion also occasionally conferred values which seemed at variance with common sense. For example, exposed late Silurian geological deposits were presumably of greater value than virgin stands of Pacific rain forests since the former were less represented in the System.

The authors of the plan seemed well aware of these problems and inconsistencies when they wrote: "Collectively among human minds, natural history becomes polydimensional and difficult to resolve into a generally acceptable rational system of categories of a nature that would be useful for purposes of evaluation and selection of representative areas."[74] Was there an implicit admission here that many square pegs were hammered into round holes and that a superficially rational plan led to conclusions which did not appear rational at all? The inconsistencies of the plan were not very troubling, however. The natural history section of the 1972 plan, was, even with its failings, an expression of, and a reconciliation with, new attitudes toward the natural world. Moreover, the plan would allow expansion of the National Park System in a manner in keeping with contemporary sensitivities and the pressures environmental groups were placing on the agency.

Many new natural areas have been added to the system since 1972, and most of them were touted as filling an empty niche in the System identified in the 1972 plan. The John Day Fossil Beds National Monument, added in 1974, had as its primary attraction the fact that five geologic epochs were represented in its exposed rock beds, including several epochs which the 1972 plan had identified as gaps in the System.[75] In the same year, Big Cyprus and Big Thicket National Preserves were authorized. The establishment of both had as primary goals the addition of forest types not previously found in the Park System. Theodore Roosevelt National Park, established in 1978, added 70,000 acres of badland topography to the

System.[76] Guadalupe Mountain National Park brought the world's most extensive Permian fossil reef into the System, while the establishment of Congaree Swamp National Monument, containing the last virgin stand of bottomland hardwoods in the Southeast, filled the nearly empty niche of eastern deciduous woodlands. The new selection criteria also formed a strong justificatory base for the vast new national parks in Alaska which grew out of the Alaska land settlement of 1981 and which doubled the area of the Park System in a single stroke.

The Alaska Lands

Park Service presence in Alaska had begun in 1917, the year after the agency was established, when Mather brought Mt. McKinley into the National Park System. In the next ten years, Katmai and Glacier Bay, each larger than any national park in the forty-eight states, came into the System as national monuments. In succeeding decades there were frequent initiatives to get large parts of Alaska protected as federal parks, forests, or monuments. For example in the 1930s, Robert Marshall of the Wilderness Society promoted the preservation of a large area of the central Brooks Range in northern Alaska by the federal government. However, it took statehood in 1959 to begin the chain of events which led to the great expansion of the Park System in Alaska.

As long ago as the mid-1960s, and perhaps even earlier, the Park Service leadership had wanted the Park System to get a share of the lands that would be retained by the federal government when the questions of Alaskan land ownership were finally settled. Parks carved out of the great Alaskan wilderness would be in the grand tradition. They would "reemphasize the primary mission of the National Park Service," that is to say, the management of spectacular natural areas.[77] A very similar idea was expressed by the agency's Northwest regional director when he said that "in Alaska the service [has] the opportunity to be doing what Steven Mather and Horace Albright did at the very start of the National Park Service."[78]

The 1972 plan strongly stated the agency's case for a large share of the Alaskan lands; thirteen of the forty-seven major gaps in the System, as identified by the double criteria of geographic and natural theme representation, would be filled by the national parks that the agency wanted to carve out of Alaska's public lands. Thus, as the agency had defined things, the proposed Alaskan parks would be a giant step toward completing the System. Moreover, the Alaska lands might be the last crown jewels added to the System. Several public officials close to the Service felt that the Alaskan additions were not only in the grand tradition but were probably the final ones in that tradition. Assistant Secretary of Interior Herbst told Congress as much.[79] An OMB officer said that: "Alaska is the only new frontier for the traditional western park, after that, expansion of the Park System has to be entirely in other directions."

It is ironic that although the 1972 plan buttressed the Service's arguments for what were perhaps the last great traditional parks, it was basically a forward-looking document of adjustment. It helped the System and Service face the closing of their own frontiers by justifying new natural area additions which would never have qualified on traditional criteria of scenic splendor or visitor appeal. It also satisfied agency critics whose credo of new environmentalism led them to believe that the grand tradition was obsolete as a basis of policy when it came to the natural wing of the Park System.

THE CONSEQUENCES OF CHANGE

Converging Agency and System Functions

Changes in management policies and selection criteria have had two important but not always obvious consequences for the National Park System and its keepers. First, the increased pressure for wilderness management brought to bear on the Park System and the Service was being felt by other public land systems and managers as well, and this introduced a new element into the long-running, strategic *pas de deux* of the National Park Service and the U.S. Forest Service. The active rivalry between the two agencies, so intense in the 1920s and 1930s, continued during the 1960s and 1970s. During these later decades, the participants had "a keen but wary respect for the ability of the other to stir up mischief." [80] The rivalry saw a brief quiescence during the "Peace of the Potomac," a truce imposed on the two agencies in the mid-1960s by Secretary of Interior Udall and Secretary of Agriculture Freeman, but according to one student of the Park Service, this only served to heighten the rivalry by making it the focus of great attention. [81] By the late 1960s, the competition was on again and it became the most celebrated interbureau rivalry in government. [82]

As we saw earlier, one important element of Mather's strategy for dealing with his agency's natural rival was to put as much functional distance as possible between the two agencies, or at least to create a popular image of a clear functional dichotomy: the Park Service as preserver, the Forest Service as exploiter. At least vestiges of this element of Park Service strategy were still extant in the early 1980s. The Park Service continued to promote a view of the Forest Service as aggressively utilitarian in philosophy and practice. For example, the author heard a young park ranger in Grand Canyon tell a tourist "our job is to protect trees; a forest ranger's job is to cut them down" when the tourist asked about the difference between the two types of rangers.

Unmitigated utilitarianism on the part of the Forest Service was no more the case in the modern era than it was during Mather's day. To be sure, in the mix of management philosophies the Forest Service embraced, a commodity-oriented utilitarianism probably predominated. As Robinson put it, "not even the timber industry, for all its recurrent complaints that the Forest Service is inadequately attentive to public timber needs, would suppose that the Forest Service is generally preservationist in its leanings." [83] Robinson's view of Forest Service sentiments was corroborated by a survey of service professionals which showed that 71 percent of them felt that the most important job of the agency was making resources available. [84] Nevertheless, dominance is not the same as exclusion, and the Forest Service's multiple use principle, officially adopted in 1960 as the basis of forest management, has allowed other attitudes to flourish over the last two decades. Fifteen percent of those polled on the above-cited survey *did* feel that preservation of the resources under the agency's management was more important than the promotion of their active use. How this attitude occasionally manifested itself in action is revealed in the way the Forest Service officially responded to a study in which rigorous techniques of economic analyses were applied to the question of the worth of wilderness areas. The response revealed an attitude not usually associated with multiple use or the general image of the Forest Service: [85]

> The Forest Service would vigorously resist any effort to set a price tag on the Voyageur's Highway of the Boundary Waters Canoe Area, the wilderness trails of the Wind River Mountains, the scenic beauty of the Northern Cascades, or the alpine meadows and the Golden Trout of the High Sierras. There are values here, and on all other wilderness type areas, that cannot be replaced at any price. These are values that simply defy comparison. In fact *they are incomparable!* The Forest Service intends to keep them that way.

The response continued, asserting that economic value was only instrumental to some other realm of value, and was not ultimate value in any sense. It argued that if pure economic value were made the criterion for land use decisions, "the superb values of snow-clad peaks, of the Grand Canyon, and of alpine lands in general sink to almost nothing."

The presence of, and official favor occasionally shown toward such attitudes within the Forest Service over the past two decades has served a very practical end; it has allowed the agency to adjust to increased pressure for preservation in land management by calling on a management strain already present. In fact, the Forest Service frequently was better able to integrate new environmental mandates into its mix of management responsibilities than agencies which were more preservation oriented by tradition. [86]

The Forest Service was initially opposed to statutory preservation of wilderness through the Wilderness Act of 1964.[87] The agency argued that it could give adequate protection to wilderness under its multiple-use management policies. Furthermore, it also felt that the proposed wilderness legislation would deprive it of flexibility in managing the national forests. The Forest Service also evaluated the wilderness proposal within the context of its interagency rivalry. The Forest Service feared that if land management were to be dictated by statutory fiat, the next step might be administrative reorganization of the federal public land systems. In such a case, the Forest Service's recreation-dominant areas might pass to the National Park Service.[88]

After the wilderness bill passed, the fears of the Forest Service seem to have been allayed, and it moved ahead rapidly in recommending the consignment of national forest lands to the wilderness system. The Park Service, on the other hand, moved slowly and it was more than five years after the passage of the wilderness bill that the first statutory wilderness in a national park was established. By 1978, national forest land comprised 89 percent of the nation's statutory wilderness.

In 1968 the National Trail System Act and the National Wild and Scenic Rivers Act were passed. As was the case with the Wilderness Act, each placed similar management restrictions on units of their respective systems, be they in national parks, national forests, national wildlife refuges, or Bureau of Land Management lands.[89]

Ironically, although the Forest Service initially feared the changes that these acts would interject into its famous rivalry, the Park Service probably had more reason to be fearful. With every public land bill which forced the Park Service and the Forest Service to manage land the same way, the distinction between the agencies' missions blurred. Reformers of the federal bureaucracy usually look first to eliminating overlap and combining similar functions when formulating their plans. For example, one of the main points of Nixon's reorganization plan was the idea of grouping bureaus with like functions into new super departments.[90] One of these super departments was to have been a department of natural resources whose charge was to include both the national parks and forests. Although such a move would not have cost the Park System or its keepers their independent identities, it would have reduced organization distance between the two agencies and the two systems, placed them under the same secretary, and considerably increased the chances of an eventual merger.

The threat of merger notwithstanding, mere functional overlap could hurt the Park System, since it increased its direct competition with the national forests for budget dollars. When this competition occurred in the past, the Forest Service, with its strong congressional support, was usually the winner. In fact, Nienaber wrote that many of the Park Service's problems could be traced directly to its "rather uneven competition" with the Forest Service when the two operated in overlapping areas of responsibility.[91]

Thus for the Park Service to encourage increased functional overlap by shifting its management of the natural parks toward strict wilderness principles put the independence, and perhaps even the separate existence of the Park System, at risk. Even if the Park Service were to put most of the great natural parks into wilderness status, the Forest Service would still manage the large majority of the national wilderness areas. The Forest Service would therefore have the strongest claims to the entire wilderness system if it were decided to place it all under one agency, either in connection with, or independent of, an overall federal reorganization. Perhaps an anonymous Park Service official feared this when he said that the wilderness system proposal would reduce all land within the wilderness system to "the low common denominator" of wilderness.[92]

Increased biocentric management and the functional convergence of the public land systems it encouraged prompted problems of identity as well as strategy. If the national parks were run with the primary aim of strict ecological preservation, what would distinguish them from mere nature preserves or national wildlife refuges? Like such areas, parks run on strict biocentric principles would require the services of scientists more than those of traditional park managers. Indeed, why should such places even be considered parks? For an agency whose collective ego was based in a good measure on its claim to being the best park manager in the world, this shifting and blurring of definition had threatening implications. Beyond questions of identity and interagency strategy, however, shifting attitudes and policies for the System's natural areas raised important questions about the Park System's base of political support.

A Narrowed Base of Support

By stressing park management for environmental ends, and shifting new unit selection criteria toward the ecological, the agency moved the Park System's base of support away from the park-using public at large and toward the environmental public interest groups. Moreover, this shift was accompanied by a willingness on the part of the agency to accept the environmental groups as legitimate sources of guidance in park management and selection and by an unwillingness to accord the same legitimacy to the public's wishes.

This has produced some disadvantages, one of which is a lessened political security. Although the value of the public as support is uncertain, so is that of the organized environmentalists. Rourke has pointed out that it is by no means certain that the goals of an agency will always be confluent with the goals of an interest group upon which it depends for political support.[93] As a matter of fact, it would seem that such a meshing of interests and goals can never be entirely the case. Those without formal power to implement policy need not take responsibility for specific courses of action.[94] So for them there is always the temptation to take an extreme moral stance, to grandstand, as the expression goes, at the expense of those who must implement detailed plans and make real-world compromises. The environmentalists certainly tend to do this; an incessant strain of self-righteousness runs through their writings on the Park System and the Park Service. Sax's preachy lecture to the Park Service on being true to its principles is a good example.[95] Neither he nor the other environmentalists, however, have to take responsibility for implementing what they propose. Moreover, such self-righteousness seldom prevents environmental groups from making themselves look good by making the Park Service look bad. As a political scientist familiar with the agency said, "Poor National Park Service, even its best friends love to kick it around in public."[96]

There is yet another point of intrinsic conflict between public agencies and interest groups. For a public agency, the best strategy for long-term prosperity is to maintain maximum administrative discretion by seeking the support of many different interest groups and by serving as many different interests as possible. In that way, it avoids capture by any one interest. An interest group, on the other hand, should strive for an exclusive relationship with an agency, which will allow it to capture the agency and bend it to its will.[97] When the environmental groups advocate quotas and reducing public access to parks, they are, in essence, telling the Park Service to knock out from under itself one of the major historical props of its support—its popularity with the general public. When Sax tells the Park Service that it must choose preserving the parks for their highest possible use over turning them into mere commodities, he is telling it to curtail its efforts at maintaining the support of the general public—the very support which could prevent its capture by environmental groups. To make such a policy palatable, environmentalists argue that the public is not much of a source of support anyway. One environmental lobbyist in Washington told the author that "public support isn't worth a damn, period."[98] Others are less extreme and simply hold that today's American public "racing through the parks in a Winnebago," as one expressed it, is not the source of support it used to be. Whether this is the case or not, it certainly serves the interests of the environmental organizations to claim it is.

Finally, there is never any guarantee that even if interests of the Park System and the organized environmentalists are presently confluent, they will always remain so, or that an interest group which today sees working on park matters as a good strategy and a wise investment of energy will continue to view it as such in the future. While it is to the advantage of a lobbying group to make an agency solely dependent on it, it is also in its strategic interest to keep open its options on whom and what it chooses to support. For the Park Service, this means that there is the danger of making an irrevocable decision which presumes the future support and political resources of the environmental groups, only to find that they have moved on to other concerns and taken their political power to other issues, thereby leaving the System out on a limb. The possibility of such a shift of interest away from the Park System by the environmental groups is a very real one, as illustrated by a comment by a Sierra Club official: "The cost of getting new national parks can be very high and seems to be getting higher. Look at [the new parks in] Alaska, or Tallgrass [Prairie National Park proposal]. There will be a point where getting new additions won't be worth our effort and we'll do other things." The number of causes embraced by the environmental movement is great; no organization need go down with what it sees as a losing one.

The Fire Island Compromise

Perhaps nothing illustrates the potential for a substantial divergence of interest between the System and environmental groups as much as recent events leading up to the passage of the Fire Island Wilderness Bill. As such they merit close review. Fire Island National Seashore, established in 1964, runs for more than 30 miles along Fire Island, the longest of the barrier islands off the southern coast of Long Island, New York. The park includes large undeveloped areas of beach and sand dunes. It also includes, as inholdings, permanent towns with both year-round and second homes, plus small hotels, bars, grocery stores, and other commercial facilities to serve residents and vacationers. At its closest point, Fire Island is only 20 miles from New York City, but except for its western and eastern tips, which can be reached by bridges, the island is roadless and accessible only by small, expensive ferrys which travel from points along the south shore of Long Island to the towns on Fire Island. The legislation which established the national seashore gave the Service two basic orders in managing it. The first was to make it accessible to the large metropolitan population which lives within an hour or so of the park. The second was to preserve the island from the private development which was likely to transform it wholesale if allowed to proceed without public intervention.[99] Toward these ends, the Park Service formulated a park development plan with four cardinal points. The agency would cap the expansion of the towns through the enforcement of zoning laws (by condemnation if necessary). It would leave some areas of the park in a natural, undeveloped state accessible only by foot. It would establish areas of concentrated visitor facilities: changing rooms, showers, snack bars, etc. Finally, it would make these facilities easily accessible to the public at large by instituting low-cost ferry service between Long Island and points along the national seashore.

For a number of reasons, discussed in chapter 6, implementation of the plan proceeded slowly. Then, in the late 1970s, a coalition of local environmental groups formed the Fire Island Wilderness Coalition with the aim of having Congress proclaim a 7-mile undeveloped stretch of the national seashore as a national wilderness area. Initially, the Park Service was ambivalent about the coalition's aim. By law, a national wilderness area has to be left free of roads and structures and is closed to motor vehicles. For the Park Service, this would mean a loss of administrative discretion, since future management decisions would be taken out of its hands. This is something which under most circumstances does not please a bureaucracy. On the other hand, the agency's own development plan for the national seashore called for leaving the area in question undeveloped anyway. This meant that supporting the national wilderness designation was an opportunity to please an active group of preservationists at no real cost to itself. The latter argument prevailed, and the Service came out in support of the wilderness designation.

The coalition soon found that, in spite of Park Service approval, there was not sufficient backing for the proposal to give it momentum in Washington.[100] Park Service support was real but tepid, and the agency did little to push the proposal through Congress. Some Long Island and Fire Island residents, especially beach buggy users, actively opposed the proposal because it would close the beaches to them; motor vehicles were anathema to the backers of the national wilderness system, and they were prohibited in wilderness areas.[101] Most residents of Fire Island, however, were simply indifferent to the proposal. Very few of them had ever used the area under consideration, and few saw its change in status from de facto to statutory wilderness as something which would affect them. Certainly it was not enough to inspire political action, in any case.

There the matter rested for several years: politically inert. A proposal had been advanced by a small group of active backers, many of whom lived outside the local congressional districts. Arrayed against it was another small group of active opponents. The two groups sparred against a backdrop of general local indifference. Given this line-up, the local congressmen were not inclined to sponsor the proposal, much less work actively for it. Then, with a stroke of skillful politics, the wilderness coalition put momentum behind their proposal. First, they neutralized opposition from the beach buggy users by shifting the boundaries of the proposed wilderness from the ocean itself back to the base of the dunes.

This would leave the beach out of the wilderness area and therefore not automatically closed to beach buggies. To sweeten the deal, the coalition added a proviso to the proposed bill that specifically prohibited the Park Service from banning beach buggies from the beach.

The coalition next gained the support of many island residents and that of the important Mid-Island Homeowners Association by adding a rider to their wilderness proposal which would prohibit the Park Service from spending more than half a million dollars on their development plans for the seashore, including the plans for the ferry and the terminal. Although they were largely indifferent to the statutory wilderness designation, this tapped an issue of considerable concern to the homeowners, for they feared the increase in day trippers that the agency's development plan would bring. The coalition rationalized its deal with the island residents by asserting that the development cap actually improved the bill. The coalition's newsletter stated that with the spending restriction, "the law thus protects not only the wilderness but the entire seashore." [102] From the point of view of the local congressmen, these changes were great improvements; they neutralized opposition to the bill while broadening its base of support. Representative Carney became an active sponsor of the proposal in its revised form, and so did Representative Burton. It now had momentum.

The Park Service also saw these changes as important, but from the agency's point of view, they were hardly improvements. The development cap of half a million dollars would allow it to spend only 10 percent of what its development plans called for, and this meant the end of plans for the ferry terminals and most of the beach facilities. Proponents of the wilderness proposal argued that the cap did not really kill the development plan since the Park Service could always go back to Congress and ask it to remove the development cap. The argument struck agency leadership as specious since, in effect, it was telling the Park Service that if it wanted to implement its plan, all it had to do was get Congress to reverse itself on the issue.

The Park Service seemed more concerned about the beach buggy clause, however, which would have ended its plans to gradually phase them out at the national seashore. [103] First, there was a question of exactly how much wilderness would be left if beach buggies were given free access to the beach in front of it. The area under consideration was 7 miles long, but in few places was it more than small fraction of a mile wide. Therefore permitting buggies on the beach meant that only a small percentage of the wilderness would be more than two or three hundreds yards beyond the permitted vehicle areas, and all of it would be easily accessible to those with these vehicles. Thus the bill seemed to have the net effect of decreasing rather increasing the wilderness qualities of the area. Second, if the law passed, this would be the first time that Congress had explicitly told the agency that it could not prohibit offroad vehicles in a particular area. To the Service this seemed like a very dangerous precedent. For these reasons it opposed the wilderness bill as amended.

Most environmental groups in Washington, including the Wilderness Society, continued to back it, however. Unlike the Park Service, they could live with the changes. No interest group is merely a lobby for a fixed set of ideals or specific goals. Such a group faces problems of maintenance that are similar to other bureaucratically structured organizations.[104] In modern environmental organizations, which depend on the support of a large membership that feels involved in the cause but which is not politically sophisticated, there is great pressure on leadership to feed that membership tangible victories and easily perceived accomplishments. The pressure is heightened by an awareness that environmental groups are competing for at least a partly overlapping pool of potential members. Thus, the establishment of Fire Island Wilderness had value as organizational sustenance. The price of compromise, the development cap and the beach buggies, was a discountable externality that the Park Service (and, arguably, the public at large), not the environmental groups, would have to live with in the future. As a high agency official close to the issue saw it, "The Fire Island wilderness issue was used by certain groups, local and national, in their own interests. The whole question of whether the bill really gave us a wilderness got lost in the process." Another agency official, when informed of the bill's passage while in the author's presence, rolled his eyes, cupped his hand behind his ear and asked, "Is that old Zhaniser [the prime mover behind the 1964 Wilderness Act] I hear turning in his grave?"

Ironically, precedent seemed as important in the decision of the Wilderness Society to support the bill as in that of the Park Service to oppose it. The environmental group, although initially concerned mainly with the great wilderness areas of the West, turned its attention eastward in the 1970s, and for many of the same reasons the Park Service had done so decades earlier. In 1975 it successfully guided the Eastern National Wilderness Bill through Congress, a bill which, among other things, eased the requirements of size and primordial condition on areas in the East proposed for statutory wilderness status.[105] Supporting the Fire Island proposal fit into the group's campaign to take advantage of that law. Since the law had been little used in New York or New England, the precedent value of a wilderness on Fire Island was great, great enough to allow the Wilderness Society, unlike the Park Service, to dismiss the question of whether the bill would in fact degrade the wilderness it claimed to be protecting. As a precedent, as the opening shot of a campaign, a flawed statutory wilderness had considerable value. A pristine but de facto wilderness had none.

Although Fire Island is hardly one of the great natural areas of the System, and in fact, given its location, it is more properly considered an urban park, what happened there does show that the interests of the National Park System can diverge considerably from those of the environmental groups which often presume moral authority to tell the System's keepers what is right and wrong. The Park Service's policies toward nature and popular access to nature in the parks have weakened its capacity to defend the System when such divergence occurs.

The Fruits of Accommodation

To quickly summarize this chapter, we have seen how, against a background of increasing public concern for the quality of the environment, the old mix of use and preservation in park management came to be seen as unsatisfactory by the Service's environmentalist allies in the 1960s and 1970s. The traditional stress on direct scenic appeal when evaluating natural areas for possible addition to the System also became progressively less acceptable. The Park Service responded to these changes by accenting resource preservation and depressing the importance of visitor access in its park development planning. In judging possible new units, the agency devalued scenic appeal and made geological and biological representativeness the primary formal criterion of evaluation. By doing so, it accommodated many of the environmentalists' wishes for the System, while at the same time it opened new avenues for System expansion. Within agency ranks, however, there was considerable dissent from official policy. Moreover, the policy shifts appear to have had their costs in a narrowed and less trustworthy base of support for the System, and in a blurred identity and a lessened clarity of mission for the System and its keepers.

When we turn to the agency's policy with regard to history, we first see problems of definition and management which were similar to those it encountered in dealing with natural areas. We see similar solutions, as well, i.e., ones which relied on increased professionalism and attempts to bring the agency's responsibilities in line with modern notions of history. The forces brought to bear on historic preservation policy were very different than those operating in the arena of natural area policy, however, and so were the agency's own attitudes toward, and interests in, historic preservation. As a result, historic preservation policy took on different hues, served different ends, and, ultimately, raised very different questions about mission and purpose in the Park Service.

5 History Policy

History and the National Park System Before 1972

America's interest in preservation and commemoration of the past has been traced back to the early days of the republic when monuments commemorating George Washington and his Revolutionary War companions were established.[1] The commemorative impulse intensified with the Civil War. Less than a year after the battle of Gettysburg and many months before Appomattox, Pennsylvania chartered the Gettysburg Memorial Association to oversee the preservation of the battlefield and the national cemetery established on it. In the next decades, the major battlefields of the Civil War became the objects of organized preservation efforts and near the end of the nineteenth century, the Chickamauga–Chattanooga, Shiloh, Gettysburg, and Vicksburg battlefields became National Military Parks under the management of the U.S. Army.

These preserved battlefields carried great emotional and symbolic weight in late nineteenth-century America, commemorating as they did, the deaths and personal sacrifices of the nation's most traumatic military conflict.[2] Veterans from every state returned to them by the thousands for massive reunion encampments on the anniversaries of the battles fought there. Monument associations in both northern and southern states turned the battlefields into vast gardens of bronze and stone with their profusion of statues and commemorative columns.

Although the National Park Service had been incidentally involved in historic preservation from its inception (Mesa Verde, Colorado, an Indian site of archaeological interest, had been included in its original holdings), it was not until these Civil War battlefields came into the National Park System in 1933 that historic preservation became more than marginally important to the agency. These battlefields were shortly followed into the Park System by the great monuments of Washington, D.C. With these additions, the importance of historic sites to the System not only increased, but the Park Service took on a sacerdotal function. It became a keeper of the flame of patriotism, the overseer of the sacred American places.[3]

Engendering patriotism was not the only possible role for the Park Service when it came to history, however. Nor was the commemoration of great events or people the only form that its involvement in history could take. There were more dispassionate voices calling for a different approach to the past. From the late nineteenth century onward, archaeological and other professional societies had worked to secure federal protection for the archaeological sites of the West.[4] For them, preservation meant protecting the sources of scientific and scholarly research into the past. Educators also saw historic sites as important, but as educational resources. They were important in teaching about the past, but not necessarily in glorifying it. Such a perspective made accuracy and breadth of resource preservation, not inspiration, the primary goals. This dispassionate view was expressed in a definitive form by Clark Wissler, an eminent historian who was asked by the Park Service to prepare a report on the role of the agency in the field of history. His 1929 report read in part:[5]

> It is recommended that the National Parks and Monuments containing, primarily, archaeological and historical material should be selected to serve as indices of periods in the historical sequence of human life in America. At each such monument the particular event represented should be viewed in its immediate historical perspective, thus not only developing a specific narrative but presenting the event in its historical background. Further, a selection should be made of a number of existing monuments which in their totality may, as points of reference, define the general outline of man's career on this continent.

This was the history of the scholar, not the patriot. It demanded a high degree of professionalism and even scientific expertise, and it was far different in spirit from the role associated with preserving the sites of battles or monuments to heroes. Thus, there evolved two quite different but not necessarily incompatible views of the role the Service could play in historic preservation.

In the decades of the 1930s, 40s, and 50s, the agency played both roles as the historical wing of the Park System steadily expanded into a truly disparate collection of historic sites and monuments. More battlefields were added, and there was a great penchant for acquiring forts of the old West. Indian sites and those which commemorated European exploration were added. The System's collection of boyhood homes of the presidents was begun. Sites not easily classifiable, such as the Truro Synagogue in Rhode Island and the Vanderbilt Mansion in New York, entered the System. If most of these sites were commemorative in purpose, some which were primarily educational or, to use the agency's preferred term, interpretative, were also added. Among these were the Salem Maritime Historic Site in Massachusetts and Hopewell Village in Pennsylvania. These two were selected more for their general representativeness than their association with great events. Salem was an early nineteenth century seaport, while Hopewell was an iron-making village which flourished around the same time.

Just as the matter of commemoration versus a more detached, scholarly preservation remained open, so did questions of significance, or more precisely, what historic sites were worthy of inclusion in the Park System. And beyond that, what made sites worthy of inclusion? We have seen how in the natural areas of the System, immediate, direct appeal of landscapes was the central criterion (at least officially). With historic sites the matter was not that simple. There were both the events and the material remnants to consider. What if a great event occurred but almost nothing remained of it on the landscape? Should such a site be added to the System? Or what of the reverse: What if the remains were well preserved but the events with which they were associated were not of prime importance? Should interpretability be a factor? What if the event was important but did not readily lend itself to interpretation? These matters appear to have been little discussed in the Service before the 1960s, and an unselfconscious eclecticism characterized the System's historic holdings before then.

From the 1930s onward, the Park Service also had "extramural" historical responsibilities, i.e., those which involved tasks beyond Park System management. Such responsibilities were conferred on the agency by the Historic Sites Act of 1935, which assigned broad powers and duties to the National Park Service.[6] The agency was to make a national survey of historic and archaeological sites, buildings, and objects to determine which of them had "exceptional value as commemorating or illustrating the history of the United States." The act also directed the Park Service to make cooperative agreements with states, municipalities, corporations, private associations, and individuals to preserve historic properties which were not under federal ownership. Discharging these tasks as well as minding its own sites kept the agency deeply involved in historic preservation.

In spite of these responsibilities and the steady increase in the number of historic sites from the 1930s through the 1960s, the historical wing of the Park System played second fiddle to the natural wing. In chapter 2 we saw how Albright, although a self-proclaimed history buff, looked at the acquisition of historical responsibilities largely in strategic terms; they secured the agency against easy assimilation by its rival, the U.S. Forest Service.

Moreover, the historical wing also had its political advantages at appropriations time since, according to Ise, "Congress showed a more generous disposition toward historic monuments than toward [natural] parks and monuments."[7] (Or at least this was the case before the rise of environmentalism in the 1960s.) The fact that the historical wing conferred strategic advantages did not mean that the Park Service had to treat it as a full partner of the natural wing, however. And it did not. In the words of an agency official who remembered the period, "history was a backwater."

For three complementary reasons historic preservation was at a disadvantage in intraagency competition for status and resources. First, the career ladder of the agency was firmly planted in what Ise called the scenic parks. The superintendencies of the great natural areas were the most important field postings in the Service and were viewed as the training ground of top agency management. Regional directors (except in the Northeast where the importance of historic parks outweighed that of natural parks) usually came from a background of natural area management. Second, most agency personnel, including agency leadership, had their professional training in such fields as forestry or other areas of natural resource management.[8] Third, outside interest groups concerned with natural areas tended to be more active, less narrowly focused in their interests, and less ephemeral than those outside groups concerned with the agency's historical responsibilities.[9]

The New Concern for Historic Preservation

An important change occurred when, in the mid-1960s, the Historic Preservation Act was passed.[10] The act was prompted in part by the wholesale destruction of historic sites during the great construction boom of the 1950s and early 60s. Urban commercial development, the interstate system, slum clearance, and large new areas of suburban development all took their toll on historic sites. This raised historic preservation to a new level of importance as an area of federal concern. The Historic Preservation Act aimed at stopping, or at least slowing, this destruction and authorized the secretary of the interior to maintain a National Register of districts, sites, buildings, and objects which were significant in American history, architecture, archaeology, and culture. The act established matching grants-in-aid for the states to help them prepare state historic preservation surveys and plans. It also established matching grants to support acquisition and preservation of state historic sites. Matching grants were also made available to the National Trust for Historic Preservation. The act established an Advisory Council on Historic Preservation, which was to count among its members the secretaries of the Interior, Commerce, Treasury, and Housing and Urban Development, plus the Attorney General, the chairman of the National Trust, and ten private citizens. Finally, the act established procedures to protect registered sites and buildings from federal or federally assisted undertakings. The act was part of President Johnson's overall program to improve the aesthetic quality of the American environment, which in turn fit into his broad Great Society vision. It also fit into Udall's notions of the New Conservation, which included aesthetic and historical preservation as much as it did conservation of natural resources, the more traditional concern of the conservation movement.[11] Although the Park Service had for a long time slighted its historical wing, this upward revision of the importance of historic preservation on the federal agenda, coming as it did on the heels of the crisis surrounding the removal of Director Wirth, and indeed springing from many of the same causes, could not help but influence the thinking of agency leadership.

Taking his cues from top administration leadership, Hartzog fought to establish a place for the Park Service on the historic preservation side of the New Conservation and the Great Society.[12] He gave the National Historic Preservation Act his strong backing, and, due in large measure to his efforts, the agency was assigned an important role in carrying out its provisions. The Park Service was given responsibility for maintaining the National Register and for evaluating sites for possible inclusion on the register. It was to administer the grants-in-aid to the states and to act as an advisor to the states in their historic preservation programs. It was to be the National Register's watchdog, protecting sites on it from threats stemming from federal projects or programs. Finally, the director of the Park Service (or his designee) was to serve as the executive director of the Advisory Council on Historic Preservation.

In addition to these new "extramural" responsibilities, the mid-1960s saw a massive entry of new historical units into the National Park System; twenty-three were added between 1964 and 1968. The group contained several of those old standbys, western forts. It also contained sites associated with artists, such as Carl Sandburg's home, and those associated with people who changed the way we view the world around us, such as the John Muir National Historic Site.

Questions of Standards

This upward revision of the importance of historic preservation to America, the federal government, and the Park Service brought several thorny problems with it. Perhaps first among them was the long-unsettled question of the agency's philosophy toward the American history it was to be preserving. The new importance given history brought new pressure for a definitive policy. While historic preservation was a minor part of the agency's mission, and while most people did not think history extended past battlefields, this was not a pressing problem; a policy of casual, inconsistent eclecticism sufficed. Now, however, with historic preservation increasingly the concern of the politically powerful, the historical wing of the System might get beyond the agency's control unless a consistent policy were thought through and standards of historical significance were developed. The Park Service had had several warnings that its control was slipping. One came in connection with Lyndon Johnson's boyhood home.

The debate over what was significant and therefore deserving of preservation at the Johnson boyhood home recalled nothing so much as the stories of Jorge Luis Borges and their concern with layered levels of reality and illusion. It also showed how bizarre debate on historical questions could become if it did not take place within established parameters of significance. When the Park Service acquired the boyhood home of Lyndon Johnson, it inherited a bit of a problem with it.[13] Park Service historians felt that in the interest of straightforward accuracy, the house should be "restored" to the condition it was in when Johnson lived in it as a boy, which is to say, dilapidated. Incumbent President Johnson, on the other hand, had different plans for the house. These included window boxes full of flowers, a fresh coat of paint on the clapboard exterior, and a well-maintained yard and fence—none of which the house had seen when Johnson actually lived there. In the end, the Service went along with the president's wishes; not to have done so would have been impolitic to say the least (figure 5-1). A former agency historian argued that "the reconstruction according to Johnson," i.e., the projection of the past that Johnson wanted, was what was significant, "and preserving that projection was the agency's chief responsibility at the boyhood home." If the Park Service had managed to establish clear guidelines for itself on what was and was not historically significant, perhaps such a retreat into the labyrinths of rationalization would not have been necessary.

Several years earlier the agency's historical wing had been involved in another project in which both the agency's freedom of action and the integrity of history appeared to come up short. In the early 1960s, President Kennedy was offended by the degree to which Pennsylvania Avenue between the Capitol and the While House had become a blighted corridor of marginal businesses and decaying, abandoned buildings. He set the wheels of improvement in motion, and Udall's Department of the Interior was given responsibility for conducting studies on how to best redevelop the corridor. The Park Service in turn was assigned the job of determining the historical significance of the buildings along the avenue. The thinking in the secretary's office was that the greater the historical value of the avenue, the easier it would be to justify spending large sums of money in upgrading it. If it could be declared a national historic site, so much the better. Accordingly, the Park Service was ordered to find a lot of history. The agency carried out the evaluation as requested, and, as ordered, it dutifully discovered that, in the former chief historian's words, "every

Figure 5-1.
The boyhood home of Lyndon Johnson as restored by the National Park
Service.

square foot [of the Pennsylvania Avenue corridor] was oozing history." [14]
The Park Service had no choice but to find it so. In 1965 the Pennsylvania
Avenue National Historic Site was established.

The agency also had little choice with regard to the Kosciuszko house,
in reality the boarding house where the Revolutionary War hero, Thaddeus
Kosciuszko, stayed for a couple of months when he returned to North
America briefly in the 1790s. The agency was not particularly interested in
the site. Agency historians felt that the site's connection to the hero was
tenuous at best and that there was already a glut of Revolutionary War
memorials and sites in the System. [15] Its addition to the National Park
System became a point of ethnic pride to Philadelphia's Polish community,
however, and once this occurred, the agency could not stop the proposal's
passage through Congress. All it could do was tastefully ignore the site
once it was in the System.

Finally, there was the matter of Ellis Island. After the vast immigrant
processing complex in New York harbor had been abandoned, it came into
the Park System as a national monument. The Park Service did not need to
maintain the entire complex to interpret it, and it certainly could not afford
to restore all of it. However, those outside the agency who supported the
establishment of the national monument on Ellis Island insisted that the
Park Service could not tear down any of it because it all had the same
historical significance. [16] The impasse pointed out the need for some policy

guidelines which would designate significance in managerially reasonable terms.

All of these incidents took place against a backdrop of increasing congressional demand for pork barrel parks, so there was now more pressure on the gates to the System. If the agency was to maintain a good measure of control over those gates, reasonable, consistent criteria for opening and closing them seemed very necessary. Accordingly, Hartzog decided that the Park Service ought to develop a comprehensive plan for the entire historical wing of the National Park System. Such a plan would allow the agency to take stock of the current state of the historical wing, and it would serve as a guide to its future expansion. It could settle matters of significance criteria and the question of commemoration versus a more scholarly preservation.

Hartzog put his staff to work on the project, and in 1972, his last year as director, the plan was complete. It appeared as volume one of a comprehensive, system-wide plan, with volume two, discussed in the preceding chapter, treating the System's natural areas. The way the set of guidelines for historical holdings was promulgated, as a companion to the natural areas guidelines, seemed a definite, final confirmation of the increased importance with which the agency viewed its historical responsibilities. Indeed, it suggested that the agency had at last accorded historic sites full parity with natural areas. The plan was intended to be a Bible for the agency in historic preservation, and during the past decade it has been one of the agency's most important policy documents. For the student of the National Park Service, it is a rich source of insights into agency thinking. As such, it deserves careful scrutiny.

PLANNING FOR HISTORY

The 1972 Plan

The 1972 plan for history reflected the comprehensive, categorical approach recommended to the Park Service by Wissler in 1929. It attempted to carry out Wissler's suggested goal of preserving and illuminating "the general outline of man's career on this continent."[17] The degree to which Wissler's spirit molded the 1972 enterprise was reflected in an evaluation of the plan by Ronald Lee, formerly the agency's chief historian. He saw the plan's ultimate aim as first, "providing the historic preservation program with an underlying framework which embraces the entire history of man on the North American continent"[18] and second, allowing the Park System to preserve and interpret "through carefully selected monuments a noble panorama of the full sweep of that history for the benefit and inspiration of the United States." Lee's very choice of words, "the entire history of man on the North American continent" seemed to be a paraphrase of Wissler's "general outline of man's career on the continent." Scientific, balanced preservation and broad representation had triumphed over the glorification of the sacred and the special.

In the plan, the history (and prehistory) of the nation was subjected to a dispassionate categorical rigor that approached that of biological taxonomy. American history was broken into nine major "themes:"[19]

(1) The original inhabitants.

(2) European exploration and settlement.

(3) The development of the English colonies.

(4) Major American wars.

(5) Political and military affairs.

(6) Westward expansion.

(7) America at work.

(8) The contemplative society, and

(9) Society and social conscience.

These were in turn divided into forty-three "subthemes."[20] For example, the theme America at work was divided into six subthemes according to type of work:

(a) Agriculture

(b) Commerce and industry

(c) Science and invention

(d) Transportation and communication

(e) Architecture

(f) Engineering

Each of these subthemes was further divided into smaller niches called "facets." For example, subtheme 7c, science and invention, defined by the plan as covering:[21]

> the discovery or invention of significant concepts, phenomena, practices, and devices in the United States, applied in such fields as commerce, industry, agriculture, transportation, communication, and medicine,

was broken down into the following "facets," each covering a separate area of American science and invention:

7c1. Agriculture

2. Anthropology and ethnology

3. Commerce and industry

4. Communication
5. Medicine
6. Natural science
7. Scientific exploration
8. Transportation

In some cases, facets were further broken down into "subfacets."[22] For example, theme 9, society and social conscience, had as one of its subthemes, 9c, environmental conservation. This was in turn divided into two facets, 9c1, the natural environment, and 9c2, the cultural environment. Each of these two facets was further divided by time periods to produce subfacets. For example, acts of preservation of the natural environment (9c1) were divided into those occurring before 1865 (subfacet 9c1a), those occurring between 1865 and 1900 (subfacet 9c1b), and those occurring after 1900 (subfacet 9c1c).

Having established these categories, the Service fitted the historical units in the Park System as of 1972 into them. In some cases, such as western forts, the fit (in this case into the theme of western expansion, subtheme of military-Indian conflicts) was effortless. Elsewhere things were not so easy. The fact that the historical wing's growth had been incremental and disjointed, responding to politics and passing urges to commemorate rather than guided by an overall plan of representative acquisition, meant that, as was the case with natural areas, some shoehorning had to be done. It showed. For example, Mount Rushmore, a tourism boosting project initiated by the Black Hills tourist industry and foisted upon the Park System in 1925, was made to stand as the System's representative of the modern sculpture facet. Piscataway Park, where the Park Service had acquired a little property and some development rights to protect the views from Mount Vernon, was classified as a historical site and placed in subfacet 9c2b, history of the preservation of the cultural environment, post–1906.

The taxonomic rigor was even extended to the point of extruding the sacred from historic commemoration. Nothing could have been more in the sacerdotal tradition of commemoration than the Statue of Liberty; it is a monument to such abstract ideals as freedom and the American spirit, and beyond that, it is one of the great altars of American patriotism. But it too was squeezed into the scheme of comprehensive yet particularistic representation. Specifically, it was made to serve as an example of a particular socioeconomic process by its placement in subtheme 9a, American way of life; facet 9a1, the melting pot. One suspects that some categories were developed to accommodate and, at the same time, ennoble holdings with scant justification. For example, the Johnstown flood memorial had a neat little niche of its own in facet 9b2, humanitarian movements; subfacet 9b2f, emergency aid. Mar-a-lago, a garish 1920s beaux-arts mansion in Florida, willed to the federal government in 1969 and turned over to the National Park Service when no good use could be found for it, was fit into the plan under theme 9, society and social conscience; subtheme 9a, American ways of life; facet 9a5, economic classes.

History and Strategy

One might justifiably ask why the Service felt it was necessary to fit every holding into a categorical scheme. The answer might be that such rigor, even to the point of disallowing the sacerdotal past of much of the Park System, was eminently useful as an instrument of agency strategy. In particular, underplaying the incremental heritage of the historical wing of the System by emphasizing (one might even say inventing) a comprehensive logic to the historical holdings was very important to the agency in its dealings with the powerful forces in its political environment. First, by denying the slightly seedy past and truly eclectic nature of the historical wing, the Service hoped to encourage respect for the System as a whole. As we saw in chapter 3, by the early 1970s the Service was losing control over its relationship with many important actors in its affairs. This was especially the case with regard to Congress, which was growing more appreciative of the pork barrel potential of national parks. The possibility that Congress would capture complete control of the historical wing was a matter of real concern for the agency. According to an agency administrator responsible for legislative liaison with Congress, "if the Service doesn't have a plan, if it just drifts, the System will become just the sum of Congress's proposals [for new national parks]." By making it appear that there was a niche for everything in the System, even the less worthy units were legitimized. Their questionable pasts were made respectable, and the true circumstances surrounding their entry into the System were obscured. With this done, they would no longer encourage similar congressional pork. Nor would it be quite as easy for cabinet officials or those in the executive office to use the historical wing of the System toward their own ends, the way Udall did with Pennsylvania Avenue. In essence, such a systemization, even if inappropriate, illogical, or simply silly, might buy the respect of powerful outsiders.

Because the 1972 plan was long and detailed (in keeping with its overt goal of covering all American history), it gave the impression that the agency was serious about its responsibilities toward history, and that it possessed the skills befitting the federal government's lead agency for historic preservation. The plan appeared to be well thought out, and it was bound in book form, with high-quality graphics and layout. It bespoke confidence, detachment, and, above all, professionalism.

Striving for Professionalism

This impression of professionalism, like the taxonomic rigor, could gain respect for the historical wing of the System, and for its managing agency. It has long been recognized that assertions of expertise and claims of professionalism are two of the public bureaucrat's strongest weapons in his battles with elected representatives for a share of power in government, and that these can be very powerful weapons.[23] Many political scientists have seen legislators as almost completely dependent on bureaucrats for solutions to public problems, and even for identification and description of the problems themselves, in areas where professional judgment counts for much.[24] Over time, this de facto dependence has even been reinforced by theories which tend to legitimize it. For example, Patterson spoke of "sapiential authority," in which the right to make the decision went to he who was most knowledgeable about the issue in question, i.e., the technical or professional expert.[25]

Natural resource agencies, including the Forest Service, have always played the card of sapient authority. In fact, some have seen the ascendency of the professional, trained bureaucrat armed with sapient authority at the very heart of the progressive conservation movement.[26] However, not all bureaus were lucky enough to have responsibilities which easily lent themselves to professionalism.[27] Some responsibilities involved decisions which were blatantly subjective or which gave little play to scientific procedure or technical expertise.

The Park Service was not among the advantaged in all respects. Although it aped the professional stance of the Forest Service in many ways and was recognized by some as "the best park manager in the world," when it came to rejecting or accepting parks, the Service had little claim to sapient authority. This was especially the case with historic sites, and doubly so when commemoration was involved. Commemoration provides little leeway for the exercise of professional expertise, for who was to say that one man's judgment was not as good as another's when it came to which of two heroes should be revered or which of the fallen should be remembered? This was exactly the problem the Service faced; claims of expertise could not be staked in the unsteady ground of subjectivity. The National Park Service was never respected as an expert, according to one writer, and when it came to history, this was a pressing problem.[28]

By shifting from commemoration to balanced representation within an overarching plan, the agency eliminated the problem of inevitable subjectivity and value judgment associated with commemoration. It introduced a strong element of detachment. Once a scheme of systematic representation was set up, underrepresentation could be determined in a simple, straightforward and seemingly value-free manner.[29] An empty niche meant insufficient representation in the System, and the issue was closed. The determination was formal and free of emotion and politics. With such a plan in place, the agency could invoke the doctrine of neutral competence, i.e., a claim to authority based on a stance above politics.[30] It had drawn up the plan with political detachment; it had no axes to grind. Even the plan's appearance of finality could serve the Service's ends. An agency cannot continually change its policies according to the whim of political sentiment. To do so amounts to an admission that the role of expert is not really important in government.[31] By implying a rejection of short-term political expedience as a modus operandi in its historical holdings, the agency enhanced its credibility and its image of professionalism. Perhaps this would set the System's historic wing above the grasp of those who would casually tamper with it for their own political gains.

One might suspect that such a structured, rigid plan would have restricted the Service's freedom of action and would have reduced thereby that which all bureaucrats value: administrative discretion. If so, this would have been a high price to pay for whatever benefits the plan conferred. Closer examination reveals that this was probably not the case, however. The plan added rather than reduced policy options, and it broadened agency horizons. It was also an ambitious instrument of expansion. In the words of an administrator close to the agency's historic preservation planning process, the 1972 plan was "blatantly a vehicle for expansion. It identified the holes in the system, it even created some." Shortly after the plan was released, Lee wrote approvingly that it would loosen the restrictions on the growth of the System's historical wing.[32]

Reductio ad Absurdum

Administrators prize the flexibility afforded by generalities and nebulous phrases. According to one student of bureaucracies, "the value of flexibility in uncertain environments is much too precious to surrender to the unbending tyranny of formulas [which] tend to freeze the political process to allocations based on measurable data."[33] In the right place, however, skillfully arranged specificity and conceptual clarity can serve equally as well as instruments of flexibility. With the 1972 history plan, the Park Service created such a vast taxonomic system that its actual holdings were almost lost among the hundreds of very specific, and usually empty, niches. For example, the Park System's fine collection of American Indian sites was reduced to insignificance when it was placed in the array of niches drawn up for aboriginal America. Table 5-1 shows theme 1, original inhabitants, fully ramified to its subthemes, facets, and subfacets. Table 5-2 shows the aboriginal sites in the System assigned to their appropriate niches. Fifteen of the twenty-four sites went into one of the thirty-two niches, while the others were spread out among another six. This meant that twenty-five of the thirty-two categories were unrepresented in

Table 5-1. National Park System History Plan,
Theme 1–Aboriginal Americans

1a. Earliest Americans
    1a1. Migrations from Asia
      2. Early Peoples of Alaska
      3. Paleo-Indian Horizon (Eastern and Western United States)
      4. Archaic Indian Horizon (Eastern and Western United States)

1b. Native Villages and Communities
    1b1. Southwestern United States
      2. Eastern United States
      3. The Great Plains
      4. Western United States
      5. Alaska
      6. Puerto Rico and the Virgin Islands

1c. Indian Meets European
    1c1. Indian Life at Time of Contact with the European
      2. Changes in Native Life Due to Contact
        a. Changes in social and political organization
        b. Native migrations and warfare due to contact
        c. Changes in economic base
      3. Native Influence on the European

1d. Living Remnant
    1d1. Government Policy toward the Conquered Aboriginal
      2. Indians of the East
      3. Indians of the Plains
      4. Indians of the Southwest
      5. Indians of the West
      6. Indians and Eskimos of Alaska
      7. Natives of Puerto Rico and the Virgin Islands

1e. Native Cultures of the Pacific
    1e1. Hawaiian
        a. Prehistory
        b. Contact with the Europeans
        c. Aboriginal history
      2. Other Pacific Cultures
        a. Prehistory
        b. Contact with the Europeans
        c. Aboriginal history

1f. Aboriginal Technology
    1f1. Hunting and Fishing Techniques
      2. Quarrying
      3. Trade
      4. Arts and Ceremonialism

Source: National Park Service, *National Park Plan,* Part I, pp. 2–5.

Table 5-2. Adequacy of Theme 1 Representation

Facet	Representative unit	Facet	Representative unit
1a1.	None	1b5.	None
1a2.	None	1b6.	None
1a3.	None	1c1.	None
1a4.	Russell Cave	1c2a.	None
1b1.	Aztec Ruins	1c2b.	None
	Bandelier	1c2c.	None
	Canyon de Chelly	1c3.	None
	Chaco Canyon	1d1.	Nez Perce
	Casa Grande Ruins	1d2.	None
	Gila Cliff Dwellings	1d3.	None
	Hovenweep	1d4.	Hubbell Trading Post
	Mesa Verde	1d5.	None
	Montezuma Castle	1d6.	None
	Navajo	1d7.	None
	Tonto	1e1a.	City of Refuge
	Tuzigoot	1e1b.	None
	Walnut Canyon	1e1c.	None
	Wapatki	1e2a.	None
	Yucca House	1e2b.	None
1b2.	Effigy Mounds	1e2c.	None
	Mound City Group	1f1.	None
	Ocmulgee	1f2.	Alibates Pipestone
1b3.	None	1f3.	None
1b4.	None	1f4.	None

Source: National Park Service, *National Park Plan*, Part I, pp. 2–5.

the System, and were therefore gaps in the collection which should be filled. Even the Civil War, generally a well-represented theme, was capable of accommodating more sites, since its facets of naval history and diplomacy were empty. Overall, of the 281 facets, eighty-five were represented by at least one unit of the Park System. This left twice as many empty.

Even the quantification of history could serve agency purposes. One might question the worth of an approach which reduced Gettysburg and Piscataway Park to equally weighted plugs in a matrix. In fact, one might suspect that the entire plan was an exercise in inappropriate exactitude or what Whitehead called misplaced concreteness.[34] Yet, numbers have an appeal and a persuasiveness of their own, if only because they are "precise quantities in a sea of uncertainty."[35]

In sum, with the 1972 history plan, the Park Service implied that the System, far from being nearly rounded out where historic preservation was concerned, was in fact little more than the beginnings of what it should have been. The agency also implied that it had measured exactly how far the historical wing had to go before it was what it should have been. (In both these respects the 1972 history plan was similar to the natural area plan.) The plan even attempted to circumvent whatever limits and drawbacks the comprehensive, representative approach entailed when it suggested that "regardless of the percentage of representation, no theme or subtheme is well represented so long as a prime site, such as Mount Vernon or Valley Forge, remains outside the National Park System." [36] In other words, the agency could have its cake and eat it too; the plan was to confer all the advantages of quanta and category, yet allow the agency to avoid the disadvantages which might attend them.

The Commitment to Expansion

It might be asked why the Park Service should commit itself to a plan which allowed and even promoted major expansion of its historical holdings. This is a very real question; we cannot simply say that organizations tend to expand and be done with it. As scholars have pointed out, expansion is seldom an unmitigated good for an organization. And it is seldom a course an agency will pursue if the liabilities are seen to outweigh the advantages. [37] The rapid expansion of its historical holdings could have brought the agency serious problems. First, there might have been staffing and management difficulties. If expansion of responsibilities stretched thin the agency's administrative cadre, mistakes might have become frequent, and its reputation for superb park management would have suffered. Second, the addition of new historic sites would have increased the fixed expenses of the agency in managing the System, and thereby reduced its ability to weather periods of fiscal retrenchment. Moreover, in the event of cutbacks, the demands the new units would have made on agency resources might have hindered its ability to manage the truly great units of the Park System. [38] Although most agencies contemplating expansion must face such risks in one form or another, this did not make them any less real or serious for the Park Service.

On the other hand, expansion, in and of itself, would have conferred advantages on the agency. More parks would have meant more management slots and greater opportunities for advancement on the career ladder. [39] This in turn would have encouraged good managers to remain in the Service rather than seek their fortunes elsewhere. Expansion of the System would have also provided the means for increased internal cohesion, by increasing the rewards for loyalty (in the form of new jobs) at the disposal of agency leadership. If it was under agency control, expansion also would have allowed the Park Service to accumulate a stock of political I.O.U.'s among appreciative congressmen. Finally, a major factor which led

Mather and Albright to encourage System expansion still operated, i.e., an expanding Park System would have meant more clients served.

The plan, however, aimed at conferring more than the general advantages of expansion. The agency's claim to professionalism and sapient authority, which the 1972 plan made, was also a claim to the role of the System's gatekeeper. As gatekeeper, the agency could narrow the broad, if self-bestowed, expansionist mandate into the specific course of expansion which most benefited the System and its keepers. The 1972 plan laid the groundwork for such a narrowing by encouraging expansion in such a way as to give the Park System a relevance and even a certain trendiness, while permitting its keepers to maintain the all-important stance of neutral competence.

The Past in Service of the Present

The use of the past to make comments on present conditions or to justify and legitimize present interests has long been a feature of American historic commemoration. Selective reverence for the past and stands on contemporary social issues often seem to go together. For example, the connections between the Washington Bicentennial, a "colonial heritage" celebration in 1932, and the nativist movement of the late 1920s and early 1930s were strong. Many of those most active in promoting the bicentennial celebration were also leading supporters of the bills to restrict immigration.[40]

Agency leadership was not blind to the advantages to be derived from linking the System, through historic preservation, to contemporary social forces whose importance appeared to be increasing (or in which agency leadership genuinely believed). Whalen and his immediate predecessors worked hard to add sites commemorating civil rights and black history. Whalen actively supported the establishment of Women's Rights National Historical Park in Seneca Falls, New York—the site of an 1853 women's rights convention. The former director said of that support: "I believed in the cause [of women's rights] and getting involved in it made good political sense, too. Think of the visibility; there is the National Park Service on the ten o'clock news, hosting annual women's rights conventions."[41]

For Whalen, the need to use history to comment on the present far outweighed the need to apply standards of national significance to the intrinsic qualities of sites being acquired. He told the author, "National significance was never a test of mine. I never got a clear definition of what it was anyway."[42] Whalen and his predecessors saw parks with the right historical themes as agents of both social redress and the historical revisionism which accompanied the heightened ethnic consciousness of the

1960s and 1970s. The opportunities for the agency in this regard were clearly seen by the NPFF's study group.[43] It suggested that the Service should undertake a "tough-minded reevaluation of what is 'officially' said in all historical interpretative programs." It judged the agency's view of history to be "narrowly conceived, and out of touch with contemporary scholarship," and it advised that interpretative material for historic sites be rewritten to remove old historical and cultural biases. The NPFF group suggested that "minority historians—blacks, orientals, Indians, and citizens of Hispanic origin" be employed as consultants to aid in this reinterpretation and to suggest additional sites of cultural and historical significance.[44]

The structure of the 1972 plan allowed the agency to pursue this relevance by investing sites with the ex officio significance which came from filling niches the plan created. For example, the theme of society and social conscience was set up to include women's rights as one of its facets. With the plan in place, the establishment of the Women's Rights National Historical Park became a professional act filling a predetermined need, rather than an act of opportunism.

With regard to black Americans, the Service gave itself considerable freedom by establishing several facets through which they could be accorded recognition: the slavery facet of American ways of life, the civil rights facet of social movements, and the abolitionism facet of humanitarian movements. By setting up separate niches for civil rights and abolitionism and then (perhaps logically, given the separation) placing the Frederick Douglass Home into abolitionism, the politically important civil rights niche was left empty. The Park Service could then move to fill it with great enthusiasm and fanfare, which it did. Former agency director Whalen took personal credit for instituting the drive which led to the establishment of a monument to Martin Luther King.[45] Ultimately, the Park System wound up not only with a site but with an entire historic district. In 1980, President Carter signed the law establishing the Martin Luther King National Historic Site. It covered 23.5 acres in downtown Atlanta, and on it the National Park Service planned to develop picnic areas, playgrounds, and parking lots. It would renovate existing structures associated with King, including the Ebenezer Baptist Church where he preached. It would establish and maintain the Martin Luther King, Jr. Center for Social Change, and it would care for King's birthplace and grave.[46] The impression that blacks previously had been underrepresented was furthered by burying sites associated with blacks in obscure niches. The 1972 plan placed the George Washington Carver National Historic Site in the agriculture facet of the science and invention subtheme. Tuskegee Institute was placed in the history of American education facet. This gave the agency the freedom to add more units commemorating black achievement as they became available.

In other cases, niches not explicitly connected with social issues were filled in such a way as to identify the Service with prominent or ascendant social forces. For example, the Georgia O'Keeffe home in northern New Mexico was recently brought into the System as a national historic site. The site ostensibly filled the niche for twentieth-century American painting. It received its most active backing on a national level from women's groups, however, and was seen within the Service as a recognition of the achievements of American women. It was this undercurrent of relevance to women that prompted the agency leadership to strongly support its entry, even to the point of bending the agency's own rules. According to agency policy, "properties achieving historical importance within the last 50 years will not, as a general rule, be considered. . . ."[47] The policy manual allowed for an exception, however, when the site under consideration was associated with "persons or events of transcendent importance." The transcendent importance clause could be used without reservation to justify the Martin Luther King site, but applying it to an admittedly good but quite contemporary artist like Georgia O'Keeffe was another matter. The agency solved the problem by determining that O'Keeffe did some of her best work more than fifty years ago, and, having determined this, it could (and did) argue that she achieved historical importance in time to beat the half-century rule.

Putting such facets as landscape architecture and urban design (both under the architecture subtheme) and commerce and industry (under the theme of Americans at work) into the plan allowed the Service to take a hand in urban renewal through historic preservation. The Lowell National Historic Park is a 450-acre site in downtown Lowell, Massachusetts whose primary purpose is, according to the Park Service, to preserve and interpret a site from the earliest days of America's industrial revolution.[48] The 1972 plan attached a value to preserving exactly what Lowell had plenty of, derelict factory buildings, while it encouraged thinking about historic preservation as applicable to an entire urban landscape, a scale necessary for an historic preservation project to have a significant impact on a city.[49]

The Lowell park's development plan called for the Service to acquire and restore several of the derelict industrial buildings in the area and to coordinate planning for the considerable amount of private land which was to remain within park boundaries. The sponsors of the bill, Massachusetts Senators Kennedy and Brooke and Congressman Tsongas, looked at the park as a means of stimulating the economy of the decaying manufacturing city within their northeastern state.[50] The importance Lowell might play in promoting urban redevelopment was also recognized by Park Service leadership, and this made it politically attractive. According to one of the Service's most highly placed historians, "Lowell was a great idea. It was a catalyst for historic preservation but it also turned a whole town around. [This gave it] a useful social role." The base of the agency's argument for the national significance of the Lowell district was the 1972 history plan.

We see, then, that the expansion of the historical wing of the Park System on the Park Service's own terms, in directions of its own choosing, and in an atmosphere of professional decorum, could have paid large dividends in respect for the agency. It also might have bestowed a new sense of relevance on the System in an era when the impression of relevance counted for much. The plan was not to have the desired results, however.

OUTCOMES OF HISTORY POLICY

The Many Uses of Historical Parks

For all its ambitions, historic preservation planning has not gained the Park Service much respect for its professional expertise from outsiders. Nor has the Service settled for itself the question of what history it should be preserving or how important history and its professional historians are to the agency. Moreover, the plan has not allowed the agency to wrest control of the gate to the National Park System from Congress.

Taking the last point first, the legislators have little respect for agency opinion, and the significance criteria which the plan promulgated have meant little to Congress. Opinion within the agency's historical wing is that Congress listens to the Park Service when convenient and ignores it otherwise. There were several reasons why the agency, even armed with its 1972 plan, failed to gain congressional respect with regard to history. First, when it came to nature, there were national environmental organizations which had to, and did, make comparative judgments about the worth of new park proposals when deciding where to allocate their lobbying energies. There was no parallel activity for the historical wing, however. Here support for new units tended to be local, and the mosaic of local support for individual projects did not transcend parochialism often enough to encourage the kind of comparative, system-wide thinking which went on where natural areas were concerned. According to an outsider close to the historical wing of the Service, "people in a city will be all fired up about saving the old bank building, but they won't do anything for the historic bank in the next city." With support being such a patchwork of idiosyncrasy, there was little external pressure on Congress to weigh the relative merits of historic sites.[51]

Second, in the 1970s, demands that the agency adopt policies aimed at ameliorating the nation's pressing social problems sometimes focused on historic sites, and this too undercut any objective criteria of historical significance. For example, the NPFF group, searching for a way to make the National Park System more meaningful in the era of the urban crisis, wrote that:[52]

> with urban park sites that are anachronistic for whatever reason—whether it be Grant's Tomb or something of this type, there is no reason why programs couldn't be developed which would get the masses out simply for a good time, to use park sites, particularly those in urban locations and to develop programs that may not celebrate the history of U.S. Grant, but very well serve local community needs.

Recommendations of this sort were not likely to encourage Congress to treat professional judgment as an important factor in its deliberations on historical parks, even if it did concede the Park Service a monopoly on historical professionalism—which in fact, it did not.

Pork Barrels

More than anything, however, this disrespect grew out of the nature of congressional politics and Congress's ceaseless search for items to put into its pork barrels. Because of their characteristics, historical parks were frequently the best kind of parks to deliver to constituents. Historic sites tended to be small, and, except for the few great ones, they did not attract large numbers of visitors to disturb local peace and traffic patterns.[53] Moreover, the visitors historic sites did attract were not likely to be disruptive types, which those in search of outdoor recreation facilities sometimes were. Unlike Army Corp of Engineers "improvements," historical parks were not environmentally damaging, nor did they usually involve much, if any, condemnation of land. In fact, the establishment of a historical park frequently relieved a hard-pressed local historical society of the burden of maintaining the site. As a result of these advantages, congressmen were usually willing to sponsor and press proposals for historical parks in their districts simply to please a few constituents. But there was often more to it. Congressmen tend to think in local terms and this local thinking could bestow exaggerated historical importance on events which occurred in a congressman's district, causing him to add personal enthusiasm to his sponsorship of bills to put historic sites into the Park System.[54]

Given the attractiveness of historical parks to individual congressmen, not only did Congress have very little interest in upholding the agency's claims to sapient authority, but it even seemed to have an interest in undercutting the agency's efforts at establishing such authority when it came to history. According to one agency administrator responsible for a liaison with Congress, the history plan "was cited [by Congress] when convenient, ignored when convenient." Because the plan was so ambitious in the terrain it staked out as the manifest destiny of the historical wing of the System, it was relatively easy for congressmen to cite it to justify units they wanted in the System.

Flaws in the 1972 Plan

The 1972 plan worked against the agency in its relationship with Congress in yet another way. Because of the representational aspect of the plan, the capacity to fill a niche in American history made a site or a structure nationally significant, *ipso facto*. But the notion of national significance was undermined by the plan even as it attempted to define it. By making niche filling and representativeness such important criteria for entry, it depressed (but did not eliminate) the unique qualities of the site as a point of evaluation. For example, the Susan B. Anthony home in Rochester seemed like the best site to represent the women's rights facet, since it was the home of the woman who most symbolized the women's rights movement. The site was unavailable, however, and the Seneca Falls convention site became available.[55] When this happened, the key question the Park Service asked itself was whether the Seneca Falls site would be presentable enough to fill the vacant niche, not whether the truly important sites associated with women's rights were being protected. Thus the search for relevance and currency tended to devalue consideration of historical importance based on inherent site qualities. Once devalued, they were of little use in preventing pork barrel parks from entering the System.

Moreover, through its sheer ambition and comprehensiveness, the history plan tended to make historical significance a very flexible term, useful to justify almost anything. For example, one service administrator defended the Indiana Dunes National Recreation Area, the site of early studies of plant succession, by asserting that not only was the site important ecologically but because early studies in plant succession were carried out there, it was a milestone in the history of ecology as a science. A service historian reflected on the colonial Williamsburg restoration, whose entry into the National Park System he foresaw as inevitable, "Williamsburg is inaccurate, and it is not a true historic site . . . but if we're forced to, we can accommodate it, probably by treating it as a landmark in the history of historic preservation." There was a Kafka-like quality to this involution, the preservation of commemoration and the commemoration of preservation, and the 1972 plan encouraged it. These rationales were not much different from those the Park Service adopted when it was forced to accommodate President Johnson's wishes for his boyhood home.

Overall, one might say that the 1972 plan was too powerful an instrument for the agency to handle. The instrument slipped out of its grasp and was used against it in the fight to control entry into the Park System. Thus the Service found itself operating with weak defenses against a constant background pressure to add new historical parks, some of which had sponsors with the power to get what they wanted. This forced the agency, in practice, to take many things beyond its own plan into consideration before taking a stand on park proposals initiated by Congress. To be sure, it usually asked itself if the site was nationally significant, and it conducted an official survey of the site and the resources it contained.[56] Unofficially, it asked itself more. For example, the present agency director decided that a recently proposed park was not a worthy addition to this system, but he gave his official support to the proposal after it became clear that his opposition would not make a difference in the final outcome. In explaining his support of the bill, he said "you don't gain anything by fighting battles you're going to lose," except enemies, that is. According to a Park Service administrator responsible for legislative liaison, "On a new historical park the initiative is inevitably with Congress; it's a political call. The question 'Do we want it?' is kicked around [by the Park Service], but so is 'Can we avoid it if we don't want it?' "

When the Service saw it could not avoid a proposal, its only recourse was what an agency historian called "creative reaction," i.e., turning an unpalatable proposal into one which was at least marginally acceptable. An example of such creative reaction occurred when Representative Bennett of Florida decided that the National Park System should include the southernmost battlefield of the Civil War.[57] According to an agency historian, "The site was a real loser, the place where two incompetent commanders met by accident," and where nothing of consequence occurred as a result. Bennett was insistent, however, and he could make serious trouble for the agency should it oppose him. Although Service historians considered the battlefield militarily unimportant, they conceded that the region of northern Florida in which the battle took place was interesting from a cultural point of view. Accordingly, the agency proposed a large park composed of many scattered sites. One of the sites would be the battlefield, but the others would illustrate the diverse cultural history of the area. Thus, by burying the battlefield in a larger park proposal, the agency made it acceptable from a professional point of view and in that form it could support the proposal before Congress.[58]

The net result of congressional interest in historical parks was that the Service's determinations of national significance, carefully laid out in the 1972 plan, usually counted for less than whatever impromptu notions of significance were advanced by congressmen. According to an agency administrator with long experience working with Congress, "the reality is that it's nationally significant when Congress tells you it is." And congressmen fishing around for justification for the park legislation they wanted or needed were very creative indeed. For example, a very imaginative use of historical significance was employed by New York's Representative Carney when he supported the Fire Island National Wilderness Area. He suggested that by preserving Fire Island we preserved the landscape that our forefathers first saw when they arrived in the New World.[59] Presumably their very glance bestowed historical significance on all they beheld. Although it is hard to believe that Representative Carney made such an argument in all seriousness, it does point out the inherent logical elasticity of historical significance and, therefore, its convenience in justifying the expedient or the necessary. In the same spirit of elastic justification, a supporter of Gateway wrote, "Besides its obvious recreation possibilities, the national park preserves the scenery through which millions of immigrants passed when they first came to these shores."[60]

Influence exercised by Congress did not always stop with the addition of a unit to the Park System; it could have a strong impact on the design of the historic park once it was authorized. An example of such influence occurred at the Roger Williams National Historic Site which was, according to one of the agency's chief planners, "a purely political park on a site of dubious connection to Roger Williams." Once in the System, in the words of the same planner:

> [the Park Service] came up with good [development] designs that the local people opposed because they included too many facilities for out-of-towners. We kept getting letters from Congressmen backing local wishes so we knew there was a line we could not cross in our planning. It wasn't spelled out to us in so many words, but it didn't have to be, it was pretty well understood.

Visitor accommodations were scaled down accordingly. Another example was at Lowell National Historical Park, where the Park Service and the local proponents of the park differed on its proper components. Agency historians, interested above all in historic preservation, favored incorporating the most historically important buildings in Lowell into the park, even if this meant a park made up of noncontiguous units. The local supporters, on the other hand, were more interested in the urban development potential of a contiguous downtown historical district, so they wanted a compact group of buildings even if it meant that the sum historic value of the park was less. In the end, the local supporters got their way when the Massachusetts congressional delegation persuaded Congress to establish a commission, made up largely of local partisans of the concentrated approach, to select the buildings to be included in the park. Not surprisingly, the concentrated approach prevailed.

Slighting History

Not only has the 1972 plan not led to congressional respect for the historical wing of the Service, the plan has not led to a role for historical preservation that is equal to that of natural preservation within the Service. An administrator whose work ranged over most agency functions told the author, "the historians in the Service are still second-class citizens." An administrator in the natural wing said that "historical areas are not very important; there isn't much of a sense of urgency connected with them. There is a System-wide bias toward natural areas [and against historical ones]." Coopers and Lybrand, in their management report to agency leadership, suggested that the historical wing of the Service, because of its second-class status, had not been able to "achieve resource allocations commensurate with [its] program demands."[61]

There were several identifiable reasons why this subordinate status continued in fact, in spite of the official parity granted history by the 1972 plan. First, very few members of Park Service leadership have come up through the ranks within the historical wing of the Service, something that has not changed since the days of Wirth's directorship. Because the historical parks tended to be small in area and staff, managing them was not considered a great responsibility, and their superintendents tended to be low ranking within the civil service. This meant that although a young manager's first superintendency might very well be in a historical park, this wing of the System did not give him or her much room to grow. If ambitious, a manager would do best to look toward the natural area wing, where higher ranking management jobs were more plentiful. Thus, a common career track would have a successful manager spending the first few years in a historical park. Then with advancement, he or she would move across the aisle for the rest of a career. The more promising a manager was, the more quickly came promotions, and the less time was spent in historical parks. If a person was not successful as a manager, however, his or her entire career might be spent managing small historical parks. Of course this is a generalization, and there were exceptions. For example, recent superintendents of Gettysburg and Lowell were widely considered to be rising management stars in the Service. Nevertheless, the great natural areas, the Great Smoky Mountains, Yosemite, Yellowstone, Glacier, and the others were the traditional breeding grounds of top agency leadership.[62]

Second, the fact that there were no public interest organizations watching over and supporting the entire historical wing meant not only that there was no one to prod Congress into respecting historical integrity, but also that there was no one to support the Service if it did decide to pay more respect to this area of its responsibility. Third, several Park Service administrators in the agency's Office of Legislative Liaison, and therefore close to the process of creating new parks, were former environmental activists whose primary interest was adding new natural areas to the System. Moreover, even those in legislative liaison who did not have this background were brought into frequent contact with environmental public interest groups by their work, and this encouraged them to share the view of these groups that the most valuable thing the agency could do was preserve nature.[63] Accordingly, in the words of one agency legislative officer, "legislative people . . . in the National Park Service . . . were most concerned with environmental matters." Finally, there was the matter of internal perception and the inherent values of agency leadership. As Utley expressed it: "The public identifies us with mountains, forests, canyons, and bears. We are the people in Stetsons who take care of superlative scenery and the natural features and wildlife associated with it. Our dominant 'establishment' . . . tends to see the Service in the same image as does the public."[64]

Since environmental groups, agency leadership, and legislative liaison tended to value preservation of natural areas while legislators keenly appreciated the down-home benefits of historical sites, whatever their historical value, the stage was set for a relationship of what Downs calls "functional interdependence" between the two wings of the System.[65] By the mid-1970s, the process of log rolling natural parks (and to a lesser extent, urban parks) into the System by using historic sites as bargaining chips was in full swing. Sometimes the process involved simple one-for-one deals. For example, the agency and its environmental supporters were very interested in expanding the Congaree Swamp National Monument in South Carolina in such a way as to give it more manageable boundaries. The local congressional delegation was indifferent or perhaps even mildly opposed to this because it would have involved some condemnation of private land. The delegation was interested in the establishment of the Camden Historical District within their state, however—a site that a Park Service evaluation team had studied and found of little merit.[66] The Park Service's office of legislation and the House subcommittee on parks, acting in concert, struck a deal with the local congressmen whereby both sides would realize their primary goals. A joint proposal incorporating both the natural site and the historic site was formulated. According to one agency official close to the process, "you don't get a good expansion, or even a good exclusion, without paying a price" and historical units of questionable value were often the currency.

More important than these one-for-one deals were the omnibus bills which combined individual park proposals into large legislative packages and which in one stroke brought numerous units into the System. Creating an omnibus bill involved lumping together all the park proposals that environmentalists wanted into a core package and then adding enough pork barrel park proposals onto this core package to ensure sufficient support for the entire bill. The omnibus bill add-ons tended to be historic sites. According to one of the agency's historians, "the omnibus bills [had] units in them that [were] purely political and most of those [were] historic sites." One purely political unit on the Omnibus Bill of 1978 was the home of Thomas Stone, whose claim to fame began and apparently ended with his signature on the Declaration of Independence.[67] The National Park Service was not looking forward to getting the site, but it was in Representative Bowman's (R-Maryland) district. The plan was to put the omnibus bill through on an eleventh hour unanimous consent request. According to a member of the House Interior committee staff, however, "Bowman had the habit of jumping up and objecting to [such bills]. The Thomas Stone House was [added to the bill] to keep him quiet." Agency leadership usually went along with such deals because their values, and indeed the values of agency rank and file, tended to emphasize the protection of nature over the integrity of the historical wing.

One should hesitate to condemn agency involvement and this process as purely cynical. Although there was undoubtedly a certain element of cynicism in this kind of policy, it also smacked of the normal reciprocity of legislator–administrator interaction.[68] The process did, however, consistently slight the historical wing of the System. It did not slight it as far as growth went, but in just the opposite way, by encouraging the kind of growth that undercut the historical wing's integrity and ultimately, the agency's control of it.

An Ambiguous Attitude Toward History

Thus, it is apparent that the 1972 plan has been largely ineffective as an instrument of strategy. It has not allowed the Park Service to claim sapient authority over federal historic preservation policy. It has encouraged the expansion of the historical wing of the System, but it has not allowed history professionals within the agency, or even agency leadership, to be the arbiters of that expansion. Its breadth and ambition have given the agency room to maneuver in seeking a trendy relevance, but the same qualities have made it easy for those outside the agency seeking parks for their own ends to justify their proposals. Perhaps the most important shortcoming of the agency's planning with regard to historic preservation, however, has been its failure to promote a consensus within itself on how seriously it wants to treat history.

If the true intention of the 1972 plan was to establish the parity of the natural and historical wings of the Service, mere proclamation was insufficient. Organizational planning which fails to take into account the different inherent capacities of an organization's subunits to defend their functions is simply unrealistic.[69] The historical wing was at a disadvantage when it came to defending its function, and, as we have seen, the disadvantage went far deeper than official policy. By failing to address the root causes of these disadvantages, the 1972 plan left historic preservation a second-class concern. In raising expectations and opening a broad gap between the formal and the de facto, however, the plan engendered what are probably the strongest animosities within the Park Service. Perhaps nowhere else in the agency is the conflict between the professional and the politician, between rigid adherence to principles and indulgence in opportunistic behavior so clear-cut, extreme, and acrimonious.

Professionalism vs. Pragmatism

The professional historians within the agency defend strict standards of historical significance, promote the comprehensive, rational approach to history as it is embodied in the 1972 plan, and insist that professional judgment be central in determining the future of the historical wing of the System.[70] Not surprisingly, given this stance, the historians are loud in their objections to the use of the historical wing as a bargaining chip to attain other agency goals, and they dislike those who would use history in this manner. Also not surprisingly, this sets them at odds with much of the rest of the agency. In part, this conflict might be viewed within the context of the broad, intrinsic one between the professional, on one hand, and the administrator and manager on the other. Many students of large organizations have observed that the professional specialist has an inherent tendency to see things through the prism of his own training, and therefore he will usually have an extreme concern for those things which his profession has taught him to value.[71] This puts him in conflict with an organization's leadership, which must look at policy from a broader perspective. As a result, there is a constant tendency for leadership to rein in the professional and to direct his actions toward broader agency goals, even at the cost, quite often, of forcing him to violate his professional norms. The professional, on the other hand, strives for the autonomy necessary to pursue the goals dictated by his training and sense of professionalism.

Students of public administration have also suggested that in a modern bureaucracy, with its division of line and staff functions, the professional frequently finds himself in conflict with line management as well as top leadership. For example, the professional puts great emphasis on standards of quality and methods in his work.[72] The manager, however, must take into account additional factors such as cost, time, and public relations, and sometimes these factors demand compromise with professional standards. For the professional, the trick is to convince others in the organization that allowing him to uphold his own professional standards is worth the cost involved, and that his professional goals are, at least in the long run, confluent with the broadest of organizational goals. Redford suggests that in the professional's ideal organization, "a ring of protection [is] run around the specialist so his function qua specialist may be performed with fidelity to his trust."[73] From the professional historian's point of view, the Park Service is far from the ideal organization. Although it gives the professionally trained historian great formal authority in evaluating the intrinsic significance of proposed sites, and the agency officially gives great weight to his evaluation in its internal decision process, his informal power in such matters seldom matches his formal authority.[74]

Few organizations enjoy the luxury of perfectly correlated ideas and action.[75] As surely as certain circumstances will inform an organization's official policy, other circumstances will prompt it to circumvent that policy. Just as there were reasons to maintain the appearance of professional decision making with regard to history, there were strong reasons for agency leadership to disregard that professionalism in practice.[76] Agency leadership was not inclined to value highly the wing of the Service over which the historians claimed professional jurisdiction. In fact, it appears that agency leadership viewed historic sites as "loss leaders," i.e., responsibilities which "represent a loss or at least small profit from the point of view of the agency's major goals, but that simultaneously widen the base of political support for the objectives that are more significant to it."[77] Given this, allowing the historians real power to exercise their professional judgment on the intrinsic worth of historical areas would restrict the freedom of agency leadership to use historic park proposals of little value as bargaining chips.

Lack of Support

Agency historians also lacked a base of professional support outside the Service with which to gain leverage on agency leadership. Associations of professional historians showed no interest in defending the integrity of the agency's historical wing, and few, if any, agency historians had large enough reputations among historians in general to command the entire profession as an audience when they spoke. Thus, agency historians were cut off from what is an important source of support for other professionals in bureaucracies, their wider professions.

This left agency historians trapped in the limbo between their official and actual power, and they were well aware of their position. One of them told the author:

> [the] party line is that we gather information on an historic resource and then decide about its preservation. In fact, a decision is made then information is gathered to justify it . . . significance decisions are never made by themselves. They're always rolled in with other criteria—and it's usually significance that suffers.

Not surprisingly, historians were deeply dissatisfied with their lot. According to a former chief historian, "Ideally the historian is important in the decision process but [in reality] the system is so unsatisfactory from the point of view of the historian that he [has] become disgruntled and alienated."[78] Much of the historians' resentment was directed at the agency administrators responsible for new park legislation, who are considered extreme opportunists. According to one historian:

> There is zero respect for scholarship in the legislative office. Objective reality, i.e. that one element is more important than another, or one site is more important than another, means nothing to them. . . . [The legislative office] short circuits us whenever possible. For example, they routinely answer letters of inquiry on proposed historical parks, not because they know anything about history but because they know about politics.

Another professional historian said of those responsible for new park legislation, "They only come to us when they get in trouble, then they want us to act as expert witnesses and say what they want to hear."

For their part, those agency administrators who work on new park legislation view the professional historians as naive. One told the author that "to get what you want, you have to be willing to cut deals, everything has a price. The historians will never understand this." They also see the historians as unrealistic. "The historical people were not supportive of [the] Martin Luther King site [because of] some technicalities, but there was no way in hell that anyone could have opposed that one." They view the proper role of the agency historians as a circumscribed one. The same administrator said that "the professional historian is important in identifying the real dog." He added that "the question posed to the historian shouldn't be 'is it the very best site?' or even 'is it a premier or just acceptable site?'"

In many agencies where professionals abound, administrators not from the professions see themselves as capable of formulating policy "with a realism and breadth of perspectives that the professionals themselves could never provide."[79] In this case, those responsible for new park legislation saw themselves in the role of the realists who must contain the narrow-minded professional historians, professionals who would be dangerous to the agency's health if they were ever permitted free rein. Thus, the resentment of the historians seems well directed. In the agency's organizational politics, professional historians and those responsible for new park legislation truly seemed locked into zero sum combat. A gain in real decision-making authority by the historians would have meant a loss of flexibility in deal making by the legislative office.[80]

Conflicts Over Standards

The historians' sense of professional responsibility locked them into similar contests with park planners and historic site managers. The planning for the War in the Pacific National Historical Park illustrates the type of conflict historians and planners got into over principles and pragmatics. For many years, the idea of a national park on one of the Pacific territories of the United States, a park designed around the theme of the indigenous cultures of the Pacific, was afloat in the Service. Albright had considered establishing one on American Samoa but ultimately rejected the idea when no area meeting his strict standards of cultural integrity could be found.[81] Much later, in the early 1970s, the idea was revived for the island of Guam, and at some point the park's supporters hit upon the idea of naming it War in the Pacific National Park. By doing so, they could argue that it filled a big gap identified by the 1972 plan in the System's historical holdings, that is to say, historic sites associated with the Second World War. Furthermore, such a name would make it a motherhood issue akin to veterans' benefits; Congress could not say no. Nevertheless, the name was largely a red herring. The park's backers still viewed it in cultural, rather than military terms. An administrator close to the park proposal said that "our idea of bringing in World War Two was showing its effects on the islanders—cargo cults and that sort of thing."

At the time of its authorization in 1978, the agency planners responsible for developing a park design found themselves with a three-part mandate which had grown out of the politics of its authorization. The first was to interpret Pacific island culture in accordance with the original idea of the park. The second was to provide recreational facilities for the use of the islanders. Representative Burton, whose subcommittee had jurisdiction over not only national parks but Interior's overseas charges as well, saw the park as an opportunity to provide recreation facilities for Guam's natives, and he made his wishes in this regard known to the Park Service. Third, the park was to pay at least some attention to the military commemoration theme. Once development planning started, however, agency historians argued that since the overt purpose of the park was to commemorate the war in the Pacific, the park planning should be under the direction of a military historian. Moreover, they argued that other uses of the park, i.e., cultural interpretation and local recreation, should be strictly subservient to war commemoration. The planners, who had already begun the design planning, objected to what they saw both as an unwarranted intrusion into their area of responsibility and an attempt by the historians to distort the proper mix of planning goals for the park. One planner pointed out a specific example of what he considered an improper intrusion by historians in park planning. "We had a conflict with the historians on the park boundary. They wanted to include Hill Forty because a bloody battle was fought there. I didn't want to because it added nothing to the park." The planner continued, "When we got to the island and started working with the local people, we realized how much they needed recreational facilities so we shifted our plans further in that direction. This got the historians even more upset." As a result of the conflict, hard feeling arose between the planners working on the project and the historians.

Elsewhere, historians clashed with park planners over what they saw as too much rather than too little historic preservation in park development plans. As we shall see later, planners at the Breezy Point unit of Gateway National Recreation Area fell back on the preservation and interpretation of the abandoned military facilities that the Service inherited with the site to justify the continued existence of the unit when the original goal of developing large-scale public beach facilities became politically unattainable. The historians opposed interpreting what one of them called "little more than surplus property." They felt that this preservation of the inconsequential was as much a violation of historical integrity as was ignoring truly significant sites. In their eyes, historic preservation had to be protected by professional historians from both types of sins.

From the viewpoint of the Service planner, who is well aware of the fact that design options are circumscribed by politics as well as by site characteristics, the historian's attitude, if allowed to deeply affect planning, would have only constituted another circumscription and thereby would have reduced the possibility of coming up with a plan acceptable to all concerned. In fact, there was a tendency on the part of some planners to see the agency's professional historians as obstructionists by nature and as people who would prefer to exercise their power in a negative rather than a positive manner. One agency planner told the author that "cultural resource people assert their formal authority to the fullest extent possible; for example they insist on a full 106 clearance on every project, no matter how much time and money it costs." [82] He compared them unfavorably with the natural scientists in the Service. "The natural resources people have the same responsibilities, but they're much more reasonable." For the difference in attitude and behavior, he offered as a possible explanation that "the natural scientists feel like insiders [in the Park Service] and the historians don't."

The professional historians also occasionally found themselves in conflict with those who managed the parks once they were established. These conflicts usually centered on "living history," i.e., the reenactment of past events and the use of historical parks for ends only indirectly connected with what the parks preserved or commemorated. The historians took exception to living history because it tended to draw attention away from the original purpose of the site and because, in its inherent gimmickry, it demeaned the history they felt was theirs to protect. A former chief historian wrote a warning about living history in the agency's historical newsletter which touched on these points:[83]

> Site managers all over the country are dying to see who can put together the most spectacular, and hence presumably the most "living," production. Costumed performers are firing muskets and cannon, dipping candles, forging horseshoes and cooking every variety of food by every variety of means known to our forebears.

Elsewhere he criticized the tendency of park managers to emphasize the portrayal of everyday life at the expense of the events the park commemorated. "We are obsessed with showing what everyday life was like in the past, surely a valid purpose. But most of our historic places are not preserved because of everyday life that occurred there."[84] He objected most vehemently to this tendency on the part of park managers to promote the prosaic through living history at two places:[85]

> At Appomattox the superintendent didn't like the country store as it was preserved, it was musty and the interior was behind glass, so he set up a living country store. It was popular with visitors and it was even a moneymaker, but I objected. First there was a question of accuracy—it isn't likely that there was much of anything on the store's shelf in April, 1865. Secondly, there was the matter of visitor impression. They were leaving with trinkets, not an understanding of what happened at Appomattox. . . . At Chickamauga, the superintendent set up a working farm because he wanted to show people that the site wasn't always a bloody battlefield. Well what the hell was the park set up for in the first place—to show people how to make sorghum?

Another service historian simply dismissed living history as a contradiction in terms: "Living history is inane, you can't live history!"

Historians were also sensitive to what they considered uses of historical sites that were in bad taste. One told the author, "the old superintendent of Fort Smith encouraged things like oyster eating contests at the site, it was the worst kind of bread and circuses." They viewed the temptations to manage historical sites for recreation or even for aesthetics as ones to be strongly resisted. A historian at Gettysburg objected to keeping much of the battlefield parklike, with wide expanses of lawn. "We're not in the lawn business. We're in the business of preserving a battlefield." He also objected to lawns for a second reason: "With lawns you get a problem; they become recreational areas for the town, and this creates the wrong atmosphere." Sometimes this notion of propriety on the part of the historians approaches the mortuarial. One told the author that the public "shouldn't be flying kites along [Antietam National Battlefield's] Bloody Lane anymore than there should be tennis courts at Yosemite. The history should be enough, recreation profanes it." But on the issue of historic site management, historians in the Service felt that they were bucking the tide, just as they felt they were bucking it with regard to evaluation and planning of new units. To them, disrespect for history, as one historian said, was "woven into the very fabric of the Service."

Historical Significance and Historical Parks

Perhaps this was so, but there were solid reasons growing out of bureaucratic *realpolitik* for several powerful groups in the agency to ignore the decisions of historical experts and to undercut their authority whenever possible. In the face of these reasons to disrespect their judgment, the historians did not seem capable of making a convincing argument that either the System or the Service would have derived real benefits from taking them seriously. Perhaps such arguments could not be made. Beneath all the rhetoric to the contrary, historical significance as judged by the professional historian did seem rather unimportant to the well-being of the System and its keepers.

One Service historian, protesting the recent authorization of what he considered an historical park of no value, observed that "it costs no more to run a good park than a bad one." The reverse of this, however, might have been more pertinent. The historical value of a site (or lack of it) did not seem to have much bearing on how successful the park developed on the site would be, or for that matter, on how easy it would be to manage. Indeed, one might suspect that the less historical significance the site possessed, the more challenged the managers and interpretative planners felt, and the more their can-do attitude rose to the occasion.

"Creative" History

Nor could the historians argue that historical significance is important to park visitors. Allegheny Portage Railroad National Historic Site is a case in point.[86] It was established on the site of a cog railroad which briefly hauled freight over the east-facing slope of the Allegheny Plateau (the Allegheny Front) in the mid-nineteenth century. Although the park is pleasant enough, there is not much there. The old locks and inclined planes are mostly gone, obliterated by U.S. Highway 22. The highway runs the length of the park, following the same route up the front as did the railroad the park commemorates. There is a visitor center on the ridge, tastefully installed in the stone nineteenth-century hotel that served the railroad. The visitor center contains a working scale model plus interpretative displays of the railroad. It also exhibits tools, furniture, and clothing from the railroad's era. Most of the park visits are local or in-state in origin. Bus tours bring senior citizens and school children. To these visits are added picnickers in the summer, cross-country skiers in the winter, and occasional railroad or canal buffs whenever they are passing through.[87] In sum, the park has all the earmarks of a good county or perhaps state park.

Perhaps Allegheny Portage will someday be viewed by the American public as a trivialization of the past that the Park Service is charged with preserving and as a cheapening of the truly great sites, but this does not appear to be happening now. Allegheny Portage is visited by local residents who want to picnic or cross-country ski, not bask in history. For those on bus tours, it is just one stop on a circuit of many stops. Those railway and canal buffs who do go out of their way to visit the park think it is fine that the Park Service is preserving what is left of the cog railroad. The lack of great historical value is not even viewed as a problem by park interpretative personnel. In response to a question about site significance posed by the author, one of the park's interpretative specialists responded: "We don't worry about it, we just try to put [the cog railroad] into its historical context. In all the time I have been here, no one has ever come up to me and asked 'why are you spending my tax dollars on this?' Maybe sooner or later somebody will, but it hasn't happened yet." The paucity of material remains is circumvented by creative management and planning (figure 5-2). According to the park interpreter quoted above, "There really isn't a whole lot left here, which is an interpretative problem—you can't go out and see much. But we get around it with a slide show and a museum."

Hopewell Village National Park Site illustrates how well inherent weaknesses can be covered by creativity and interpretative skills. In 1935 the Park Service acquired Hopewell, an abandoned iron-making village in southeastern Pennsylvania, and restored it as a CCC project. Three years later it was designated a national historic site. Like many CCC projects in the Park System, the need to find work for the unemployed was more important in prompting the park's creation than any inherent site qualities. There was an abandoned village which could be restored and doing so would absorb idle labor. That was enough; it mattered little that the park's theme was amorphous and universal rather than particular and American. It also mattered little that it did not fit in with the era's spirit of commemoration—or even the American history themes of Wissler's 1929 report. According to the interpretative prospectus for the park that the agency prepared in 1972, "the basic message of Hopewell [is] the dependence of

Figure 5-2.
The Allegheny Portage Railroad National Historic Site. Note the condition
of the remains and the use of interpretative materials.

man on iron and the heritage of this ancient dependence." [88] This lack of
significance by any of the agency's criteria in no way inhibited interpretative plans, which stressed such themes as "blowing the furnace, the labor
of filling it, the frustrations of clotheswashing in a soot-smudged environment." [89] The interpretation would also show "the sociological self-suffi-
ciency of the close-knit paternalistic society, the financial dependence
upon the company store, the character of recreation, and the function of
education." Just as interpretation was ambitious in its ends, it was dramatic
in its means. The interpretative prospectus suggested making a movie
which would parallel the lives of eighteenth-century steel workers with
their twentieth-century counterparts. The prospectus added that:

> The film should end on a powerful note that will draw attention to the invest-
> ment of individual lives in the making of iron and steel, and to the thunderous,
> dominating power of the process. A sequence of workers' faces at the time of
> the opening of the furnaces, both historic and contemporary, might be ex-
> tremely effective in bringing this message home.

For good interpretation, measuring up to criteria of national significance
was not important; the potential for a good show was.

Activities such as craft demonstrations, fairs, and musical events usually
reflect an entire historical period or the history of an entire region. Therefore an historic site's lack of significance need not detract from the value of

those events to their viewers or participants. Moreover, living history demonstrations are an attractive means of building a clientele for a park. Demonstrations and historical recreations are exciting and memorable to visitors, and children love them. A good example of Park Service living history programs was that at Turkey Run Farm which was, according to the agency, a "re-creation of a low income north Virginia farm in the 1770s."[90] The park relied largely on volunteers to create "the experience of reliving the past."[91] As the agency described the living history program:

[The volunteers] dress in period clothing and work alongside the farm staff in performing the chores typical of the sons and daughters of our Colonial forefathers. Their tasks include helping to cultivate, weed and harvest the tobacco and corn fields and the large kitchen garden areas, tending the farm animals, and helping with domestic chores primarily related to preparing the midday meal.

Such a program does more than attract and amuse visitors. By providing an outlet for the skills and enthusiasm of the performers involved, it turns them into special park clients and supporters. Park managers are well aware of this reciprocal relationship between the park and the volunteer.[92]

The Attractiveness of Postings in Historical Parks

From the point of view of agency rank and file, qualities such as size and location are more important than any intrinsic historical significance in determining their attractiveness as postings. For example, historical parks tend to be small and to have small, relatively unspecialized staffs. This makes them attractive to those who are by inclination generalists or to those whose ambitions lead them to seek exposure to many aspects of park management. Also, historical parks are more likely to be in densely populated areas than are natural parks. For the family person who values proximity to schools and the possibility of professional work for a spouse, for the employee who does not want his social contacts limited to other agency personnel, or for the employee who prefers to live in a town, these historical sites are very attractive postings. They are not attractions which are affected in the slightest by historical value.

Likewise, the historical importance of a park is not likely to affect the careers of its management personnel. There are historical parks of indisputable worth and significance, for example, the Gettysburg and Chickamauga–Chattanooga battlefields, Independence National Historical Park in Philadelphia, and Colonial National Historical Park in Virginia. Superintendent jobs in these crown jewels of the historic wing, as they are sometimes referred to, are considered plums. However, this is because of the size of these parks in staff and budget, and the high level of management skills they demand. In some cases historical parks without a patina of great significance, Lowell National Historical Park, for example, are equally if not more valued as career steps because of the administrative challenges they present. One is far more likely to establish a reputation as a young manager on the way up by successfully handling a difficult management task in a park with little historical significance than by acquitting oneself well in a park which presents only minimal problems and straightforward management demands, no matter how strong the latter's claim to historical worth might be.[93]

Historians' Responses to Predation

Thus we see that the goals and values of agency historians, as well as their official powers, isolate them within the agency. Their response to their isolation and to their lack of real power takes several forms. One is support for a separate agency to manage the System's historical sites or, failing this, a Park Service in which separate career ladders, staffs, and responsibilities come together organizationally only at the level of the agency director. The author found nearly universal support among agency historians for one or the other of these rearrangements, for either would insulate historians from competition with other Service interests and responsibilities.

Another response of the historians is to draw the cloak of professionalism tightly around themselves by denying or underplaying the commemoration aspect of their roles whenever possible. Commemoration is open to value judgments and nonprofessional decision making. Moreover, there are limits on how much professional skill is necessary if one is simply commemorating a past event. Preservation, on the other hand, has an infinite capacity to absorb professional skills—those of the military historian in reconstructing the battle with exactitude, those of the architectural historian in restoring buildings to their original condition and so forth. This tendency toward professional exactitude and the psychic comforts it offers is illustrated at Gettysburg. There, while a strip of fast food establishments was springing up along the street leading to the main entrance of the park and encroaching on the site of Pickett's Charge, historians were studying how accurately the battlefield monuments were located.

The historians also tend to play up the august lineage of the historic wing of the System, sometimes even at the expense of historical accuracy. For example, a historian called on to lecture on the origins of the Park System stressed the evolutionary symmetry of the historical and natural wings. His lecture contained the following:

The National Park Service is supported by two foundations that have their origins in the 19th Century. These foundations eventually led to the development of the National Park Service in 1916. We can illustrate this in the following chart.

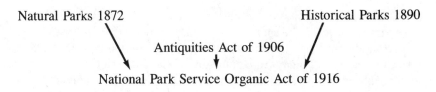

Natural Parks 1872   Historical Parks 1890

Antiquities Act of 1906

National Park Service Organic Act of 1916

He seemed to imply the equal pedigree of the natural and historic branches of the System and therefore, presumably, strong vested rights for the historians. In fact, the diagram is not accurate. Of the agency's patrimony, only Mesa Verde, an Indian archaeological site, had what might be called historical overtones. The historical parks of 1890 were not placed under Service management by the Organic Act of 1916. It was to be almost two decades before they came into the Park System and when they finally were added, it was at least in part for the strategic reasons discussed in chapter 2. But the diagram did support the historians' claims to parity on grounds of lineage.

Some historians become disillusioned by their situation and leave the Service. Some stay on, embittered by their lack of power. Some adjust to circumstances and try to exert whatever influence they can on decisions by using their credibility carefully and sparingly, like precious ammunition. As one who stayed and adjusted explained it: "You can't become a [nuisance], if you do you are turned off. You've got to appear reasonable. Otherwise you get to the point of being a gadfly. I've cautioned [agency historians] not to get into that role. Make [agency leadership] value your opinion, even if they can't or won't use it." Although such a strategy seems reasonable and based on an accurate reading of the position of the professional historian in the Service, it is nonetheless a strategy pressed from a point of weakness. Therefore it, as much as the other reactions of the historians, reflects their isolation and powerlessness, and beyond that, the agency's tendency to sacrifice the professional view of historical importance and historical values to other agency goals when tradeoffs have to be made.

The professional historians argue that by disregarding their objections to sites whose intrinsic value is questionable and by disregarding their objections to plans and programs which demean history as they see it, the agency has also disregarded restraint and perspective. There are two views on everything, however, and what looks like restraint to one side can look like obstructionism to the other. It is fair to say that agency leadership has taken the latter view of the historians (in fact, if not on record). Whatever the agency's view of the intrinsic importance of historic preservation, its search for relevance led it to see history as a means of identifying with causes and themes emerging in the interplay of social forces. It also saw historic preservation as a base for claims of sapient authority, and as an area where its responsibilities could be expanded to its advantage. To do these things, however, the agency needed historical credibility. That meant respecting the professional historian. But politics and the pursuit of other goals prevented the agency from doing that. This circularity has given rise to a double bind and to great internal tension. Because it cannot completely ignore the professional historian nor truly respect his judgments, the agency cannot answer once and for all those questions which, if answered, would define the extent and nature of its mission with regard to the nation's history.

6 Urban National Parks

The Issue Develops

As we saw earlier, the Park Service's basic responsibility for the preservation of the nation's history and its nature was never seriously questioned. Questions arose only when it became necessary to resolve this general responsibility into the specifics of policy, priorities, and management decisions so as to give shape and definition to the expanding Park System in the 1960s and 1970s. With urban national parks, there were the specific questions of management and policy, to be sure, but beyond these questions was the more basic one of whether the Park Service should be involved with cities in the first place. In fact, perhaps no question has been so debated within the National Park Service as that of the appropriateness of urban parks in the National Park System. Aside from the use versus preservation question, no issue has been its equal in forcing the agency to ask itself questions about its basic mission or in involving it in controversy. Moreover, it is a policy area in which definitive commitments, rather than settling things, seem only to launch new rounds of dissension and indecision.

Although the agency's assumption of urban responsibilities is commonly viewed as a very recent thing, the Park Service has been involved with parks in and near cities for a long time.[1] In 1933, it was given responsibility for managing the parks of Washington, D.C., and the following years saw historic sites in several cities added to the National Park System. For example, the agency took over responsibility for managing Federal Hall in lower Manhattan in 1939. Independence National Historical Park in center city Philadelphia was established in 1948, and Jefferson

169

National Expansion Monument, located on the waterfront in downtown St. Louis, was authorized in 1954.

As far back as the 1930s an element within the Park Service felt that the agency should establish and manage recreation sites close to the nation's largest population concentrations. This idea took a concrete form when, in the national seashore studies of the 1930s, the agency suggested that beach areas near Atlantic and Gulf coast cities be acquired by the federal government and managed by the Park Service for mass recreation.[2] With this proposal, the agency asserted that the federal government shared with the cities themselves a responsibility for placing recreation facilities within the easy reach of urban dwellers. Considerations of urban access also had a significant effect on park development planning. For example, Shenandoah National Park's proximity to Washington and Baltimore led agency leadership to assign a high priority to its completion during the 1930s.[3] The Great Smoky Mountains National Park, although containing far more impressive scenery, was relegated to a lower priority because of its distance from urban areas.

Even taken together, however, these actions hardly added up to a definite commitment to urban national parks or to an important urban responsibility for the Park Service. Even managing the parks of the District of Columbia was considered a chore and very incidental to the main goals of the agency. According to an agency administrator, "the Service felt that the capital parks were a horse of a different color, separate from the rest of the System. Even the personnel were different."[4] The agency's organization table reflected this isolation; until 1960, the capital parks administration was entirely outside the agency's normal regional chain of command. The agency viewed Federal Hall, the Jefferson Expansion Monument, and Independence National Historical Park as historic sites that happened to be in urban areas through accidents of history. They were not in the National Park System because they were in urban areas but, if anything, in spite of it. Shenandoah National Park, although within a morning's drive of two major cities, was still well beyond the outermost edges of suburban development. Cape Hatteras, the only national seashore established in the 1930s, was about as remote from population centers as a stretch of the Atlantic coastline could be. The more general recommendation of the national seashore studies, that the federal government should become directly involved in providing recreation sites easily accessible to urban populations, came to nothing. The 1930s ended before anything could be done about urban national recreation areas, and the times were not right for such an involvement during the war. After the war neither Drury, with his custodial view of the Park Service, or Wirth, concerned as he was with Mission 66, paid much attention to urban recreation.

In the late 1950s, federal interest in open space in and near major urban places quickened. In 1955, the Mellon family sponsored an update of the national seashore studies which the agency had conducted in the 1930s.[5] The principal recommendation of the update was the same as that of the original studies: that the national seashores should be established near the nation's population centers wherever possible. Now, however, there was a great note of urgency in the recommendation. Metropolitan areas had grown rapidly in the twenty years since the original studies, and there was a fear that many of the beach areas close to the cities would soon undergo private development and that rapid price escalation would put the rest of them out of the public's reach. At least in part, the same concern for vanishing open space in or near urban areas prompted Congress in 1958 to establish the Outdoor Recreation Resources Review Commission (ORRRC) to examine the entire question of recreation in the United States.[6] When ORRRC's final report was released, it reflected the same sense of urgency in the face of closing opportunities as did the Mellon report. One of its principal conclusions read that "outdoor opportunities are most urgently needed near metropolitan areas." These areas "have the fewest facilities (per capita) and the sharpest competition for land use."[7] In what would turn out to be an important corollary of this conclusion, ORRRC found that:[8]

> Over a quarter billion acres are public designated outdoor recreation areas; however, either the location of the land, or restrictive management policies, or both, greatly reduce the effectiveness of the land for recreation use by the bulk of the population. [This land is] of little use to most Americans looking for a place in the sun for their families on a weekend, when the demand is overwhelming. The problem is not one of total acres but of effective acres.

During the early 1960s, federal interest in urban open space further increased. Udall effected the conceptual nexus between conservation and the whole complex of urban problems subsumed under the term "urban crisis," and then he called for a policy of federal activism to man the "battle line of conservation in our cities."[9] Part of the battle was to be for public open space. Kennedy, Johnson, and much of Washington accepted the New Conservation's call for federal involvement in the provision of urban open space, in part because they realized that the benefits of programs designed to bring parks closer to urban people did not stop at the end of the grass. For a federal government, now steering by the referents of the Great Society and newly concerned with a whole gamut of urban problems—uncontrolled growth on the urban periphery, decay in center cities, overwhelming financial burdens, and a general decline in the quality of urban life—public open space programs had considerable appeal.

First of all, public open space programs fit into the theme of maldistribution that ran through Great Society thinking. The establishment of urban parks could be viewed as righting wrongs and as ending the old inequalities identified in ORRRC's final report, inequalities which had resulted from past public policies. President Johnson was undoubtedly thinking in such terms when he wrote that "a park, however splendid, has little appeal to a family that cannot reach it. The magnificent areas preserved in the early days of conservation were remote from the cities—and many Americans had to travel half a continent to visit them. The new conservation is built on a new promise—to bring parks closer to the people." [10]

Second, programs to provide open space were often exempt from the criticisms of inappropriateness, ineffectiveness, and subversiveness leveled at programs designed to alleviate directly other urban problems, such as income maldistribution or poor housing. This was something Smith recognized when he asserted open space, as an urban need, was inherently "less controversial than such other urban concerns as housing, urban decay, and income maintenance." [11]

Third, the provision of public open space was a relatively straightforward task. It was a matter of setting aside acres, putting up swings and slides, constructing restrooms, and so forth. Success was easily measured (or at least easily claimed). On the other hand, federal programs aimed at establishing orderly urban growth, arresting urban decay, or controlling crime needed to be far more complex, and they were therefore uncertain undertakings. Finally, as New Deal administrators had known, federal open space programs could take forms which threatened no one. When President Johnson advocated using antipoverty funds to train city park workers, he could assure his listeners that such training would not rely on the forced involvement of businessmen, would not take business away from them, and would have no direct costs to them. He could also point out that once trained, the program's graduates would enter jobs made for them in the public sector, rather than compete for existing jobs.

In the early 1960s, in accord with its increased interest, the federal government enlarged its involvement in urban public open space with several significant moves. The Housing Act of 1961 was, according to Foss, a "major breakthrough in open space planning and acquisition." [12] Title VII of this wide-ranging act had as its avowed purpose helping state and local governments in their efforts to provide open space. Such open space was, according to the act, "essential to the proper long-range development and welfare of the Nation's urban areas." [13] Specifically, the act allowed federal housing money to be used for the acquisition and development of public open space in conjunction with housing projects. In 1964, Congress, in keeping with an ORRRC recommendation, established the

Land and Water Conservation Fund (LWCF).[14] The LWCF was an earmarked federal fund out of which, first, federal land-managing agencies could take money to acquire recreation land and, second, states and cities could be given grants for their open space purchases. ORRRC had hoped that the creation of such a fund would help correct the rural bias of public open spaces that it had identified in its final report. To make sure this would happen, several pro-urban stipulations were written into the act. For example, not more than 15 percent of the acreage added to the National Forest System through the program could be west of the 100th meridian.[15] In other words, 85 percent of the land the Forest Service bought with LWCF money had to be in the populous East. The LWCF also gave priority to state projects designed to make their recreational facilities more accessible to urban populations. In 1965 another ORRRC recommendation was implemented when a new bureau within Interior, the Bureau of Outdoor Recreation (BOR), was established and given responsibility for coordinating federal open space and outdoor recreation activities. One of the new bureau's major mandates was providing urban open space.

Thus by the mid-1960s, several things—the New Conservation, the Urban Crisis, and the Great Society—had come together to place urban open space high on the federal agenda. This immediately made a previously neglected policy area an attractive piece of bureaucratic real estate, one which promised great benefits to agencies which could claim a part of it.[16] Moreover, the field was an open one. Although ORRRC had contributed mightily to making urban space a federal issue, for fear of giving offense to entrenched bureaucratic interests, it did not spell out explicitly how the various federal agencies with recreation experience should involve themselves in urban open space. Nor did it offer explicit recommendations on the question of direct federal land management in urban areas versus more indirect involvement through subsidies and other forms of aid to the cities.[17]

The Park System as an Instrument of Urban Policy

George Hartzog, the man Udall picked to succeed Conrad Wirth as director of the National Park Service in 1964, had a reputation in the Service as an innovator. Among the accomplishments he brought to the directorship was his successful development of one of the System's most important parks in an urban area, the Jefferson Expansion National Monument in St. Louis. Hartzog was, if nothing else, sensitive to the winds of change.[18] Coming in to pick up the pieces after Wirth's departure, he realized that innovation and adaptation were being demanded of the Park System.

There were compelling reasons for Hartzog to attempt to bring the Park System into the cities in more than an incidental way. Above all, an urban commitment seemed very much in keeping with the times. A Park System with more equalized public access would tap a main current of Great Society thinking, and perhaps it would also tap the wellsprings of rapidly increasing federal spending.[19] Moreover, the agency's own past policies and performance had to bear at least some of the blame for any maldistribution of open space. Although ORRRC was hesitant to place blame for the state of things squarely at the feet of the National Park Service, others were not.[20] After all, since 1936 the Park Service had been the officially designated lead federal agency in outdoor recreation. This blame could be turned to the agency's advantage, however. If the Park Service caused the problem, it might follow that it bore the responsibility for finding a solution. Part of that solution could be urban national parks.

Furthermore, land for urban national parks was already in federal hands in many cities. For example, in the San Francisco area there were seven unneeded military installations and one former federal prison, Alcatraz, which could be put to recreational use by the Park Service. The New York area was also speckled with military posts of declining defense value, including Floyd Bennett Field and Fort Tilden in New York City and Fort Hancock across the bay on New Jersey's Sandy Hook. Although New York and San Francisco were exceptional in the amount of surplus federal land available, many other cities had an unused fort, airfield, or federal institution waiting to be put to other uses. The federal government had a long-standing procedure for transferring such land to the cities and states, but the Park Service realized that if it were retained in federal hands, such surplus property could constitute the inexpensive nucleus of a federally managed urban Park System.

Finally, the agency, out of favor with its department and its preservationist allies, might find political support in the cities—and urban support was becoming increasingly attractive. With the advent of the civil rights movement, the minorities, who were largely urban dwellers and not very valuable constituents, suddenly became worth courting. Also, the cities became "class conscious" in the 1960s, and they united in an active National Council of Mayors, with New York's dynamic John Lindsay at their head.[21] Moreover, the Supreme Court's one man–one vote rulings and the legislative reapportionments they required meant that metropolitan political representation would finally come to reflect the large populations of cities. Hartzog was well aware of the constituency-building uses of urban parks. According to one agency administrator, "Hartzog saw what reapportionment would do and this made urban parks a good bet." Another recalled that "Hartzog was desperate to expand the political base of the system. He knew that a major urban park would bring the [active support] of a couple dozen representatives while a new crown jewel would only affect one, and probably make him an enemy of the System at that."

Drawbacks

There were serious drawbacks to an urban commitment on a large scale, however, and agency leadership was well aware of them. First, there was the question of the Park Service's ability to assume the role of lead agency in any urban open space policy thrust by the federal government. Its western, nature-oriented image, which had served it so well in the past, made it seem to many like an inappropriate and unqualified agency for an urban mission.[22] There was undoubtedly something to this. The management problems in large urban parks, especially ones accessible to ghetto areas, would present the agency with problems, constituents, and demands it had not previously encountered. They would be ones which, in the words of an agency administrator, "would blow the mind of your traditional ranger." Moreover, the agency had no proven criteria for urban park selection. It did not have a clear idea of which parks would be good ones from the standpoints of management ease or political support. It did not even know which ones would be able to meet their stated objectives.[23] One thing was certain, however; the urban parks would require large staffs, large development outlays, and large budgets, perhaps large enough to starve the other parks in the System. These questions and facts meant that a major urban commitment would be a very large gamble at unknown odds.

There might also be the problem of resistance in the ranks. Urban parks would run counter to the image of tall trees, clean lakes, and big mountains. It was an image within which agency employees found comfort and from which the Service derived its values and sense of worth. From the point of view of a bureaucracy's rank and file, its leadership is a protector of agency values above anything else.[24] A commitment which went so completely against mystique and tradition was bound to be viewed as a betrayal in many corners of the Park Service. There was more than mystique and abstract values involved for agency personnel, however. Hertzog had made it clear that an urban commitment would mean a large infusion of new kinds of professionals into the agency: sociologists, psychologists, urban recreation and design specialists, and so on. Urban parks would therefore mean that the traditional rangers and managers would have to share their agency, and its rewards, with persons with very different skills and values.[25]

From the perspective of agency leadership, this resistance could be more than an annoyance. If not handled properly, it could undermine and ultimately bring down an urban commitment. Even if this internal resistance were handled right, and the necessary infusion of new professional skills and perspectives took place, there were still major risks. Any originality of vision that a newcomer brings to an organization will almost certainly be balanced by inexperience in dealing with day-to-day problems.[26] The possibility of errors, including serious blunders, due to inexperience in managing urban parks was great, and this then was yet another source of uncertainty and danger. For an agency already under a cloud of disfavor, an undertaking with such large unknowns might have fatal consequences.

Finally there was the problem of interagency competition. A National Park Service bid for the role of federal provider of urban open space would not go unchallenged, and the primary challenger would be the agency's newly created sister agency, the Bureau of Outdoor Recreation.

The Bureau of Outdoor Recreation

As discussed above, the very establishment of BOR was a blow to the Park Service. The ORRRC study was a thinly veiled condemnation of the agency for failing to discharge its recreation responsibilities, and BOR was established to assume those responsibilities and carry them out properly. Not surprisingly, the Park Service and the new bureau were at each others' throats from the very beginning. BOR argued that it, rather than the Park Service, was the appropriate instrument for federal urban recreation policy because it was concerned with "the recreation needs of people rather than the utilization of resources."[27] The implication was that the Park Service was primarily a resource management agency and only incidentally concerned with serving people.[28]

Although coordinating the outdoor recreation activities of other federal agencies was part of BOR's official charge, the new bureau, a small one with support in the administration but with no power base in Congress and no articulate body of supporters in society at large, was no match for the older agencies like the Forest Service or the Park Service, agencies whose policies it was supposed to coordinate. BOR quickly recognized the obvious, that is to say, since it could not impose its will on them, it would have little real say in recreation policies for federal lands. Accordingly, it abandoned efforts in this direction under a rhetoric of compromise and conciliation and put its energies into its responsibilities for federal liaison with lower levels of government.[29]

In pursuing this strategy, BOR played up the aid and advice side of federal open space policy as much as possible, and tried to convince state and local governments that their interests lay in a national recreation policy for the cities which relied on grants rather than direct management and therefore on BOR rather than another federal agency.[30] From the local viewpoint, however, there were advantages to direct federal management of open space. First, there was the axiom that held that the higher the level of government, the more professional the management was likely to be.[31] The axiom seemed to be confirmed by the National Park Service, which had a reputation as an excellent manager of parks. Therefore recreation facilities managed by the National Park Service could be counted on to be top quality and to please the public. Second, although direct federal management would inevitably mean some loss of local prerogatives, it would also relieve the local government of all expenses connected with running a park after it was established, something grants-in-aid from the LWCF would not do.[32]

An Urban National Park System Is Established

Through the 1960s, the extent to which the National Park System would develop an urban wing remained unclear; administrations and agency leadership seemed uncertain as to whether the benefits would outweigh the costs. During the decade, however, the National Park System expanded rapidly. Hartzog's directorship (1964–72) saw the System expand from 226 to 284 units.[33] The new parks of the 1960s included several on the periphery of metropolitan areas, units for which access to large urban populations figured as an important rationale. The decade opened with the authorization of Cape Cod National Seashore, a park which had its genesis in the 1955 national seashore study. In 1962, Point Reyes, just north of San Francisco on the Pacific Coast, was added to the System and two years later Fire Island, to the east of New York City, became a national seashore. Like Cape Cod, the Fire Island park had been recommended in the national seashore study. In 1965, Delaware Water Gap National Recreation Area was authorized in conjunction with an Army Corps of Engineers' plan for a dam and water storage project at Tocks Island on the Delaware River. Although this cooperation between the Park Service and another agency on a combined dam and recreation project was not new, this was the first such project in the populous Northeast, and it was touted as a recreation facility for metropolitan New York. The following year, Indiana Dunes National Lakeshore was established. The park was close to Chicago, from which it derived considerable public and private support, and it too was considered a recreation facility for a major metropolitan area.

None of these new units of the system represented a distinct break from past policy. The seashores fleshed out a plan advanced in the 1930s and vitalized in the 1950s. The Delaware Water Gap park was the latest manifestation of an old alliance, albeit in an unfamiliar part of the country. For Indiana Dunes, urban recreation was but one of many rationales. Thus, even taken as a whole, these new parks, while marking a drift toward a more urban system, hardly represented a conscious, coherent commitment to urban recreation or preservation of urban open space. Their addition was too disjointed, and, perhaps above all, none of them were thought of primarily as instruments with which to combat urban problems. They did not bring the agency into, to use Udall's phrase, "the battle lines of conservation in the cities."

Then, almost by accident it seemed, the agency was definitely drawn in. In early 1968, the Park Service tentatively instituted its Summer in the Parks program in the Washington, D.C. parks it managed. The program was designed to get the people of Washington, D.C., particularly the young, into the parks in order to increase environmental awareness, teach outdoor recreation skills, and provide cultural enrichment. Not incidentally, the program might also keep ghetto youths off the streets and out of trouble during a long, hot, late 1960s summer. These goals were to be pursued through programs which went far beyond the limits of traditional, essentially passive, park programs, and which included artfests and cultural events, environmental school programs, daycamps, overnight camps, and rock concerts.[34] The program had just gotten under way when the riots following the shooting of Martin Luther King erupted. Many in the city, including the mayor, credited the Summer in the Parks program with preventing the trouble in Washington from becoming worse than it was. The city government publicly praised the Service and thanked it for saving the city. This convinced agency leadership that the Park Service could play a highly visible, effective urban role and reap great rewards from it. According to agency director Dickenson, "the whole idea [of urban parks] fused with the burning of Washington, it showed we could be activists."[35] Tentativeness became a commitment. "After that [Service leadership] backed us full guns," recalled one of the agency's urban park advocates.[36]

Searching for an Urban Mission

During the last years of the Hartzog directorship, the Park Service actively promoted itself as an agency in search of an urban mission. Hartzog was convinced of the positive net value of urban parks to the System and Service and he was willing to throw the die. According to McPhee, at the end of his directorship, one of Hartzog's two principal goals for the System was to give it "a new emphasis toward the cities."[37] (The other was to "maintain the Park System's vast existing apparatus.") This new emphasis was embodied in the agency's Parks to the People policy, which called for the establishment of urban recreation areas. This would be a major undertaking which involved more than putting a few traditional rangers in urban or semiurban areas. It was a commitment to more than recreation. Through these parks, the Service would commit itself to engendering basic social changes in the cities. Hartzog said, "I'm looking for social scientists [as well as park managers] to staff big recreational areas near ghettos."[38] The program also involved more intensive use of existing national park facilities in or near urban areas, including historic sites and natural areas. The name of the policy, Parks to the People, conveyed relevance and action. It echoed radical slogans of the past, "all power to the Soviets," "land to the tiller" and so on. It was, to use Tom

Wolfe's term, radical chic. It reflected a search for relevance that in retrospect might seem naive and perhaps even a little silly. But it also reflected the commitment of the times, and it did seem to be a practical strategy for an engagé era.

It was not until 1972, however, at the end of Hartzog's directorship, that the first of the urban parks that Hartzog envisioned were added to the National Park System. Nixon's first secretary of the interior, Walter Hickel, was enthusiastic about urban national parks. He had inherited the belief of his predecessor, Stuart Udall, that Interior had a mission in the cities. During this era "the call was for the courage to plan on a large scale," with regard to the nation's cities.[39] Hickel must have heard the call; he proposed a whole system of urban national parks.[40]

Federal open space aid to the cities fit into Nixon's expressed interest in improving the quality of urban life, which, in general terms, was very much like that of his predecessor. And like Johnson, Nixon committed himself publicly to improving the urban dweller's access to open space.[41] There were important differences in that commitment, however. Nixon took a less expansive view than did Johnson of the federal government's responsibilities in areas traditionally considered the domain of the states, and the well-being of the cities was traditionally a state concern. Nixon was also more concerned about public spending, and he made periodic efforts to reduce the federal budget. Although Nixon was well aware of the political benefits to be gained from establishing federal open space near large population centers, his sense of federalism and his concern for public spending did not favorably dispose him toward the whole system of costly urban national parks that Hickel envisioned.[42] Hickel's public utterances had gone a long way toward committing the administration to such a system, however, and considerable political momentum had developed behind two specific urban park proposals.

One proposed park was in New York. There, a plan for a national park had enlisted the support of a disparate alliance of public, semipublic, and private groups. The alliance included the Regional Plan Association, which was historically concerned with urban open space. It included minority organizations, which saw a federal park as an amenity for their constituents. There were the neighbors of Floyd Bennett Field, who feared that unless a park incorporating the nearly abandoned airfield was established, it would be used as a site for low-income housing. The proposal also had the support of Mayor Lindsay's administration, which was on the fiscal ropes and looking for any help it could get in running the city. It hoped that a federal park would relieve it of some of its recreation responsibilities. In San Francisco, an urban–suburban alliance, which had the support of the city government, promoted a park which would include islands in the San Francisco Bay, military property, city parks in San

Francisco, and large tracts of open land in Marin County to the north of the city.[43]

Here, then, was a problem for the Nixon administration. It was not in favor of a major federal move into the direct management of urban open space, yet the president had committed himself in general terms to federal aid for open space in cities. The secretary of the interior, acting largely on his own, had deepened and focused that commitment with his call for an urban national park system. Moreover, two urban national parks proposals already had gathered considerable political momentum.

Urban Parks Enter the System

The solution arrived at by the administration was a clever one at first glance. The proposed parks in New York and San Francisco were both oriented toward their respective waterfronts (not surprisingly since many of the city parks and the unused military properties available to be incorporated into the parks were on the waterfront), so this common element was played up by the federal government. The proposed New York park was named the Gateway National Recreation Area while that of San Francisco was named the Golden Gate National Recreation Area. Nixon furthered the impression of symmetry by referring to the two parks as Gateway East and Gateway West. This made the New York and the San Francisco park proposals appear to be a complementary, complete, and exclusive set of two. As a set unto themselves, the administration might be able to support them without raising demands for federal parks from other cities.

Some read cynicism into this support for only the two urban national parks which had gathered the most local momentum. For example, one writer suggests that the two gateways "were nurtured on the interest of the Nixon Administration in demonstrating good faith on urban problems without spending much money."[44] One need not be so cynical, however. Such a position was consistent with the Nixon administration's desire to rearrange responsibilities among the levels of government so as to centralize some functions at the federal level while returning others to the states and local governments.[45] One of the unquestioned responsibilities of the federal government in Nixon's view was that of playing a demonstration role in areas where its expertise would allow it to develop programs which could serve as models for the states, counties, and municipalities. The gateways could be viewed in this context. Within the Nixon administration there was a feeling that the two gateways should serve as models of what state and local governments might do in meeting urban recreational needs.[46] Whatever its motives, the Nixon administration threw its support behind the two park proposals. In 1972, the national parks for New York and San Francisco were authorized. Soon afterward, in the New York area, Floyd Bennett Field, the Jamaica Bay Wildlife Refuge, Fort Tilden on Breezy Point, and Fort Hancock on Sandy Hook were transferred from their various federal managers to the National Park Service. In addition, several city-owned properties on Breezy Point, on the eastern shoreline of Staten Island, and on the periphery of Jamaica Bay were also transferred. Together these properties formed the basis of the Gateway National Recreation Area (figure 6-1). In San Francisco, several city parks plus the federal presidio became components of Golden Gate National Recreation Area (figure 6-2). Alcatraz Island in the bay and several properties, forming an almost continuous stretch from the north shore of the Golden Gate to Point Reyes Station, some 25 miles to the north, were also included in the unit (figure 6-3). With these two new parks, the National Park System's incremental drift toward urban parks was supplemented by a conscious and substantial, if circumscribed, commitment.

As some had predicted, the symmetrical, closed-set idea of two urban national parks was not well received in other urban areas of the country, and the two gateways served more to encourage than inhibit other cities and their congressional delegations when it came to advancing proposals for their own federal parks.[47] In 1974, the Cuyahoga Valley National Recreation Area, stretching between Cleveland and Akron, Ohio was authorized by Congress. In 1978, the Chattahoochie River National Recreation Area, extending along the banks of the Chattahoochie within and to the north of Atlanta, was authorized. That same year saw the authorization of the Santa Monica National Recreation Area, a park which ran in a wedge-like shape from Griffin Park in Los Angeles westward through the Santa Monica mountains north of Malibu. The establishment of Jean LaFitte National Historical Park and Recreation Area, a park consisting of several parcels scattered in and around New Orleans, also took place in 1978. Thus by the end of the 1970s, the National Park Service was deeply and directly involved in urban recreation. The two gateways were among the most important units of the National Park System when measured by the size of their staffs and budgets or by the number of annual visits they received (table 6-1). The other urban parks showed promise of becoming their equals. The question of whether the federal government would become an important manager of urban open spaces through the National Park System was effectively settled. However, this by no means ended questions surrounding urban parks.

THE POLITICAL AND CONCEPTUAL CONTEXTS
OF THE URBAN PARKS

Preservation and Mass Access as Park Goals

Urban parks, especially those on the periphery of metropolitan areas but, also, to a lesser extent, those closer to the hearts of cities, brought the Park

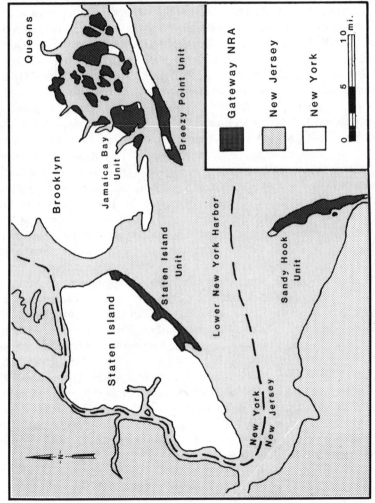

Figure 6-1.
Gateway National Recreation Area.

Figure 6-2.
Golden Gate Bridge as seen from the Marin County section of the Golden Gate National Recreation Area. (Photo by Richard Frear).

Service a major dilemma. This dilemma is perhaps most clearly illustrated at Fire Island National Seashore. Two qualities, the proximity to large urban populations and the presence of undisturbed nature, were prominently cited in the bill establishing the seashore, which read that:[48]

Table 6-1. Budgets, Visits, and Personnel Ceilings of Selected National Parks, 1979

Unit	1979 Budget allocation (in thousands of dollars)	1979 Authorized personnel ceiling[a]	1979 Visits (in thousands)
Gateway	9,362	156	9,006
Golden Gate	7,168	124	11,371
Cuyahoga[b]	1,823	15	543
Yellowstone	8,853	82	1,875
Grand Canyon	4,673	97	2,310
Great Smoky Mountains	4,498	104	11,187

Source: U.S. Department of Interior, *Budget Justifications, F.Y. 1981—National Park Service.*

[a] Full-time personnel only.

[b] Cuyahoga is still in the initial development stage; authorized operation ceilings and budgets will rise as the park moves toward full operation.

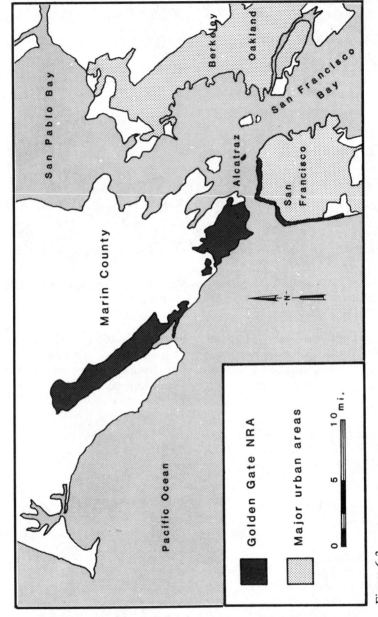

Figure 6-3.
Golden Gate National Recreation Area.

For the purpose of conserving and preserving for the use of future generations certain relatively unspoiled and undeveloped beaches, dunes, and other natural features within Suffolk County, New York, which possess high values to the Nation as examples of unspoiled areas of great natural beauty in close proximity to large concentrations of urban population, the Secretary of the Interior is authorized to establish an area to be known as the 'Fire Island National Seashore.'

When Secretary Udall visited New York's Fire Island in 1963, he was impressed both by its unspoiled quality and its proximity to New York City, and he cited both in justifying his support for the national seashore proposal. For Udall, linking nature preservation with urban access on Fire Island made sense intellectually, and it paralleled the conceptual links he had made between the traditional conservation and urban problems in *Quiet Crisis*. He would later make essentially the same arguments about the virtue of preserving nature close to the city at Indiana Dunes and elsewhere. President Johnson, like Udall, appreciated the unspoiled quality of Fire Island, for Johnson, like Udall, had his doubts about what material progress meant to urban landscapes: "Cities reach out into the countryside, destroying streams and trees and meadows as they go. A modern highway may wipe out the equivalent of a 50-acre park with every mile. And people move out from the city to get closer to nature only to find that nature has moved farther from them."[49] Yet Johnson saw Fire Island as fitting into his plans to improve access to public amenities for the nation's urban dwellers. At the close of his term, when he looked back at his administration's goals and accomplishments, Johnson viewed with pride the establishment of the national seashore on Fire Island. He wrote that "the residents of New York City can within an hour or so reach the beaches and waters of the Fire Island National Seashore."[50] Bring the people without bringing the city, that was the best of both worlds.[51] In fact it was the very unspoiled quality which would draw them to Fire Island.[52]

It was all well and good for Udall and Johnson to strive for conceptual elegance. The fact that two ideal goals could be fused into one appealing synthesis such as preserving nature for the enjoyment of millions, however, does not guarantee contradictions will not arise in the implementation. Plans without implementation "reflect the freedom of ideal choice."[53] And often it is the public agency which must cope with the implementation problems associated with ideas whose superficial appeal hides deep, perhaps inherent contradictions.[54]

Secretary of Interior Hickel, Udall's successor, speaking in support of urban national parks, said, "we have got to bring the natural world back to the people, rather than have them live in an environment where everything is paved over with concrete and loaded with frustration and violence."[55]

There is a very good reason why urban areas are paved over with concrete; the natural environment will not stand the intensity of use that urban life subjects it to. Historically, permanent concentrations of people have meant intensely altered environments. Places of intense urban leisure, like places for other types of urban activity, must be modified considerably to serve their function. Hickel could ignore this fact because, like Udall before him, he did not have to face the hard realities of implementation that park managers and planners would have to confront.

When one gets to the specifics of park planning and management, there is an incompatibility between the preservation of nature and indigenous landscapes on one hand and mass recreation on the other. No amount of artful design will cover this fact. A million people cannot walk in solitude through unspoiled nature. They have to have places to change, throw their litter, defecate. If such places are not provided, people will improvise. A million people will also need lifeguards to protect them, police to restrain and sometimes cull the inevitable lawbreakers, and sanitation men to clean up the mess they leave when they go home. Without these services and facilities, a beach is not capable of fulfilling the mass access goals which figured in the rhetoric justifying Fire Island. Yet if they are provided, the resource's unspoiled quality is destroyed. Compromise is possible, but a compromise is just that; a forsaking of the complete attainment of one goal for the partial attainment of the other.

Thus we find that there is a contradicition in the very heart of the urban national park concept. The park is to be an agent of the improvement of urban life, and it is to be an agent of preservation in the face of what urbanization must mean when it comes to parks, i.e., the alteration of their natural resources and their integration into the modern urban system. Out of this central contradiction came the ideational cross-currents and conflicting political interests which have formed the context of the Park Service's urban park policies.

Local Interests in Urban National Parks

The political campaigns that led to the establishment of many of the great national parks of the West in the late nineteenth and early twentieth centuries were nationwide. They were usually spearheaded by an elite group consisting of professionals, progressive politicians, and businessmen of a philanthropic bent.[56] Civic and conservation organizations lent support; for example, groups like the American Federation of Women's Clubs and the Audubon Society mobilized their nationwide membership in support of national park proposals. The campaigns also had the support of the national press, which aroused middle-class public opinion by extolling the virtues of the areas proposed as national parks. On the other hand, many of those in the vicinity of the proposed parks—farmers, ranchers, lumber-men, etc.—were opposed to, or at best indifferent toward the proposals. If matters were left to them, it is unlikely that any of the great parks would have been established at all.[57] Thus in the geography of their support and opposition, the campaigns for crown jewels were similar to other causes of progressive conservation, with national "progressive" forces pitted against "hidebound" localism.[58]

The urban parks established over the past two decades were set on a very different political landscape. The impetus for establishing them was inevi-tably local or regional, not national. On the periphery of metropolitan areas, local citizens were usually prodded into support for park proposals by the desire to thwart the changes which accompanied metropolitan ex-pansion, or at least to channel that expansion into what they considered benign forms. It was Robert Moses's plan to link Fire Island to the wider New York metropolis as an extension of Jones Beach that coalesced local support behind a national seashore proposal that had been languishing for almost a decade.[59] Residents feared that if the national seashore were not established, the island would become the site of a major state beach and uncontrolled private development, and so they acted. The Cuyahoga Valley National Recreation Area grew out of "a great local fear of general devel-opment" in the words of the park's superintendent.[60] Local residents were driven by fear of suburbanization in the Santa Monica Mountains and in Marin County, California. They were moved to action by fear of low-income housing at Floyd Bennett Field and a fear of heavy industry at Indiana Dunes. According to Platt, at Indiana Dunes the most active members of the key Save-the-Dunes Council "resided in or near the dunes, and undeniably had an immediate personal stake in the outcome."[61]

Once this local support for an urban park was activated and the local congressmen had declared in favor of it, national environmental groups might put their energies and lobbying skills behind the proposal. Local activism and local commitment, however, were usually the *sine qua non* of parks on the urban periphery.[62] National support could be weak and it still might be possible to establish the park, but without active local support, a metropolitan park proposal could go nowhere. The national conservation groups carefully gauged the amount of local support behind a project that they were thinking of actively supporting. They knew that if sufficient local commitment were not there, their own efforts at the national level probably would be wasted. Just how important this local activism was in giving momentum to an urban park proposal was illustrated by Udall's insistence that before he "stuck out his neck" to support the proposal for a Fire Island National Seashore, the local coalition had to prove to him that it was willing to commit itself without reservation to the park.[63] Then, only when he was sure of local support, did he come out publicly in favor of it.

For the local citizens who organized to establish urban national parks in the centers of cities and on the periphery of metropolitan areas, the task did not end with the passage of authorizing legislation. These citizens had a great stake in how the parks would be developed. As mentioned earlier, local activists usually supported a park proposal because it was the least of possible evils. Its establishment was seen as a means of precluding highly undesirable change. If an urban national park meant massive public recreation facilities, it would be little better than the development options it precluded.

For those who lived on the periphery of metropolises, the arrival of the urban public in search of recreation had always been viewed with horror. In the 1920s and 1930s, the estate owners on the north shore of Long Island and the farmers and fisherman on the south shore fought Moses's recreation plans with equal vigor (and, in the end, with equal lack of success).[64] Within the cities, however, there had been important changes in what recreation facilities for the general public meant to their neighbors by the time the urban national parks were established in the 1970s. For example, in New York City the presence of large public recreation facilities near a middle-class neighborhood would have been viewed as a mixed blessing in an earlier era.[65] Added noise and perhaps some congestion were balanced by the convenience of having such facilities nearby or by the knowledge that there were many forms of urban development worse than recreation facilities, no matter how large the scale. But the prospect of such facilities had become less attractive in the decades after Robert Moses had built large city parks like Jacob Riis Park and Orchard Beach.

Although New York City had always prided itself on its social liberalism and, indeed, formal racial discrimination was never prominent in the city, informal means of segregation had long been practiced at public facilities.[66] In the late 1930s and 1940s the water in public pools near black neighborhoods was purposefully kept cold on the theory that blacks did not like cold water and therefore would not use the pools.[67] Exclusively white lifeguards and pool attendants at such facilities would make blacks, and later Hispanics, feel distinctly unwelcome, hoping by their actions to persuade those who came once that they did not want to return to such uncongenial surroundings. In other cities, the means of segregation varied, but the effect was the same.

From the viewpoint of their white, working or middle-class neighbors, a public recreation facility with a relatively homogeneous, white public, although causing some inconvenience, did not pose much of a threat to neighborhood security or social order. By the early 1970s, however, such a facility was a threat. In New York, the aggressive commitment of the Wagner and Lindsay administrations to integration and racial equality had ended segregating practices such as those mentioned above. Minorities were encouraged to use public facilities and minority group members were actively recruited for employment by the city's parks and recreation department. As a result, the user public of a major city facility was now a multiracial one.

To those in Brooklyn and Queens neighborhoods near Gateway, who had watched as the city's changing racial pattern and the sequence of housing filtration changed white neighborhoods into minority ones all around them, the prospect of being outflanked by a massively used, multiracial national recreation area was not a pleasant one. They feared that a general public which included a large number of blacks and Hispanics would bring with it high levels of litter, violence, and vandalism. They would fiercely fight the development of Gateway as a truly regional facility, because to them the very survival of their neighborhoods was at stake. In other cities, white neighborhoods came to feel similar paranoia as they saw the political power of minorities increase, while city hall, sometimes under black mayors, came to be viewed as unsympathetic to their needs.

Because they had such a stake in park development, the local organizations which were active in the establishment of the urban parks usually remained active in overseeing their development. In some cases, important battles were won in the early days of park planning. For example, the Breezy Point Co-op, a summer home community situated in the middle of what was to be Gateway's most intensely developed unit, Breezy Point, won exemption from condemnation. The park planners would have to plan around the community and its zealously guarded privileges. In several other cases, the ability of the Park Service to resort to condemnation was severely curtailed when Congress, prompted by the wishes of park neighbors, imposed low limits on the amount of land the Service could acquire. For example, Chattahoochie National Recreation Area was restricted to a maximum purchase of 5,000 acres. In other urban and semiurban parks, including Fire Island, the agency's planning was severely restricted by low ceilings on the amount of money it could spend to develop recreation facilities, ceilings imposed by Congress at the prompting of local groups.

Deterrents to Mass Access

Several large-scale social and economic trends worked in favor of local wishes and against the development of urban parks into facilities for the public en mass. First, the days of cavalierly running expressways through city and country were over by the time the first urban parks were established. The local opposition to, and a political disaffection with, road building as it was practiced in the 1950s and 1960s, plus the general impoverishment of the public fisc, meant that it was no longer possible to plan a facility for a metropolitan public and assume that once the facility was established, the means of getting there would follow.[68] If anything, the reverse was now the case; the availability of roads, bridges, public transportation, etc., was a severe limiting factor in recreation planning. By the 1970s planners were asking themselves if it was wise to plan regional recreation facilities if regional access did not already exist. They had the example of Fire Island; more than a decade after the establishment of the national seashore as a metropolitan facility, getting to it was still an ordeal of delay and expense. Although Johnson pointed with pride to Fire Island's accessibility to the "urban working man and his family," as the Park Service described the trip, it did not sound like much to be proud of: [69]

> Visitors who do not have access to private boats may have to endure many frustrations before reaching Fire Island National Seashore. Those who arrive by car may have to battle several hours of traffic on Long Island highways, only to become lost en route to the ferry slip because routes are poorly marked. At the ferry slip, public parking costs $2 to $3 per day, and the insufficient parking area is often full. Visitors must then leave their cars some distance away, occasionally at the risk of an illegal parking fine. Their frustration may be magnified when they see that the parking lot contains many empty stalls, which are reserved for the exclusive use of local residents. If visitors miss a ferry connection, they may have to wait an hour or more for the next run. Once on the ferry, visitors have about a 30-minute ride to the island; round-trip ferry rides cost $3.00 to $3.75 per person.

Moreover, the situation seemed likely to get worse rather than better. The congestion on Long Island's highways increased every year, the deterioration of the Long Island Railroad meant longer and less comfortable rides for those who wished to take it to Fire Island, and both the railroad and the ferries were raising their fares steeply. In 1978, planners on Fire Island evaluated the regional accessibility of their park and the possibility of improving it. They concluded that the National Park Service would do well to give up thinking that the national seashore could become a metropolitan facility in the foreseeable future: "For several reasons the regional population doesn't represent the potential visitors to the national seashore. . . . The provision of an inexpensive, convenient means of public transportation is the only way by which the national seashore is likely to become more accessible, and the possibility of such a system appears relatively remote." [70]

At Golden Gate, access was less of a problem. Many of its units were already served by public transit, so increasing access involved little more than putting extra buses on established routes. Serving New York's Gateway was another matter. The units of Gateway were not along major, established lines of public transportation, and the roads leading to it were often jammed. The Sandy Hook unit in New Jersey had no public transportation to it, period. Improving access to the units of Gateway would have required a major commitment by the entire region,[71] but as the expense of such systems went up and governments moved further into the era of fiscal austerity, ambitious public transit programs and ambitious road building programs looked more and more like pipedreams everywhere. In short, the entire urban transit mess of the 1970s worked against urban national parks which were truly metropolitan in character and use.

A second trend over which the agency had little control and which encouraged the development of the urban parks along local, neighborhood lines was the rise of participatory, or open, planning. In the late 1960s and early 1970s, the idea that government agencies had become too removed from popular sentiment and too insensitive to popular wishes in their planning gained wide currency.[72] Highways had been rammed through stable neighborhoods, destroying them. Large public facilities had been located with little thought to their impact on surrounding areas. The Army Corps of Engineers was tearing up whole landscapes to keep its battalions occupied and its congressional supporters happy. As discontent found its way into the political process, federal and other government agencies were required to open their planning to public advice and scrutiny through such devices as public meetings, educational workshops, and wide circulation of plans for their proposed projects. Sometimes agency proposals had to be accompanied by a discussion of considered and rejected alternatives. In many cases, advisory councils with representatives from many sectors of society were established to oversee agency planning.

The way the National Park Service planned its parks was very much affected by the open planning movement. Most of the large new parks had citizens' advisory councils established for them. In some cases, the agency only had to take the recommendations of the council under consideration, but in other cases, the council's recommendations were made binding on the agency.[73] With open planning, the Service was also required to circulate draft versions of its park management plans to other public agencies, to place them in libraries, hold public meetings on them, and to make copies available upon request.

Those citizens who took advantage of participatory planning were usually those who felt they had the largest personal stake in the parks, those who lived close to them and wanted to see their development produce as few changes and disturbances in their lives as possible. Those who already used the areas in their present form and who, therefore, did not favor changes which would threaten their enjoyment of them, were also likely to take advantage of the avenues of direct public participation. When the general management plan for Gateway was circulated publicly for comments, almost all the responses the agency received were from organizations whose members lived near the park and who objected to plans to increase access to the park or to expand facilities to accommodate those who did not already use it.[74] The mandated public meetings on the general management plan turned into raucous, hostile encounters between the Park Service and citizens who accused it of wanting to destroy their neighborhoods. The other side of the argument, that inconvenience for a few might greatly improve recreation opportunities for the many, was not well articulated through the participatory planning process. Citizens who might have benefited from ambitious planning for urban parks lived far away, were diffused throughout the metropolitan area, and undoubtedly had other things closer to home to worry about.

The hand of the park's neighbors in influencing Park Service planning was further strengthened by the importance the agency placed on getting along with a park's neighbors when it evaluated a manager's performance (and therefore his career potential), and by the high value it placed on the ability to produce development plans which moved smoothly through the public participation process when it evaluated a planner's worth. Managers in the field were well aware of this. Several invidiously compared the Forest Service, which they said stuck by its managers when they got into trouble with the local citizens for upholding agency principles, to their own agency, which, in the words of one park manager, left its managers "twisting in the wind" in similar circumstances. Other Park Service managers complained of two levels of instructions which agency leadership communicated to its personnel in the parks. The first, the overt one, was high minded and principled: "Manage the parks for all the people of this and future generations." On the covert level, the message was "oil the squeaking wheel." In the case of message conflict, the managers were sure the covert one took precedence.

Planners too were well aware of the importance of not arousing local opposition with their plans; they had had plenty of lessons. For example, the task of developing a general management plan for the Upper Delaware National River was initially given to the planners of the agency's regional office in Philadelphia. However, their efforts aroused considerable hard feelings among those living near the river, who felt that their interests were being sacrificed for the benefit of outsiders, especially recreationists from the wider region. When this resentment broke into the open, the project was taken out of the hands of the region's planners and assigned to the agency's central planning unit in Denver. The planners who inherited the project felt that because their predecessors had aroused the locals, they had "botched the job" and it was now up to them, the Denver-based professionals, to smooth local feathers so a management plan for the river could be completed. The whole affair was seen within the Service as reflecting poorly on the Philadelphia region's planners and on the regional director (whose blame was widely felt to have been lessened by his inexperience in the job). Conversely, agency planners had seen positive reputations built by, and choice planning jobs assigned to, those who were good at getting along with locals and at developing management plans which met with local approval.

Thus, the combination of local forces brought to bear on the urban parks was a powerful one. It was not, however, entirely unopposed.

The Weakness of Wider Interests

There were several voices which spoke for the inclusion of more than local considerations in park planning. First, there were political interests within the metropolitan areas whose designs on urban national parks ran counter to those of the local community groups, and who wanted to see park access made as wide as possible. For example, in New York there was the Regional Plan Association, a venerable organization which had promoted integrated regional planning for almost fifty years and which had first proposed Gateway in the early 1960s. There was also the Gateway Citizens Committee, a "citizens" organization with individual and corporate members. The committee had a decidedly upper-class cast and a philanthropic interest in seeing that the park was made available to the city's poor.[75] The committee has long advocated that the Park Service, the city, or the federal government subsidize bus transportation to the park. It has also campaigned for subsidized ferry service to Gateway's Sandy Hook and Breezy Point units from waterfront points close to the city's poorest neighborhoods.

Organizations representing minority groups and politicians from minority districts insisted that the park should serve their members and constituents—and this meant increased access to, and expanded facilities in the units that made up Gateway. Although the voices of the minorities hardly spoke for the entire New York region, they did represent political power on the side of wider accessibility and large-scale facilities. Thus one found in New York what has been called the Manhattan Alliance, a coalition of the unquestionably poor and the clearly privileged, both putting their weight behind Gateway in the service of the entire region. There were similar alliances in Cleveland, Atlanta, San Francisco, and Los Angeles speaking for increased access to their cities' urban national parks. The voices speaking for the broad access and large-scale development were not always well articulated, however, and by and large they were ineffectual. To have had an impact, they needed the support of the metropolitan governments for their ambitious plans.

One might have expected the city governments themselves to have been strong advocates of urban national parks which were truly regional in their development. They tended not to be, however. When the city governments of New York and San Francisco came out in support of the national parks in their cities, their rhetoric was lofty, usually along the lines of bringing the taste of the great national parks to all their citizens.[76] But there were more prosaic reasons for them to support the proposals. The cities, especially New York, were in desperate financial straits, and they needed all the help they could get from the federal government just to maintain their existing recreation facilities.

In the two years before Gateway was established, New York had been forced by its public impoverishment to reduce its park maintenance force from 5,620 to 4,700, and more cuts looked inevitable. Neighborhood parks in the outer boroughs were becoming ad hoc dumps, littered with mattresses, old refrigerators, and abandoned autos. Nets had disappeared from tennis courts, so players brought their own or did without. Resodding programs were cancelled, and brown patches of beaten ground expanded at the expense of the grass around them.[77] In San Francisco, the situation, while not as extreme, was similar. Because park and recreation facilities were deemed to be among the less essential city services, they were among the first to be cut when the money ran out. Nevertheless, parks were highly visible public goods, and their deterioration was a constant reminder to citizens of their city's plight and of the inability of current officeholders to deal successfully with it.

Bailing Out the Cities

One suspects that what the city governments really wanted the National Park Service to do was to take over the day-to-day management of the city parks, remove the junked cars, put nets on the tennis courts and, in essence, make things look right by restoring the status quo ante.[78] Both New York and San Francisco urged the National Park Service to take over as many city-run facilities as possible, regardless of how blatantly lacking in national park qualities they were. Ironically, in New York, the Lindsay administration even wanted to include Coney Island, a faded seaside amusement park, in Gateway. Perhaps it was unmindful of the fact that years before, Mather had cited Coney Island as a specific example of the kind of facility the National Park System should not include.[79]

For New York's Mayor Lindsay, there was a further reason to support Gateway. The Navy's Floyd Bennett airfield, situated on a large island in the Jamaica Bay, was obsolete, due to be closed, and shortly to be declared surplus property by the federal government. Governor Rockefeller had revealed plans to use the island as a site of a massive, low-income housing project and had interested the federal government in supporting it. Lindsay opposed the plan.[80] Such a project would require the city to provide services—sewer hookups, roads, schools, additional firehouses, and police precincts—for which it did not have the money. The housing proposal also had enraged Brooklyn and the Queens communities near the airfield, who did not relish the possibility of tens of thousands of new low-income neighbors. A Gateway which incorporated Floyd Bennett Field would preclude the housing project and all the problems for the city administration that it would entail.

Thus it was undoubtedly as much to bail their cities out of financial straits (and their administrations out of political ones) as anything else that caused the mayors of New York and San Francisco to support the gateways. Ideally, the National Recreation Areas would mean federal aid without federal control. In fact, once he felt that the Nixon administration was irrevocably committed to Golden Gate, San Francisco's Mayor Alioto indulged in a bit of coquetry. Although he wanted Golden Gate to include many city parks, he wanted the parks to remain under the jurisdiction of the San Francisco Recreation and Park Commission. In other words, he wanted the federal government to provide the money to maintain and operate the parks but to allow control of them to remain with the city. In a letter to Secretary of Interior Morton, he asserted that such an arrangement was "a condition of the city's necessary participation" in the project.[81] His bluff was called with a curt rebuff from the secretary, and he was forced to back off. There was no getting around it, the price of federal support would be a good deal of federal control.

In the meantime, New York City's Mayor Lindsay played a nervy gambit of his own in the hope of squeezing as much federal aid as possible out of the park plan. He declared that since the recreation area was being touted by the federal government as a regional park for all metropolitan New Yorkers, the federal government should not only maintain the park but should also subsidize New York City's mass transit system as a means of getting to it. Mayor Lindsay hinted darkly that unless such transit subsidies were included, he could not support the plan for a national recreation area at all. Later, when it became clear that his reach had exceeded his grasp and that his position on transit aid was threatening the entire project, he backed off. The chairman of the city's planning commission meekly assured Congress that transit aid was not necessary; if the federal government established Gateway, New York City would provide the means of reaching it.[82]

Although the Lindsay administration touted Gateway as a park for the metropolitan region, one suspects that a Gateway which actually lived up to its rhetoric about being a regional facility would have been as much of a threat as a blessing to the city government. The political and social ecology of New York City was and still is a fragile one. Its maintenance depends on reconciling the frequently conflicting demands of different groups and different neighborhoods within the city. White, middle- and working-class neighborhoods look to the city government as a protector of community integrity, as a provider of those services which would allow stable, culturally homogeneous areas to retain their character and security. They also expect the city government to be a shield against any changes in public facilities, housing stock, transportation patterns or whatever would threaten these neighborhood characteristics. On the other hand, minority groups look to city hall to provide increased residential options and access to public service and facilities.[83] Given the intrinsic conflict implicit in these differing goals and the broad diffusion of political power within the city, the most expedient solution to any problem involving major changes in the city's spatial ecology would usually be that which maintained the appearance of impartiality while engendering only slow, incremental change.[84]

A Gateway which did little more than refurbish existing parks for their current users would deliver benefits to some without threatening anyone. A Gateway which lived up to its high promises of becoming a major regional facility was another matter altogether. It would involve major new public works. It would change the entire city's recreation habits and its patterns of weekend movement. It would have unforeseen and probably negative consequences for those who lived near it, and perhaps dangerous ones for a city administration which was party to such an undertaking.[85]

The Rhetoric of Urban Parks

A final factor which might have been expected to prompt development of the urban national parks into truly metropolitan parks was the force of the rhetoric used to justify them. The rhetoric which attended the founding of the urban national parks emphasized that they were to be special places in the service of entire metropolitan regions. They were to do something or provide something that the regions or the municipalities within the regions could not do or provide for themselves. In short, the urban national parks were to be special, unique places. For example, Assistant Secretary Herbst, testifying before the House Appropriations Committee, reassured the congressmen that the National Park Service was not merely managing urban parks, rather it was building national parks in urban areas.[86] Hartzog had made similar points in his campaigns to get the two gateways established. In fact, rhetoric about serving whole metropolitan regions in extraordinary ways is in the authorizing legislation of all the urban national parks which have followed the gateways.

Unfortunately, urban national parks which were to be more than merely local facilities demanded far greater creativity than did national parks in spectacular natural areas. In fact, perhaps they demanded too much creativity. Great natural features were the supreme referents of the crown jewels; they were the reasons for the creation of parks like Zion and Grand Canyon, and, to a large extent, these features provided the guidelines of park planning. Unlike the great natural areas, urban parks could not take their planning cues from the site resources, since the sites were little more than land on which a park could be developed. Sax wrote, and rightly so, that the "new urban parks cannot feed on the traditional symbolism of wilderness."[87] One suspects that efforts to create Yellowstone-style parks in New York City or in the valley of the Cuyahoga would have fallen of their own weight. Sax went on to assert that the urban parks, nevertheless, did have a rightful place in the contemporary national parks system.[88] According to Sax, they would only find their rightful place, however, if the symbolism of great national parks were brought closer to the public. But if the symbolism of the great parks was urbanized in the process, the new urban natural parks would fail. In a way, this is reminiscent of an episode in *Candide* in which Voltaire got his hero and his companions out of a fix by transporting them across the Andes in a marvelous flying device which remained conspicuously undescribed. It was all well and good to say what should be done in general terms, like creating urban national parks which would bring the symbolism of parks closer to the public. Describing the specifics of how to go about it was the trick. What exactly was the symbolism of the parks? How was it to be brought to the city without urbanizing it? Neither Sax nor the Park Service seemed to know. Thus when the National Park Service took the metropolitan mandate seriously, it found itself facing the task of designing urban national parks which could be neither contemporary urban parks nor traditional national parks, and it had no clear idea of what was in between.

Forms of Urban National Parks

In the face of this problem, the agency's promotion of urban parks as something special, something beyond merely pleasant, local open space, took two specific forms. First, the urban national parks were promoted as prototypes of future urban open space. By its perspicacity and length of foresight, the Service would accommodate and help shape the city of the next century. Secretary of the Interior Andrus called Cuyahoga Valley National Recreation Area, the urban park of the twenty-first century.[89] Cuyahoga's superintendent made a similar point when he said that "we might well be the Central Park of the future." Second, urban national parks were touted as special because they would fit at the very apex of a hierarchy of open spaces which had evolved in urban areas over the past century. For example, a truly regional Cuyahoga Valley National Recreation Area would fit at the top of the hierarchy of open spaces in the northeast Ohio metropolitan area. As Cuyahoga's superintendent expressed it, "first there are the local parks, then the metro parks, and we're the third phase."[90]

At the first level there were local, municipally run parks of Cleveland, Akron, and the suburbs of these cities. They were the small, intensively used parks, usually drawing a strictly local clientele for an hour or two of use. Next were the parks of the metropolitan park systems. In the early twentieth century, a multicounty metropolitan park board was established for Cleveland and the surrounding area, and soon it was acquiring park land in the scenic areas to the south of the city. Olmsted himself was brought in by the park board to draw up plans for the new system. Planned as a large semicircle of greenery around Cleveland, the system was called the "emerald necklace," and by the 1970s it had grown to include 18,000 acres of parkland.[91] Several of its largest units were in the Cuyahoga Valley, directly to the south of Cleveland. In the meantime, less than 30 miles south of Cleveland, Akron was developing a metro park system of its own, and several of its units were in the Cuyahoga Valley to the north of the city. Unlike the city parks, the larger size of these metro parks meant that they could be used for land-extensive activities such as hiking and cross-country skiing as well as for large organized events like fairs and craft demonstrations. Logically, the Cuyahoga Valley National Recreation Area, encompassing most of the Cuyahoga Valley between the two cities, was the next step up; its open space was to serve an even wider region and to provide opportunities not present in the metro parks.

Behind the logic of both the urban national park as the city park of the future and as the apex open space lurked serious conceptual and practical problems. When Frederick Law Olmsted designed New York's Central Park in the middle of the nineteenth century, he was criticized for laying out the park on what was then the periphery of the city. His answer to such criticism was that he was building a park for the city of the next century.[92] By the next century, New York's growth had placed Central Park near the center of the metropolis, and Olmsted's answer looked, in retrospect, like the height of vision. But perhaps it was not such a visionary coup after all. The street plan for the entire island of Manhattan, down to its numbered avenues and cross streets, had been drawn up in 1811, and by the time Olmsted started planning Central Park, forty years of urban growth had been pushing the periphery of the city northward as expected, fleshing out the grid plan.[93] The same process of fleshing out the outline would continue unabated through the nineteenth century. Therefore the Manhattan of the early twentieth century could have been, and was, predicted from the robust, city-forming processes of economic growth and physical expansion which were so apparent in the nineteenth. Moreover, there was little controversy in Olmsted's time over whether such growth was a good thing. It was widely accepted as something to be encouraged and accommodated. Nor was there much question of what, in essence, an urban park should be. Like many men of both genius and action, Olmsted was frequently embroiled in controversy over his plans, but usually it was over such things as whether the city could afford the park, over elements of design, or over particular facilities to be included. There was little objection to his basic vision of what a large urban park should be, how it should fit into the fabric of the city, or what its role in the life of the city should be.[94] This is not to deny Olmsted's genius; he planned open spaces on a scale previously unknown in this country and with a degree of creativity perhaps unmatched before or since. But it was virtuosity within an established precedent of urban parks and within a long-established and approved vision of the urban future.

The late twentieth century is not like the late nineteenth in its certainty about the city of the future. The things which in the past acted like strong, clear directional signals to the future have disappeared or become ambiguous. The urban growth on which the last century based its predictions—constant population increase and constant economic expansion, is no more. The deconcentration of the city during the early and midtwentieth century was fueled by cheap energy, and perhaps the era of cheap energy is ending. Who can say? Moreover, even given a public choice in the form of the future city, it is not clear what would be picked. Aversion to high density urban life is countered by disenchantment with suburban living. As a result, those who plan the great urban parks of the future have little in the way of predictors or even norms to build on. How can the Central Park of the city of the future be planned in ignorance of the morphology of the city of the future?

The idea of the urban national park at the apex of a metropolitan open space hierarchy, the "third phase" in the words of Cuyahoga's superintendent, also carried with it much debilitating ambiguity. As is the case with the urban park of the twentieth century, it is not clear what the "third phase" means in practice, or what specific services it is to provide that the metro parks did not already offer. Cuyahoga was larger than the units of the metro systems; in fact, figure 6-4 shows how it encompassed many units of the two metro park systems like a matrix. But metro parks already included the most attractive natural areas. The lands added by the national recreation area were therefore usually of lesser rather than greater scenic value. The new land could be used for activities that needed a lot of space, such as hiking and cross-country skiing, but the metro parks already accommodated these activities. Thus, the Cuyahoga Valley park might expand opportunities for certain activities, but it did not create them. What it offered were changes in quantity, not in kind.

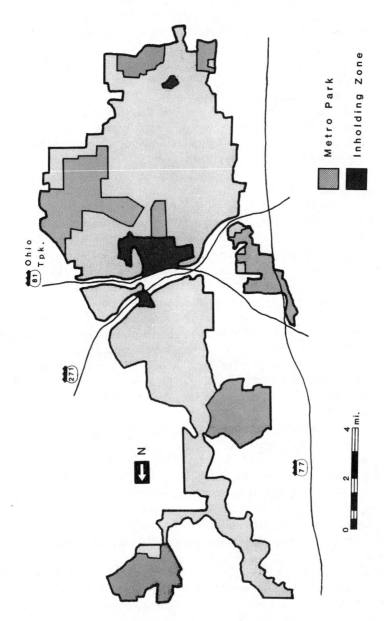

Figure 6-4.
Inholdings of private land and preexisting metro parks in Cuyahoga Valley National Recreation Area. (From Cuyahoga Valley National Recreation Area, undated map published by the Standard Oil Company, Ohio.)

These conceptual problems confronted both agency leadership and managers in the field. For example, in spite of all the pronouncements about Cuyahoga Valley's place in the present and future Cleveland–Akron metropolitan area, when the park was proposed, agency leadership wavered. It was unsure of whether to support it or not because it did not know what to do with the relatively undeveloped but already settled valley running between the two Ohio cities. According to one agency administrator who opposed the proposal initially, "I did a 180 degree turn on Cuyahoga; I didn't think it was national park material. But then I thought what the hell, let's save it, then work out what to do with it." It was left to those who drew up the first general management plan for Cuyahoga Valley to wrestle with the problem of what to do with the park. They lost. The first superintendent, pressed to show progress, condemned and acquired land without having definite plans for it once acquired, stirring up considerable local resentment in the process. His successor was considered one of the most resourceful and skilled managers in the Service, but he too faced the problem of lack of direction. "I never get pressure [from headquarters] on how to run things. I have a free hand because they haven't got the slightest idea of what to do with it."

On one point, however, the superintendent was firm, "I'm not here to try to make a Yellowstone in Ohio." But this was a negative point, a denial of the traditional model, and it was not accompanied by anything to put in its place. "I have no real model in mind," he told the author. By and large, he has relied on eclecticism and opportunism to make up for a lack of clear purpose, "picking up ideas from all over." For example, he refurbished an old train which now runs from downtown Cleveland to points within the park and has cooperated with the Cleveland Metro Park System in offering joint programs. He has also promoted the park's negative virtue, i.e., the way it has saved the valley from unregulated private development. That one of the most creative managers in the System was forced to rely on opportunism and fall back on what were essentially negative accomplishments shows the depth of the problem that conceptual ambiguity visited on agency ambitions in one of its major urban parks. The National Park Service's urban park planning was limited by demands of those living near the parks that they be neither threatened nor inconvenienced by park use. These were demands to which the Park Service had to listen because they were backed by citizen participation in the planning process, by political power, and, increasingly, by the reality of an impoverished public fisc. The agency was also subjected to the demands of those who spoke for wide access and ambitious planning for mass use. Such ambitions, however, were not backed by political will or conceptual clarity, which greatly weakened them.

One of Gateway's planners felt that conflicting regional and local forces operating on Gateway were roughly in balance and that this allowed the agency the freedom to take both views into account in its development planning. Another agency planner thought this was not the case and that at Gateway, and at all the urban parks for that matter, the agency had caved in to local pressure for inaccessible, exclusive parks. According to him, when the Park Service planned, it was "screw the people every time." There was clearly conflict between regionalizing and localizing forces. But was there balance? And what was the net impact of these two forces on the development of the urban parks? Let us first look at agency attitudes toward their development.

DEVELOPING THE URBAN PARKS

Conflicting Guidelines

In the face of this contest of political power and points of view, agency tradition offered conflicting guides. Traditions could be used to support either a mass recreation or a preservation role, either a policy in which the Park Service tried to incorporate its parks in or near cities into the metropolitan systems or one in which the Park Service actively resisted such incorporation.

Since the early National Park Service was very much a part of the conservation movement in its moral outlook, Pinchot's rhetorical question about selfishness and generosity, "for whose benefit shall [the nation's natural resources] be conserved . . . the benefits of the many or the use and profit of the few?", served as a policy referent for the agency.[95] "The many" in Pinchot's rhetoric meant serving as broad a range of beneficiaries as possible, and for the Park Service this meant serving a national clientele. That there might be legitimate local interests in the parks was not often acknowledged in the early days of the agency. Concessions to such interests were made only when political power was brought to bear on agency leadership, and then only grudgingly. The tradition of seeking a mass constituency, when brought to the urban parks, told the Service to be integrationists, to bring the parks fully into the urban system, to make them as accessible as possible to as wide a range of tastes as possible.[96]

Tradition could also be invoked, however, to support an urban park policy which preserved both the cultural and natural elements of the landscape from the changes associated with urban growth, and which thus acceded to local demands. Although the Park Service began life as a preserver of nature, its ambitions led it to extend its preservation mission to other things, first to remnants and reminders of the past, and, then, gradually, to living landscapes. When Albright was director, he flirted with the idea of parks on Samoa and Hawaii to preserve the indigenous cultures. From the first days of the Shenandoah and Great Smoky Mountain national parks, efforts were made to preserve their cultural as well as natural landscapes.[97] Hartzog explored the idea of using the National Park System to preserve special cultural areas like California's Sonoma country and the Amish region of southeastern Pennsylvania. The agency's experiments with less-than-fee land acquisition and its occasional reliance on regulation rather than purchase acknowledged that under certain circumstances the national interest demanded the guided evolution of an existing privately owned and occupied landscape rather than its replacement by a wholly owned federal area.

Metropolitan vs. Localist Planning

Out of these traditions, two general positions emerged among those who dealt with urban parks. One agency planner, discussing urban parks, told the author that:

> there are two kinds of people in the Service, the resource types and the people types. The former say manage the people to preserve the resource and the latter say manage the resource to serve the people. The Service looks at some of its parks, and it can't figure out whether it ought to be preserving resources or serving people.

The view held by the "people types" was considered the traditional one among agency managers and planners. It was the metropolitan view which emphasized mass access, large-scale facilities, and fee-simple acquisition of parkland. Implicit in this view was the notion that the Park Service should be an instrument of the expansion and rationalization of the metropolitan system, and, through it, of progress as it is conventionally considered.[98] A corollary of this view was that the preservation of the indigenous landscape and its human and natural characteristics was clearly secondary to popular access and service to the entire region. One of the agency's senior planners, one widely identified as a traditionalist, summed up his position on popular access by saying, "I want to see people in the parks," and this meant that preservation goals had to be subservient ones.[99]

This metropolitan view identified the national interest with the interests of the entire metropolis and, like the view of the architects of early twentieth century metropolitan expansion, it tended to see opposition to what it interpreted as metropolitan ends as selfish and narrow-minded. The inferences of this view for park management (or at least the psychology of park management) could be read in the words of one urban park superintendent, whose ideas (although not necessarily his actions) leaned toward the metropolitan view: "A superintendent should be a chess player to understand his role. It's the national interest against a whole set of local ones. You have to know the rules and the plays, be prepared to give up a pawn every once in a while and go out and play the best game you can."

The complement of the metropolitan view was the localist view. Whereas the former was considered traditional, the latter was considered modern. It was associated with the dynamic younger planners and it tended to be far more local in its sympathies. It stressed preserving the indigenous landscape for the high value it had to those relatively few people whose lives were lived on it, or who took special enjoyment from it. According to one of the younger planners, considered one of the best in the Service, "The landscape [of] towns, villages, and farms is the important one. It's where people live their lives, and that's the one we've neglected—or obliterated." Reflecting a similar view, the superintendent of a major urban park, when asked about the mission of his park, told the author, "We're here to do more than provide recreation, we've got a broader . . . purpose, preventing urban development from ruining what's here." [100] The superintendent of one of the national seashores, which incorporated small towns as inholdings, reflected the localist view when he said that, "I'm not sure that there is national significance here, but if there is, it is because of the towns. Perhaps the real national interest here is in preserving them in their unique lifestyles."

There were several reasons why localist sentiments tended to be predominant among the younger planners. First, there was the natural tendency for those working at the park level and in direct contact with those living near the parks to "go native" in their outlook, and younger planners found themselves working at the park level more frequently than did the older planners, who often had moved into more senior positions at the agency's planning center in Denver or in one of its regional headquarters. One planner (who was not referring to an urban park but whose sentiments were equally applicable to them) said, "when you're in the field, far away from headquarters, you naturally think of local considerations and needs." Another planner working at Santa Monica National Recreation Area said, "you get caught up in the local enthusiasm and you figure that if something is so important to [the local residents], it should be important to you."

The changing political climate of the past decade or so was undoubtedly an important factor as well. There was widespread popular disillusionment with Washington as a source of intelligent and just policies or even as an efficient deliverer of public services. [101] This attitude was reflected in the detached, critical view some younger planners had of the federal government and even of the agency for which they worked. One told the author, "the Service has a big ego, its attitude is that you can't go wrong with the Park Service." It was reflected in what psychologists call role distancing, i.e., refusing to identify with one's role even though one continues to play it. For example, one young agency planner asserted that, "I have no time for bureaucrats, I don't like them and I'm not one of them." This attitude of detachment, combined with disillusionment with the federal bureaucracy, encouraged the younger planners to see their proper role as one of mediator between the insensitive agency for which they happened to work and people, especially the people near the parks, who were likely to be hurt by the clumsy or insensitive operation of that agency.[102]

But the localist views of the younger planners also reflected a sensitivity to the political realities of urban park planning. In his study of the TVA, Selznick found that the younger administrators, whose careers and reputations were still to be made, were far more willing to adopt "realistic" policies, that is to say, those which conformed to the wishes of the powerful interests in their task environment, than were the older, more established men.[103] Indeed, the younger planners of a localist persuasion in the Park Service stressed their realism as frequently as their idealism. According to one, "Listening to [a park's neighbors] is simple common sense. They can stop you dead in your tracks so what are you going to do? When they talk, you listen." Another planner, who had successfully handled several politically sensitive park planning assignments for the agency and who was prominently identified with the localist perspective, told the author, "First, I ask myself who has the power to stop this, and then I proceed from there." He usually proceeded by listening carefully to what those living close to the parks said they did or did not want.

They can stop you dead in your tracks so what are you going to do? When they talk, you listen." Another planner, who had successfully handled several politically sensitive park planning assignments for the agency and who was prominently identified with the localist perspective, told the author, "First, I ask myself who has the power to stop this, and then I proceed from there." He usually proceeded by listening carefully to what those living close to the parks said they did or did not want.

There was clearly an affinity between planners who had to confront strong power behind the local wishes for the parks and ideas which justified acceding to that power.[104] A moral position which emphasized their sensitivity to local demands and stressed the Park Service's role in preserving indigenous nature, the local community, or the traditional landscape allowed Park Service personnel to respect local demands without feeling that they had sold out their agency's principles. In fact, it allowed them to think they had behaved in a high-minded manner. If it could be argued that the Service's true responsibilities in urban areas were largely local, then there was no hypocritical set of double messages; upholding Service principles and not arousing the locals were confluent, not conflicting goals. The stronger the local power, the more useful to managers and planners was a localist view of the role of the urban national parks.

Gateway National Recreation Area

The degree to which local pressure and the localist view have shaped the development and management of the urban national parks can be seen in the agency's plans for these open spaces. Let us look first at Gateway, one of the two oldest urban parks, and the largest in terms of staffing and annual budget. Figure 6-5 is a rendering of the National Park Service's initial plans for Gateway's Breezy Point unit, envisioned as the recreation centerpiece of the park. The rendering comes from the Green Book, the agency's first published report on its intentions at Gateway.[105] (The report appeared even before the park was authorized by Congress.) The Parks to the People policy is evident in the plan; mass recreation was clearly to be the unit's primary goal. On Breezy Point the landscape of prior structures and use was to be wiped clean. Obsolete Fort Tilden was to be removed, and so was the Breezy Point Co-op, the mixed enclave of vacation and all-year houses. The roads that laced the areas were to be obliterated. In their places were to be ferry terminals and broad promenades leading directly across the narrow peninsula to the large public swimming beaches. Behind the beaches the agency planned golf courses, playing fields, parking lots, an amphitheater, an environmental education complex and areas of what it called creative open space. At the tip of the point was going to be a "walk and a wander" nature area. The facilities would be designed to accommodate 300,000 people on a summer weekend day and 27 million visitors a year.

Figure 6-6 shows the current development plan for Breezy Point, drawn up in 1979, almost a decade after the original Green Book Plan was developed.[106] The changes from the original plan are great and obvious. Most obviously, the slate is no longer to be wiped clean of prior structures. The two enclaves of the Breezy Point Co-op are still there, and so are the roads connecting them to the rest of New York City. The residential enclaves proved politically immovable after they enlisted the support of both the city government and the local congressmen in their cause. U.S. Army and Coast Guard facilities also remain; they too proved immovable in the face of demands that they stay. Park planners found it necessary to work around all these permanent remnants of the past. The plan for what remains is one of conceptual and physical clutter, and it reveals very different goals than those of the original Green Book plan.

Accommodation of mass recreation and ease of public access have been deemphasized. The two centrally placed ferry terminals on the original plan have been reduced to one, and it has been shifted far west of the best beaches. The promenades to take people directly across the peninsula from the ferry terminals on the bay side to the beaches on the ocean side are also now gone, and so are the playing fields and the walk and wander area. The large swimming beaches are greatly reduced while access to most of the point relies on a Rube Goldberg arrangement of interconnected shuttle bus routes. The exact degree to which mass recreation has been deemphasized can be seen in the number of visitors now being planned for.

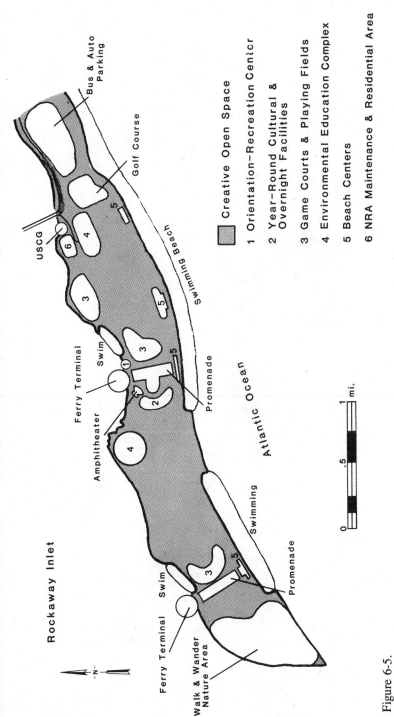

Figure 6-5.
The 1969 plan for the Breezy Point unit of Gateway National Recreation Area. (From National Park Service, "Green Book".)

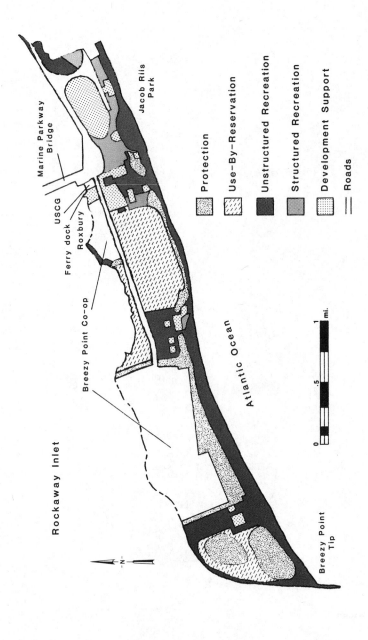

Rockaway Inlet

Marine Parkway
Bridge

USCG

Ferry dock

Roxbury

Breezy Point Co-op

Jacob Riis
Park

Atlantic Ocean

Breezy Point
Tip

N

Protection

Use-By-Reservation

Unstructured Recreation

Structured Recreation

Development Support

Roads

0   .5   1 mi.

Figure 6-6.
The 1979 plan for the Breezy Point unit of Gateway National Recreation Area. (From National Park Service, Gateway NRA General Management Plan, p.35.)

Whereas peak-day planning initially aimed at accommodating 300,000 people, that number is reduced in the current plan to 90,000. A large percentage of this will be local use and the number represents little, if any, increase over the area's recreational use before Gateway was established.[107]

This decreased emphasis on mass accommodation has been accompanied by an increased concern for, in the agency's words, "the quality of the recreation experience," a term which means low-density usage.[108] At West Beach, the removal of all existing parking lots will ensure that use remains low, since people will have to walk around Breezy Point or rely on the shuttle to get to it. According to the current plan, "it is anticipated that this westernmost segment of the beach will be a place with substantial beach area per person." The plan envisions maximum densities of one person for each 575 square feet of beach. Since most recreation standards recommend between 50 and 100 square feet per person as an acceptable minimum for an urban beach,[109] Breezy Point's tip will indeed offer a high quality recreation experience to those who can get there.

Concern for the preservation of both historic and natural resources is also much in evidence in the current plan. Fort Tilden has been declared an historic district and its ammunition storage magazines, Nike missile sites, and radar tracking stations have been declared historically significant. The plan calls for the Park Service to restore and interpret many of these features.[110] Nor has nature been slighted. Terns were found nesting at the western tip of Breezy Point, and so the walk and wander nature area was turned into a restricted access area. Small patches of "locally unique" woodland were discovered and they too were afforded the protection of restricted public access. In fact, the impulse for nature preservation is so strong that where natural resources worthy of protection do not exist, they will be created. The beach in front of the Breezy Point Co-op, identified as "the widest and best" on Breezy Point in the first plan, is to be the site of extensive dune building. Once the dunes are built, they will be, in the words of the current plan, "managed as protection zone lands and protected from random access by means of boardwalks, designated routes to the beach or other operational methods."[111]

Overall, the current plan leaves very little of Breezy Point available for unrestricted use. There are the beaches, not the broad expanses of ocean front envisioned in the Green Book, but rather the preexisting facility at Riis Beach (formerly a city park) and a small, less accessible beach to the west of West Beach. In addition, there are four widely scattered patches of land dedicated to unrestricted recreation.

While plans for unrestricted use of Breezy Point have been greatly curtailed, plans for groups of visitors now figure large. There are to be youth hostels in the Fort Tilden area, and in the area east of the Breezy

Point Co-op, supervised group campsites are planned. Much of the park is now to be accessible only by special permit, which for all practical purposes closes it to public uses except for preplanned and supervised group activity. In keeping with this shift to group use, there are now plans for large charter bus parking lots near the West Beach, Tilden Beach, and the group campsites. There is also a special program center to accommodate groups.

Education of various sorts is also emphasized in the current plan. A public education program aimed at showing how to live in a world of diminished energy resources is to be built around the park's energy-saving efforts. Accordingly, energy conservation is to be an important feature of all building design and nonfossil fuel energy sources will be used whenever possible. Much environmental education is to take place in three "gateway villages." According to the current plan, these gateway villages are envisioned as "major educational centers" and as "object lessons in the relationship of man and his environment."[112]

All of these goal shifts served to bring planning into line with political reality. By reducing mass access to Breezy Point, the agency brought its plans into conformity with the wishes of the Breezy Point Co-op, whose continued existence attested to its political potency,[113] and with the wishes of the park's neighbors in Brooklyn, who feared that mass access to Breezy Point would increase traffic congestion on their streets. The emphasis on "quality" recreation experiences covered the retreat from mass accommodation and as a cover it served well. By stressing a quality park experience rather than the total number of park visits, the agency did appear to bring one of the qualities of the great national parks—spaciousness—to Gateway.

Likewise, the emphasis on resource preservation at Breezy Point counterbalanced Gateway's failure to live up to its promise as a truly metropolitan recreation facility. While Gateway was being condemned by the current plan to remain a collection of local parks in terms of access and use, the designation of so many of its features as nationally significant historic resources justified keeping the Park Service involved. In addition, this shift of emphasis to resource preservation brought Gateway planning in line with planning trends throughout the Park System and especially, as we saw in chapter 4, with trends in managing the great natural areas. At Gateway, however, the intrinsic qualities of the resources to be protected were of little importance. It was the strategic advantages of preservation that mattered. Even invented resources, such as the reconstructed dunes in front of the co-op, seemed adequate for this purpose.

The shift in emphasis to group accommodations also served strategic ends. It allowed the agency to promote some nonlocal park use and, by doing so, to accommodate those few pressures for wider use which did exist. It was wider use in a form more palatable to the Breezy Point unit's neighbors than unstructured recreation would have been, however. In the current plan, group campsites are consistently referred to as *supervised* group campsites, while group activities are inevitably referred to as *supervised* group activities. The same emphasis on supervision of groups appears in the agency's responses to questions about crime and vandalism raised by the park's neighbors.[114] It was hoped that if the outsiders, many of whom would be ghetto youths, came to the park under supervision, they would be less likely to indulge in the mayhem that the park's neighbors feared. Also, youths coming in supervised groups would present fewer management problems for the park's own personnel than those coming individually or in informal, unsupervised groups.

Another advantage of emphasizing group access stemmed from the fact that spontaneous use of Breezy Point, or any large urban park for that matter, tended to peak strongly on summer weekends, drop on summer weekdays, and drop even more when summer ended. It was these great peaks, occurring when most people want to go to the beach and have the time to do so, that Breezy Point's neighbors objected to most strenuously. This objection forced the Park Service to agree to cap peak-day use at or near that which the constituent units of Gateway had experienced before the park was created. By stressing organized group visits, the Service gave itself control over the timing of many park visits. It could schedule groups for weekdays, and with school groups it could schedule visits in the spring, fall, or even winter. Thus, by shifting from the accommodation of the individually initiated visit to the group visit, the agency could discourage the visit it did not want, the one on a summer weekend, and encourage the one it did want, the off-season visit which increased annual visitor counts without increasing the political (and staffing) problems associated with high peak-day usage. Table 6-2 shows the degree to which the agency planned to increase nonpeak-day visits. Group visitors were a key part of its plans.

The need for off-season visits and the importance of organized groups in reaching goals of off-season use might explain the seeming incongruity of a gateway village whose ostensible goal was to teach lessons about mankind's relation to its natural environment yet whose facilities would include restaurants, cafes, stores, theaters, studios, gymnasiums, day camps, hostels, playgrounds, day-care centers, plazas, open-air markets and promenades.[115] It was well and good for the Park Service to sound high minded about things like teaching how to live in harmony with nature, but if Gateway's use during off-season and inclement weather was to increase substantially, the gateway villages had to be designed to attract and entertain that most valuable visitor, the one who came in a prescheduled group regardless of summer rains or winter snows.

Table 6-2. Breezy Point Projected Use Levels

		Projected use			
	1976 Daily visits	With ferry	% Increase over 1976 visits	Without ferry	% Increase over 1976 use
Peak summer day	90,000	100,000	11	90,000	0
Average summer weekend day	45,000	55,000	22	45,000	0
Average summer weekday	25,000	40,000	60	35,000	40
Average spring/fall weekend day	12,000	22,000	83	17,000	42
Average spring/fall weekday	5,000	9,000	80	7,000	40

Source: National Park Service, *Gateway NRA, General Management Plan*, pp. 44–45.

Thus we see that at Gateway, emphasis on preservation and education allowed the agency to shift planning away from mass access while emphasis on group accommodation allowed it to channel that little public access which remained into the most politically acceptable and managerially convenient forms. Such shifts of emphasis were by no means limited to Gateway, however; they occurred to different degrees and in different mixes in most of the urban national parks. Although all of the urban parks were initially justified by their mass recreation potential, little of their actual development planning reflected this.[116]

The Other Urban Parks

Cuyahoga Valley

Planning for Ohio's Cuyahoga Valley National Recreation Area placed much emphasis on "environmentally neutral" design, that is to say, development which would only slightly alter the present characteristics of the land or its flora and fauna.[117] Since land characteristics must be modified considerably to accommodate large numbers of people, mass access and environmentally neutral design are largely incompatible in practice (although as we have seen above, one can always pay lip service to both of them when they are ambiguous, general goals).

The Cuyahoga Valley General Management Plan, the most detailed expression of agency intentions toward the park, showed a decided preference for environmental neutralism over mass access in practice.[118] Park planning also stressed what is called the "innate development capacities" of the land within its boundaries. Park planners used these innate capacities as the basis for a map of the park's recreation potential (figure 6-7). It might be noted that much of the park, perhaps most of it, had none. On the sloping land, the danger of erosion reduced its recreation potential to near zero and made strict preservation the only reasonable management option. The hilly, rolling upland which constituted much of the rest of the park was deemed suitable only for dispersed, low-density recreation. The two types of land were identified as suitable for concentrated recreation, however. First, there were scattered patches of upland plateau, most of which were on the park's periphery and were too small for any large-scale recreation development. There was also the flood plain of the Cuyahoga River; it too was deemed capable of sustaining a high density of human use. But there was a catch here too; the constant danger of flooding prohibited permanent facilities. Thus respect for environmental neutralism was at the same time respect for the political reality, which would not condone an ambitious recreation-oriented development plan for the park.

Santa Monica

In the Santa Monica Mountains National Recreation Area, preserving the scenery was strongly emphasized as a goal. Therefore one of the major objectives of the park could be attained simply by maintaining the landscape's status quo through the purchase of development rights and scenic easements, and by encouraging the municipalities within the park boundaries to adopt and enforce strict zoning measures. Since the land valued for its scenery was not to be purchased in full fee, it was not accessible to the general public. Even in areas where outright acquisition was to take place, planning for large-scale intensive use would seldom be practical since, as a concession to local sentiments and power, few existing houses were to be acquired. The agency might purchase most of the land, but most of the private residences and a small amount of land around them would remain in private ownership.[119] This would leave the Park Service with a fragmented ownership pattern and great difficulties in developing intensive recreation facilities.

As was the case at Gateway, and probably for the same tactical reasons, planning for supervised groups took precedence over planning for individuals in search of high-density, active recreation: swimming, playing sports, and so forth. Individuals looking for low-density, resource-oriented recreation—hikers, bicyclists, bird watchers, and the like—were an exception, however. They would be well accommodated; walking trails, bike

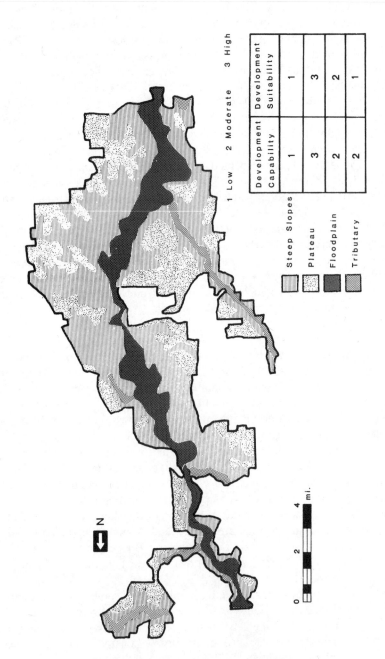

	Development Capability	Development Suitability
Steep Slopes	1	1
Plateau	3	3
Floodplain	2	2
Tributary	2	1

1 Low   2 Moderate   3 High

N

0   2   4 mi.

Figure 6-7.
Cuyahoga Valley National Recreation Area land classification for development suitability. (From National Park Service, Cuyahoga Valley NRA General Management Plan, p.23.)

paths, and natural areas were prominent in park planning. But, then, accommodating the latter types was not politically difficult. They were a well-mannered segment of the general public and therefore more acceptable to the local residents than a wider cross section of the population of Los Angeles would have been. Moreover, once areas were dedicated to such low-density recreation, they were not readily available for more intensive recreation development. The Park Service might later attempt to convert them, but if a constituency for low-density use had already been established, this would be a difficult and unpopular move.

Golden Gate

Of all the urban national parks, San Francisco's Golden Gate National Recreation Area was widely viewed as the most successful by the urban park advocates within the Park Service. It was frequently cited by agency administrators as an example of an urban national park which had lived up to its original intentions. Here, it was claimed, the agency had attained the twin goals of mass access and environmental preservation. In fact, it was partly on the strength of the park's perceived success that its superintendent, William Whalen, was appointed agency director in the mid-1970s.

Golden Gate's success in simultaneously attaining these two goals can be attributed to what might be called an accounting device, however. It has attained the two goals in two different places. There are two distinct parts to the Golden Gate National Recreation Area—the segments within the city of San Francisco and those lands to the north of it in Marin County. Each part had its own coalition of supporters. Each part was the object of a separate campaign to put its management in the hands of the federal government. Only at the last minute was a proposal advanced which joined the urban and nonurban parts into one national park.[120] Much of the National Recreation Area within San Francisco city limits consists of former city parks whose accessibility and patterns of intensive use were established long ago. Other segments in the city, mainly former military properties, were also readily accessible to the general populace via San Francisco's excellent public transit system. Here the Park Service simply built on inherited precedents of public access and use. The parts of the park lying to the north in Marin County, however, were similar to other national parks on metropolitan peripheries—Santa Monica, Fire Island, Cuyahoga Valley, and so on. Public access, especially by public transit, was difficult to begin with, and the Park Service, whatever its intentions, has done virtually nothing to improve it. Here active local groups feared mass recreation facilities and used their considerable political power and skills to thwart any agency moves in this direction. Agency planning at Golden Gate reflects the reality of these circumstances and promotes intensive use where it is physically and politically possible to do so, that is to say, in San Francisco. For the Marin County part of Golden Gate, the usual pieties about preservation, environmental harmony, and the need to be a good neighbor prevail.[121]

Fire Island

Perhaps nowhere more than on Fire Island has the goal of mass access been so thoroughly abandoned. But this is not surprising since problems of transportation are very large and nowhere in the Park System has local paranoia about, or opposition to, increased public access been so great or unyielding.[122] Twelve years after the national seashore was established on a crest of Great Society optimism, it was getting an average of 6,450 visitors on a peak-use summer day. The 1978 park development plan foresaw increasing this gradually to 9,170 peak-day visits by 1987.[123] The 1975 plan, however, had projected 17,450 peak-day visits by the late 1980s.[124] In the face of local opposition, this goal was abandoned for the more modest 9,170 visits. To put these figures into perspective, we might compare them with the combined summer weekend population of the towns on Fire Island—60,000 to 100,000 people.[125] Peak-day visitor use of nearby Jones Beach, which Fire Island National Seashore was initially intended to supplement, is on the order of 200,000 visitors. While plans to accommodate visitors to the national seashore have been scaled down continually since it was established, private development of Fire Island has increased dramatically. In the first decade of the National Seashore's existence, the number of structures on the island, mostly dwellings, increased 31 percent.[126]

Nothing more strikingly illustrates the relative strength of local and wider interests that have been brought to bear on urban national park planning than does the agency's development plan for the Talisman area of Fire Island (figure 6-8). Barrett Beach, an inholding within the national seashore reserved for the exclusive use of the residents of Islip (the Long Island town across Great South Bay) slices across the island. The inholding's boundaries are semipermeable barriers; Islip residents can legally cross them to the national seashore, but nonresidents cannot cross into the beach from the national seashore. While the national seashore next to Barrett Beach will have a small, seasonal dock, the facility for Islip residents includes a permanent marina with a bulkhead. The national seashore facility will include a small environmental education center, but it will also include a vehicle path across the dunes to allow those on Barrett Beach access to the national seashore in their vehicles. Undoubtedly the environmental center at the national seashore will teach its visitors about the need to preserve the integrity of the dune line and how destructive motorized vehicles can be to the dunes. Perhaps it will also stress how, by using a small temporary dock rather than a permanent one, the Park Service is planning with rather than against nature. One suspects, however, that for

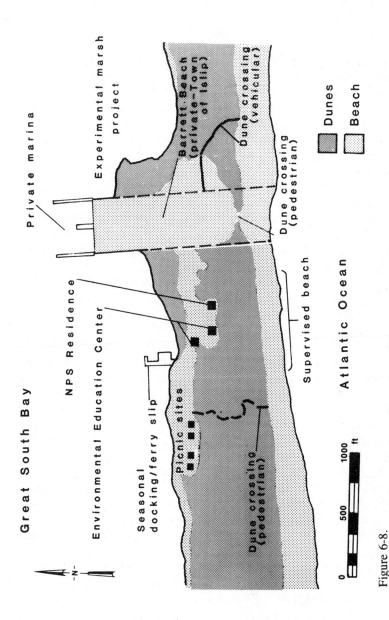

Figure 6-8.
Fire Island National Seashore, Barret Beach Section. (From National Park Service, Fire Island National Seashore Final General Management Plan, figure 7, p. 77.)

astute visitors to the Talisman area of the national seashore, the important lessons will be political rather than environmental.

The Consequences of Limited Success

Before the Gateways were established, there was considerable uncertainty over the wisdom of establishing urban national parks. Neither the problems those parks brought with them nor the solutions worked out for them have done much to dampen disagreements. Political scientists have pointed out that public administrators tend to be positive about their own programs. Administrators look for strong points, overlook weak ones, and are inclined to accept weak indicators of program success.[127] This is very much the case with Park Service officials who are by position or inclination committed to the urban parks. Even those who fought for mass access and intensive development in the urban parks are inclined to look at Gateway, Fire Island, Cuyahoga Valley, and the rest, as limited successes rather than failures. For example, an agency planner, deeply committed to the metropolitan vision of what an urban national park should be, told the author that, "at least on Fire Island we've controlled development, and we've provided the facilities for those who can get there on their own." The superintendent of Fire Island National Seashore expressed a similar rationale for his park, "If you can say nothing else about [Fire Island National Seashore] you can at least say that our presence put the brakes on private land development."

An agency administrator involved with Gateway from its earliest days told the author that the removal of the junked cars and dumped mattresses from the areas incorporated into the national recreation area went a long way toward justifying agency presence there. In other words, supporters of the urban parks have tended to use what has been accomplished as a justification for their presence, even if the accomplishments add up to just hanging on. According to former agency director Whalen, "At Gateway, survival, not staying on top of it, was the name of the game."

Whereas those in the agency with a special commitment to urban parks might be willing to accept anything accomplished as an indicator of success, there was no reason for others in the Service to do so. The unique characteristics of the urban parks, the resources they contained, and the purposes they served, made them special to their supporters, but to their detractors these characteristics simply made them inferior. According to Robert Cahn, a "cause of frustration and discontent [within the Service] is a belief among many veteran employees that national park standards have been lowered to accommodate new areas, especially places such as Cuyahoga Valley and Santa Monica Mountains recreation areas."[128] Plans which stressed maintenance of the status quo, marginal improvements, and the preservation of historical and natural resources of modest value only reinforced such attitudes. In fact, it appears that the rank and file in the Park Service is not convinced that urban parks belong in the System. After conducting extensive interviews with agency management, Coopers and Lybrand concluded that "if given their choice, National Park Service staff members would elect to stay out of urban areas." [129] Nothing succeeds like success and strong evidence of success would have convinced opponents and quieted doubts, and thereby restored internal cohesion on the issue. But there was no unambiguous success.

Some telling analogies have been applied to the urban parks by agency personnel attempting to sum up the Park Service's experience in metropolitan areas. One is the opening of Pandora's box. Park Service leadership flirted with the idea of urban parks and in the end committed the agency to it. In the cities, the Service hoped to find large, appreciative new constituencies, and a new role in keeping with the spirit of the times. In short, the agency was to find a part of its lost relevance in an urban commitment. Instead it found dilemmas, internal dissension, and severe political constraints. The Service's experiences were also compared to the story of the sorcerer's apprentice, in that a process, once initiated, could not be controlled. The agency found that, having committed itself to urban parks, it could neither control their authorization, nor gain the initiative in planning for their use. Finally, the urban parks were compared to large, grounded vessels. They had been launched on a great tide of optimism about the possibilities of effecting social redress and making metropolitan cities better places to live. When that tide ran out, the parks were stranded on the shoals of localism and other hard realities of the nations' decaying cities.

The analogies were very different from one another, and they stressed separate aspects of what was a complex and in some cases, an unclear chain of events. Nevertheless they were not happy analogies, and whether they placed the blame on personal foibles or impersonal events, they were all analogies of unrealized expectations.

In fact, critics of the urban national parks have argued that even the very success of the agency in surviving at parks like Gateway was, in a sense, a failure. They pointed out that the urban parks became even more attractive pork barrel goods for local constituents when it became clear that they would not be developed for mass access. [130] They further argued that the very possibility of getting such an attractive benefit for one's constituents encouraged congressmen to hold their support for the entire park system hostage for their own local proposals. There appeared to be some truth to this. While congressmen from the urbanized areas of the East traditionally provided the largely western (or at least remote) National Park System with a strong basis of support, arguing that such parks were national resources, the 1970s found some eastern congressmen taking a different stand. They asked why they should support a remote Park System which provided few local benefits for their constituents. Perhaps the most chilling expression of this sentiment came from Representative Seiberling, chief sponsor of the Cuyahoga Valley National Recreation Area and chairman of the House subcommittee responsible for national parks in the 97th Congress.[131]

> Year after year, members of the Ohio delegation vote money for great national parks in California, in the Rockies, along our sea coasts, and yet their constituents are beginning to say, 'Are we who put up a lot of the taxes and pay for those parks entitled to some return on our investments?' . . . If we are going to treat all of the people equitably, if we are going to have a continuing constituency for national parks and outdoor recreation, then we are going to have to provide for the people in our most populous urban areas, an outdoor recreation opportunity comparable to that we have provided for those already located near some of our great parks.

Opponents of the urban parks also argued that units like Gateway and Cuyahoga were competing for funds which could be put to better use maintaining the great natural areas of the System. In other words, a dollar spent at Cuyahoga Valley was a dollar taken from Grand Canyon.[132]

One should not ignore the accomplishments of the urban parks, however, nor should one dismiss their defense by their supporters as mere rationalizations. "Just hanging on" is an accomplishment. For if nothing else, it means that options on the future have not been foreclosed. Although the original goals of the urban parks have been thwarted, if there is ever a change in the array of forces operating on the agency so as to make planning for mass access a possibility, the agency could always respond accordingly. Perhaps the urban parks are like beached vessels, but then, tides turn. It should also not be forgotten that the urban parks are serving people right now. Planning for group access might have been a means of covering a retreat from planning for the individual, but at least groups are now using the urban parks. Natural and historical sites are being protected, even if they are not easily available for the enjoyment of many urban dwellers. Large areas which otherwise might have fallen to developers and speculators are now open to those who enjoy walking, bird watching, and other low-density activities.

Furthermore, the urban parks do not appear to be starving their more spectacular but remote siblings for funds. Table 6-3 shows the large construction projects included in the agency's 1981 fiscal year budget. Although Gateway was included, and so was Washington, D.C.'s Rock Creek Park, the largest single project was in Yellowstone. The Natchez Trace Parkway, Glen Canyon, Lowell, and Lassen Volcano all were the sites of larger projects than any undertaken in the urban parks. Moreover, the

Table 6-3. National Park Units with Construction Budget
Allocations over $1,500,000 for Fiscal Year 1981

Unit	Project	Allocation (in thousands of dollars)
Yellowstone	Concession facilities rehabilitation	6,963
	Water treatment and storage system	2,903
		Total 9,866
Natchez Trace Parkway	2.9 miles of parkway	Total 7,100
Yosemite	Final effluent disposal system	Total 4,258
Lowell	Historic structure rehabilitation	2,500
	Lowell Manufacturing Company rehabilitation	1,500
		Total 4,000
Glen Canyon	Visitor and support facilities	2,982
	Sanitary facilities	110
		Total 3,092
Lassen Volcano	Sewage facilities and power plant	Total 2,889
Rock Creek Park	L Street bridge reconstruction	Total 2,882
Haleakala	Parking lot and trail access	688
	Potable water system	1,192
	Sanitary facilities	675
		Total 2,555
Lake Mead	Sewage system	1,349
	Flood protection	786
		Total 2,135
Carlsbad Caverns	Water system modernization	1,158
	Utility line relocation	655
		Total 1,813
Gateway	Roads and parking lot resurfacing, buildings and grounds rehabilitation	1,572

Source: U.S. Department of Interior, *Budget Justifications, F.Y. 1981—National Park Service.*

project planning budget, a good indicator of future capital expenditures, revealed a similar pattern in 1981.[133] The biggest allocations here were to, in descending order, Sequoia, Yosemite, Canaveral (National Seashore), Gateway, Olympic, Yellowstone, Sleeping Bear Dunes (National Seashore), Mount McKinley, and Voyageurs. The list was dominated by the great natural areas of the system.

As to annual operating expenditures, the urban parks do not seem to be outcompeting the rest of the System. The fiscal year 1981 budget saw a small system-wide increase over 1980 in allocations to park operations, approximately 18 million dollars.[134] Of this increase, well over 60 percent went for the management of the new parks added to the System as a result of the Alaska settlement. Most of the urban national parks, like most of the System outside Alaska, showed either no increase or a very slight increase in their operating budgets.

Thus it can be seen that the Park Service's urban thrust, while failing to reach its intended goals, has had some successes. Moreover, it cannot be convincingly argued that those successes were bought at too high a price to the rest of the Park System. This confronts the policy analyst with the central problem in evaluating the agency's involvement with urban parks: Is the bucket half empty or half full? Are the agency's failures the proper starting point for recommendations on future policies, or are its successes? This question is most profitably left for the concluding chapter where it can be integrated into an overall discussion of the Park Service's current predicament. Before initiating such a discussion, however, we turn to the Park Service's efforts to guide the fate of lands it does not own, for it is a policy area with many striking similarities to that of the urban parks.

7 Beyond Park Boundaries

BACKGROUND

One of the central problems the Park Service faced in the modern era was that of establishing the limits of its responsibilities. The preceding chapters discussed how the agency dealt with this problem with regard to the nation's history, its nature, and its cities. More literal questions of responsibility also arose with regard to the land the agency did not own outright, land which was sometimes far beyond park boundaries. With new private development on the periphery of many of the older parks, the agency was forced to search for ways of ensuring that such development did not harm the parks and to ask itself how far beyond park boundaries its efforts at protecting the parks should go. As new units, Fire Island and Santa Monica among them, came into the Park System with provisions that allowed for large, permanent inholdings of private land, the agency found it necessary to turn to the use of less-than-fee rights to control private land use within the Park System. As regional planning and public control of private land development were placed on the national agenda, the Park Service found itself with an opportunity, and perhaps an obligation, to play an important role in federal land planning policy. Although only a small fraction of the Park Service's energies are absorbed by land it does not own, nevertheless it is a growing fraction. Furthermore, it has forced the agency to ask itself basic questions about its future role in American society.

Early Agency Policy

For Mather and Albright, outright land ownership was the preferred basis of park management because it meant a full measure of protection and

223

control. Lane's 1918 letter to Mather made this point: "Wherever the Federal Government has exclusive jurisdiction over national parks, it is clear that more effective measures for the protection of the parks can be taken."[1] Inholdings, that is to say, enclaves of privately held land, were present in some of the earliest parks. Usually they had resulted from concessions which allowed the parks to be established in the first place and they were viewed as necessary but temporary evils by agency leadership. The desirability of their ultimate removal was not questioned. On this point as well, Lane's letter was unambiguous: "All of [the inholdings] should be eliminated as far as it is practicable . . . in the course of time."

As much as agency leadership desired a clear separation of the public and private, it realized that it could never own all the land necessary to afford the parks complete protection. Occasionally, it would have to take a hand in the uses to which private land within or near the parks was put. The fate of Yellowstone National Park's elk herd was an early case of such involvement. In a sense, it was a misnomer to call the herd Yellowstone's since it spent only part of the year within the park. The herd's pattern of migration in search of grazing took it out of Yellowstone for much of the year, and when it was beyond park boundaries, it was subjected to heavy hunting pressure. By the 1920s, the herd was showing signs of serious depletion. Mather responded by orchestrating a publicity campaign that included a story about the elk of the old West in the *Saturday Evening Post*, and soon the national park's problem appeared to be a national problem.[2] The publicity forced Montana and Wyoming to tighten up its game law enforcement to protect the herd while it was outside the park. Eventually, the herd was restored to its former size.

The transfer of the great Civil War battlefields to the National Park System in 1933 forced the agency to rethink its unqualified preference for full acquisition. In the late nineteenth century, the army had considered full acquisition of the entire Antietam battlefield and explicitly rejected it. General Davis, in charge of developing a plan to preserve the battlefield, argued in the 1890s that the best way to preserve the site in the condition it was in when the battle was fought was to maintain the land in agriculture.[3] With what sounded like a modern argument, he reasoned that complete acquisition would be expensive. Furthermore, it would ultimately defeat the preservation of the site, since it would undermine the local agricultural community, which was keeping the land as it was when the Union and Confederate armies met on it. Accordingly, General Davis developed, and the army implemented, a battlefield preservation plan which did not involve the purchase of any large tracts, only narrow corridors along the lines of battle. Light duty roads lined with fences were then constructed along these corridors, and, where appropriate, historical markers were erected on the roadside.[4] Thus, with the Antietam battlefield, the National Park Service inherited a site where full acquisition had already been con-

sidered and rejected. In spite of its penchant for full acquisition, the ·
agency continued to manage the site according to the army's original plan.
It was to be a model for similar ventures in the future.

The 1930s also saw the agency settle for less than full acquisition when,
to protect the vistas from the Blue Ridge and the Natchez Trace parkways,
it acquired the rights to restrict the use of land bordering the parkways.
The restrictions prohibited erecting billboards, establishing dumps, cutting
trees, or constructing nonfarm buildings. The total acreage involved here
was not great, however; there were approximately 1,200 acres along the
Blue Ridge Parkway and 5,000 along the Natchez Trace.[5] Overall, the
agency's use of less-than-fee acquisition remained minor. It was used when
special conditions seemed to warrant it or where it was inherited convention, but it was not the subject of any basic policy. Through the 1950s, the
agency continued to show little inclination to involve itself in the gray area
of partial ownership of land or partial rights to determine the future of
lands it did not own outright.

In retrospect, it appears that this agency preoccupation with land it
owned outright was logical, given American attitudes toward land as they
had evolved in the nineteenth and early twentieth centuries. It has been
argued that one of the dominant characteristics of American landscape
aesthetics was its disregard for the qualities of the close-at-hand, mundane
places of ordinary life and its near reverence for the unique and exceptional
places.[6]

Lowenthal observes that the national parks fit into this American cult of
the unique. "The national parks were originally set up to enshrine the
freaks and wonders of nature, and park literature still touts the Grand
Canyon, the Grand Tetons, Yellowstone, and Yosemite as unique."[7] Lynch
too was struck by the way a great national park was thought of as an
isolated, special entity, as "something separate from other aspects of living."[8]

Participation in the sorting out of places into the preserved unique and
the abandoned mundane appears to have been an important part of the
agency's mission in the early days. Certainly Mather subscribed to the cult
of the unique and the best when evaluating potential additions to the
System. Was it the best of its type? Was it unspoiled by human alterations?
Was it unlike any other unit in the System? These were all important
considerations to him. If the places qualified, the agency was interested in
them, if not, it was not. (At least officially it was not. As we have seen
earlier, actual practice was sometimes another matter.)

Changing Attitudes and Interests

Although this emphasis on the fully owned unique was a basis of agency
and System success during the first half of the century, it was unclear to

Park Service leadership whether it still was good policy in the 1960s and the 1970s, and if so, for how much longer it would remain so. Several things eroded agency confidence in it. One was the attitude changes which seemed to be taking place in American society and the opportunities they appeared to create for the agency. The expansive context of environmentalism, with its emphasis on the interrelationships of nature, which emerged in the 1960s, undercut the cult of uniqueness. Udall explicitly rejected it when, in *The Quiet Crisis,* he wrote that too many conservation organizations "overconcentrate on a chosen holy grail," and that few of them had "entered the fight for the total environment."[9] Udall's concern for visually pleasing and unpolluted total environment led him to discover that which traditional American landscape aesthetics had dismissed as valueless—the landscape in its entirety.[10]

Within government, Udall was not alone in his comprehensive aesthetic and environmental concern. It was reflected in the concern shown by ORRRC for the lack of recreation opportunities in the landscape of everyday life. It was reflected in President Johnson's highway beautification schemes and his efforts to bring beauty and access to nature into people's daily lives. Following in Johnson's footsteps, Nixon adopted concern for the total landscape as a public position. In 1970, in his report to Congress on the state of the nation's environment, he declared that the United States was at an "historic milestone." It was "the first time in the history of nations that a people has paused consciously and systematically to take comprehensive stock of the quality of its surroundings."[11]

Land Use Controls

Embedded in this concern for the total landscape was a swell of interest in increased public control of land use and in a role for the federal government in ensuring orderly, rational land development. In 1963, a prominent land use lawyer surveyed the national land regulatory scene and pronounced it one of "total confusion."[12] Even by then, however, the status quo was a source of great dissatisfaction. Reilly, writing in the mid-1970s, asserted that a "new mood in America" had evolved over the previous decade and that the public was no longer willing to accept the landscape transformations that strictly local control of land use usually allowed.[13] According to Reilly, one of the principal reasons for the reduced public willingness to accept unregulated land development was "a rising emphasis on humanism, on the preservation of natural and cultural characteristics that make for a humanly satisfying living environment."[14] There is evidence that there was such a shift in popular attitudes and that it was a strong one.[15]

More than popular attitudes seemed at work, however. It is possible that the increasing scale of land development was ultimately behind the growing dissatisfaction with then-current land use controls.[16] According to Plotkin, strictly local land use control was favored by land developers in the first half of the century because the zoning anarchy it engendered meant few fetters on development plans and therefore few restraints on opportunities for profit. By the 1960s, however, the nature of land development had changed with the advent of great residential subdivision projects, massive industrial parks, and planned unit developments the size of whole cities. This resulted in part from the fact that large national corporations were entering the building field, one previously dominated by small, usually local entrepreneurs.

The fine-grained mosaic of local regulations existing since the 1920s came to be seen as an obstacle by the new large developers because of the coordination problems it now presented. Accordingly, they aimed to alter the arena of land use decision making by expanding it to involve state and federal governments. For those large-scale builders' interests, "centralization of power was perceived as a mechanism for smoothing prevailing patterns of investment."[17] In other words, they wished to do away with local obstructions which might not hinder the small-scale builders but which could interfere mightily with the large projects of the big corporations. Although a shift in popular attitudes might have formed a backdrop against which the politics of regulatory change were played out, Plotkin argues convincingly that profit-based interest also had a large hand in forcing the issue of public land use control onto state and national political agendas.[18]

From the 1960s onward, public officials responded to these changes in interest and attitude by advocating increased land use planning at levels of government above its traditional seat, the municipality. President Kennedy called for comprehensive public land use planning which would embrace "all major activities, both public and private which shape the community."[19] Udall, his secretary of the interior, placed a similar high value on planning that transcended municipal boundaries and treated regional problems comprehensively. In *The Quiet Crisis* he wrote that: "If we are to create life-enhancing surroundings in both cities and suburbs, the first requirement is the power to plan and to implement programs which encompass total problems. . . . As long as each city, county, township, and district can obstruct or curtail, planning for the future cannot be effective."[20] Presidents Johnson and Nixon both made similar arguments and advocated increased public control of land use as an answer to problems of environmental degradation, social justice, and economic efficiency. Their advocacy of such planning was buttressed by reports from official study organizations such as the Advisory Commission on Intergovernmental Relations and the National Commission on Urban Problems, which stressed the need for more centralized land use control.[21]

Interest and advocacy soon translated into political action. Exclusive local control was challenged and an upward drift of regulatory responsibility was set in motion.[22] Many states moved to gather back to themselves the powers they had distributed to their counties and municipalities several decades earlier. California established its Coastal Commission and gave it considerable power over land use along the entire length of the state's coast. Hawaii adopted a statewide land use plan which greatly restricted the freedom of local municipalities to approve (or reject) land use changes within their boundaries. Other states, including Vermont, New York, New Jersey, Maryland, Oregon, and Florida, determined that certain types of development projects, as well as certain types of areas, were of special concern to the entire state and would therefore be subject to a measure of state control.[23] This upward shift of land regulation authority was further encouraged by the erosion of judicial deference to home rule and by a series of expansive judicial interpretations of government police power in matters of public land use regulations.[24]

Although President Johnson had made growth management one of the cardinal points of his Great Society program, a coherent federal role in such management never emerged during his presidency. In Nixon's administration, however, support for a definitive federal commitment to land use planning in the form of legislation was widespread in Washington, and it looked like it would translate into action. Many supporters of the National Environmental Policy Act (NEPA), including NEPA's chief legislative patron, Senator Jackson, saw a federal land use bill as the next logical step. NEPA's passage encouraged them to pursue a parallel course in regulation of private land use.[25] Senator Muskie announced in favor of national land use regulation, while in the White House, John Ehrlichman, chairman of Nixon's Domestic Council, made efforts toward the same end. An issue network on federal land use regulation developed around these men and extended into such federal units as the Department of the Interior, the Department of Housing and Urban Development, the Environmental Protection Agency and the Council on Environmental Quality, as well as into nongovernment Washington-based organizations such as the American Law Institute, Resources for the Future, and the Conservation Foundation.

This, then, was the background against which the Park Service approached involvement in land use regulation in the 1960s and 1970s. While it was unclear what form federal policy would eventually take, it was clear that the issue of a federal land use policy, long dormant, was very much awake. It had become part of the world in which the Park Service operated.

New Opportunities and Demands

Capturing Regional Policy

When national land use planning found a place on the federal agenda, the Park Service seemed like a logical instrument of planning policy for several reasons. First, there were historical connections between resource management on federal lands and broader environmental, urban, and regional planning efforts. For example, in the 1930s agencies like the Tennessee Valley Authority and the U.S. Forest Service extended their concern for public land to the management of private lands and the development of entire regions. There were also historical connections between recreation and regional planning. Benton MacKaye, one of the foremost regional planning theorists of the 1920s and 1930s, saw a generative link between the two. He wrote that, "The natural tendency [is] to start regional planning with a designing of recreation areas." [26] Later, when the cities were identified as a national problem, Russell Train saw parks as a possible catalyst for ambitious urban planning efforts. He wrote that, "The planting of flowers to brighten the heart of a city may not accomplish an environmental revolution, but it may well lead to a new awareness of their surroundings on the part of many members of the public—flowers can lead to trees, and trees to public parks, and parks to comprehensive planning programs." [27] Second, concern for the parks, especially concern over a proposal to build a jetport athwart the route of the water supply for the Everglades National Park, had contributed to the political momentum behind efforts to involve the federal government in land use regulation in the first place. [28] Thus the Park Service was conceptually close to the regional planning issue when it emerged. It was organizationally close as well. The Solicitor's Office of the Department of the Interior was an important part of the network of support for federal involvement in land use planning, and Interior was viewed as the "lead agency" for land use policy by most of those involved with the issue. [29]

Finally, the agency's combination of public image, attendant interest groups, and traditional concerns gave it the freedom to move into an ambitious role in regional land use planning. According to former Director Whalen, "the Director of the National Park Service has a great pulpit for preaching wise land use and he isn't hemmed in by conflicting interests." The NPFF study took a similar view of the agency's potential role in any land use planning thrust by the federal government. "The National Park Service, because of its broad mandates and wide spectrum of holdings, is uniquely qualified to exercise national leadership with respect to land use." [30] Here, then, was opportunity and the agency sensed it. As an agency planner, considered one of the best, put it, "there is a vacuum out there, planning the landscape [of everyday life], and with some organizational commitment the Park Service can play a big role in filling it."

Environmental Planning

The opportunity for expanded responsibility seemed to extend naturally to general environmental planning. Just as MacKaye and Train saw a close connection between public open space and more comprehensive regional planning, others saw a connection between providing open space and maintaining the overall quality of the environment. For example, the Bureau of Outdoor Recreation's 1970 study, *The Recreation Imperative,* stressed the close relationship between outdoor recreation and overall environmental quality: "Outdoor recreation generally is dependent upon the quality of the environment. Much of outdoor recreation is simply enjoyment of the environment, and to the extent the environment is impaired, the quality of outdoor recreation is reduced."[31]

Moreover, organizational changes within the executive branch drew the Park Service into the environmental policy arena. The Water Pollution Control Administration was transferred from the Department of Housing and Urban Development to Interior in 1966. (This was in keeping with Udall's expansive view of Interior's area of concern and indicative of his power within Johnson's administration.) When this happened, an increased concern for the general quality of the environment diffused among Interior's agencies, including the Park Service.[32]

The creation of a separate Environmental Protection Agency (EPA) in 1970 and its designation as lead agency for environmental concerns did not distance the Park Service from the question of overall environmental quality. If anything, it drew the agency closer. First, the creation of a powerful executive agency confirmed the importance (and probably the permanence) of environmentalism as a national issue. Second, the National Environmental Policy Act of 1969,[33] although preparing the way for the establishment of EPA, also spread responsibility for maintaining environmental quality broadly among federal departments through its environmental impact statement requirements, its insistence that departmental regulations be brought into conformity with the provisions of the act, and its suggestion that federal agencies interpret their own environmental responsibilities as broadly as possible.[34] This upgrading of environmental protection as a public issue meant that the roles of the National Park System and Service had to be viewed anew in the light of their relationship to the quality of the total environment.[35] The NPFF study group saw NEPA as conferring a new "extraterritorial" responsibility on the Park Service, and with the responsibility went a great opportunity. It wrote: "[The role of] the National Park Service as environmental advocates should be strengthened. Advocacy for the values on which the Park Service has special expertise should not be limited to the confines of [the] National Park System." [36]

For the Park Service, such an expanded role in the 1970s made good strategic sense for another reason. As Nicholson observed in 1972 at the 2nd World Congress on National Parks, national parks were no longer the spearhead of the environmental movement. It had been "overtaken by the struggle against pollution, while other great causes . . . will rightly claim more and more attention." [37] Environmental activism and the extension of agency concern to land use problems beyond park borders, some of which directly affected parks and some of which did not, appeared to be the path back to the center of public environmental concern. Furthermore, the transformation of the President's Recreation Advisory Committee into the more broadly concerned and powerful Council on Environmental Quality during the Nixon administration showed that organizations could cross the conceptual bridge between recreation and wider societal concerns, and they could acquire great power in doing so.

Cultural Preservation

Moving beyond the acquisition and management of traditional, fully owned national parks might bring the agency closer to the center of new cultural concerns as well. The 1960s and 1970s saw the melting pot both attacked as a theory and questioned as an ideal. [38] Perhaps because of this, those years also saw a great concern for preserving unique cultures from the homogenizing influence of middle-class American life. Some thought that the Park Service might be an appropriate instrument for a federal policy aimed at ethnic and cultural preservation. After all, such preservation could be considered a logical extension of the agency's other preservation missions, and saving unique human lifestyles and cultures was similar to preserving special natural areas and historic sites. For example, the NPFF study group recommended that the Park Service be charged with maintaining cultural diversity, writing that "it certainly serves the national interest to involve the Park Service in cultural diversity, just as the national interest is served by sustaining ecological diversity." [39]

With Udall's encouragement, the Bureau of Outdoor Recreation had been feeling its way into an active policy of cultural preservation. In 1968, it released its *New England Heritage,* a report in which it advanced a plan to preserve the unique scenic and cultural qualities of the Connecticut River Valley while at the same time increasing the valley's capacity to provide recreation for the northeast region of the country. [40] The report made it clear that there was a big part for the Park Service in the protection of the cultural landscape as well as in the provision of recreation facilities.

Agency leadership under Hartzog was drawn to the idea of the Park Service as an agent of cultural preservation. According to an agency administrator close to Hartzog, "Hartzog and those around him had the idea of a national cultural park as a new category [of national park]. They were not going to be [centers for the performing arts] like Wolf Trap or Kennedy Center. They would be areas with special folk cultures like the Pennsylvania Dutch country, the Sonoma country in California, for example. The presence of the Service would protect them."

It was clear that simple, straightforward land acquisition was hardly an appropriate means of preserving special regional cultures. Indeed, such acquisition would probably destroy the culture it aimed at preserving by destroying the patterns of property ownership and livelihood on which it was built.[41] The unique cultural areas which Hartzog and the Bureau of Outdoor Recreation were thinking about were large, at least in comparison with historical sites, and even when compared with most natural areas in the Park System. The area included in the Connecticut River Valley study was 400 miles long and varied between 20 and 30 miles in width. Pennsylvania Dutch country covers most of several well-populated counties in southeastern Pennsylvania. The acquisition of such large areas would have been economically impossible, even if it were desirable. No, if the agency were to become involved in cultural preservation, it would have to do so through regional planning or some mix of regional planning, full-fee acquisition, and less-than-fee land ownership; traditional parks would not do.

Defending Established Parks

While agency involvement in land it did not own appeared as an opportunity to expand into new areas of responsibility, such involvement also appealed from a defensive point of view, that is, as a means of protecting the agency's traditional charges, the national parks, from depredation. It also seemed increasingly important in the pursuit of traditional agency goals such as providing new recreation opportunities and bringing new natural areas under federal protection.

While the use of neighboring lands had always caused some problems for the parks, it was only in recent years that the Park Service became acutely stressed by development close to the parks. The problems seemed to multiply with each passing year.[42] In the early years of the System, neighboring land use was not usually a problem because the early parks were surrounded and therefore protected by a huge and largely unused public domain.[43] However, the same burgeoning land development which prompted interest in federal land use control in the 1960s and 1970s forcefully and painfully showed the Park Service it could no longer count on isolation from development to protect the System.

Unfortunately such isolation was just about the only protection the parks had, and where it didn't exist, the agency was virtually powerless to protect the parks from external threats. This was illustrated at Gettysburg, where an entrepreneur planned on building a giant tower overlooking the battlefield and then charging tourists admission to it. He bought a tract of land adjoining the park and revealed his plans for the tower. The Park Service tried to stop him by condemning the proposed site, but the businessman simply found another one. Since the agency had no formal power over local land use decisions and the city fathers of Gettysburg were not particularly sensitive to the park's plight, condemnation was the only strategy open to it. Since the Service could not condemn all possible sites for the tower, it was bound to lose. In the end, it made a loser's compromise, dropping its opposition to the tower and even giving the builder a right-of-way across park land in exchange for some design changes.[44] Today, the futuristic tower hovers over the battlefield like a nosey spacecraft (figure 7-1). At night its presence is made even more obvious by its flashing red aircraft warning lights. Not far away from the tower, a chain of fast food establishments, whose construction the agency also was powerless to prevent, presses hard against the main entrance to the park.

Tacky tourist towns grew up around the entrances to some parks, while others, near growing towns, found themselves outflanked by rural industry. Park managers fought these developments as best they could, relying on persuasion and whatever little real power they had at their disposal.[45] It was perhaps in connection with the parks of southern Utah that these external development threats were most severe. Unlike parks like Yellowstone or Glacier, the parks of Southern Utah—Arches, Zion, Capital Reef, Bryce Canyon—and even Canyonlands are relatively small, although they are among the most spectacular in the System. Bryce Canyon is little more than 50 square miles in area; Arches is little more than 100; Zion is about 250. The impressive vistas they offer and the feeling of grandeur they evoke depend on the vast empty spaces of the southern Colorado Plateau which surround them. As one agency official expressed it, "Southern Utah is one experience, it isn't just a bunch of separate national parks. The real resource out there is what you can see from the parks." However, the last few years have seen mineral extraction and energy development threaten the matrix in which the parks are set, and consequently the views from the parks and sometimes the air quality within the parks.

If such direct experiences were not enough to convince the Park Service of the need to concern itself with neighboring lands, it had the exhortations of public interest groups which were also worried about the problem. By

Figure 7-1.
Privately built observation tower overshadows Gettysburg Battlefield.

the late 1960s, the National Parks Association was encouraging the agency
to concern itself more with the changes taking place beyond park borders.
Through the 1970s the NPA continually pointed out to the Service the
danger of development near the parks.[46] The article "Cutting Glacier to
Size," which appeared in the association's magazine, reflected many of the
primary concerns of the association when it asked, "Is Glacier National
Park destined to become a green island, surrounded by development and
cut off from adjoining wild lands?"[47] The article stresses the similarity of
Glacier to surrounding land and the ecosystemic relationships which
united them. It argued that the park was not unique within its setting,
rather it was "simply the highest of a lot of high country, the crown of a
vast stretch of relatively remote and primitive lands," and that "plant and
animal communities belonged neither to the park nor to adjacent properties but to overlapping ecosystems."[48] As NPA saw it, that relationship was

being undermined: "Human activities are making the distinction between park and nonpark sharper all the time." The NPA catalogued these human activities and the list was a long one. It included overintensive hunting and trapping on neighboring tribal lands, fossil fuel development, logging, water pollution, and highway improvements. The article closed by asking the rhetorical question: "Are American citizens willing to let Glacier National Park be strangled by exploitive activities surrounding it?"[49]

In 1972, the NPFF study group took a stance similar to that of the NPA on the need for the agency to involve itself in land beyond park boundaries to defend the existing parks. Like the NPA, the study group appealed to the principle of environmental unity: "It has long been known that the environment is indivisible, but the meaning of the concept to the national parks is only dawning."[50] The report asserted that because of this interdependence, there had to be a "new sharing of planning and land use control responsibility with other federal, state and local agencies." The NPFF group dispelled any notion that the problem was one of the Park Service's attitude, observing that agency officials were becoming increasingly aware of the importance of external influences on the the well-being of parks.

The NPFF study group was correct in its observation; agency leadership down to park level was acutely worried about external threats to the System in the 1970s. In 1976, agency director Everhardt declared that "the most deadly threat to the national parklands exists in the Southwest where existing electric generating plants powered by local coal supplies have already created haze and smog in the once clear desert air."[51] A regional director said that: "The Park Service is coming to realize how much air quality is related to the Organic Act of 1916. Are the plants dying? Are the resources eroding? These are questions that our enabling legislation compells us to ask, and when necessary to become a player in regional planning."

The Conservation Foundation surveyed park managers about development beyond the parks and found much concern everywhere.[52] The superintendent of Big Thicket National Preserve in eastern Texas saw water quality being lowered by nearby logging and oil drilling. Death Valley reported air pollution drifting in from Los Angeles. The staff of the Grand Tetons worried about residential subdivisions going up on adjacent ranchlands. The park staff at Rocky Mountain National Park complained that numerous obtrusive strip developments already lined the two primary approach corridors to the park, and the superintendent of the Great Smoky Mountains National Park expressed a similar concern about the park entrances.

When the Park service conducted its own study of threats to the parks in late 1979 (in large part prompted by the Conservation Foundation's study), responses from the park staffs indicated that external threats had become more prevalent than internal threats to park integrity.[53] Table 7-1 shows management concerns by category of threat. The report offered the same reason for the high incidence of external threats as did Sax and the NPA when it said that while most of the national parks "were once pristine areas surrounded and protected by vast wilderness regions, today, with their surrounding buffer zones gradually disappearing, many of these

Table 7-1. Threats to the National Park System

	Number of threats from internal sources	Number of threats from external sources	Total threats reported	External threats as percent of total
Air pollution	83	609	692	88
Water quality and quantity	142	324	466	70
Aesthetic degradation	423	662	1,075	62
Removal of resources	376	262	638	41
Exotic species encroachment	277	325	602	54
Visitor impacts	399	106	505	21
Park operations	254	103	357	29
Total	1,954	2,391	4,345	55

Source: National Park Service, *State of the Parks—1980*, p. 4.

parks are experiencing encroachment."[54] This view of the problem's origins is common among park managers in the field. Perhaps the superintendent of one of the crown jewels expressed it best when he said, "The parks used to be islands of civilization in the wilderness. Now the thing has inverted, they're islands of wilderness in a sea of civilization."

We see, then, that there was agreement within and without the Service; times had changed and the faithful discharge of the agency's original mission, the protection of the national parks, required that it move beyond the park boundaries in controlling land use. The agency's manual made it clear that bifurcated thinking would no longer suffice when it said that:[55]

The plans of outside agencies and interests affect and are affected by proposed actions within units of the National Park System. Cooperative planning is needed to integrate the park into its regional environment and to ensure that potential conflicts between interdependent actions are minimized or eliminated.

Pursuing Traditional Goals

While the Park Service was forced to cast its concern beyond land it owned outright to protect the traditional parks, outright land acquisition was becoming a less satisfactory means of acquiring new park units aimed at the traditional goals of providing recreation opportunities and preserving natural and historic sites. Most of the early units of the National Park System were either carved out of national forests, were the result of state donations, or were the gifts of wealthy philanthropists. They had entailed little outright land purchase by the federal government. But by the 1960s such sources could no longer be counted on. Few, if any, tracts of land comparable to the Rockefeller gifts—the Grand Tetons and Acadia National Park—remained in the hands of individual private owners. The U.S. Forest Service had moved to secure its prime scenic areas against loss to the Park Service by giving them special recreation or preservation status. This meant that future efforts at expanding the Park System would increasingly involve large, expensive land purchases.

In 1964, the Land and Water Conservation Fund was established in large part to enable the Park Service and other federal land managing agencies to acquire recreation land from private owners, by condemnation and at prevailing market prices if necessary. During the halcyon, recession-free days of the early and mid-1960s, this seemed reasonable. The American economy looked like a great, unflagging money machine, and to what better use could the wealth be put than expanding the National Park System? Ambitious, expensive thinking characterized open space studies conducted during the 1960s and even well into the 1970s, before new realities caught up with planning dreams. For example, in 1972, the NPFF study group concluded that too much of the nation's land had passed into private ownership. The solution? Buy it back. The cost? One hundred billion dollars.[56] The group proposed a large national bond issue for land acquisition, capital development, and improvement of the national, state, city, and county park systems. Such a scheme would allow the public to "buy back America."

Even during the Kennedy and Johnson years, however, it was apparent that direct acquisition alone would not do—too much had to be preserved. Therefore, in 1962 the Natural Landmarks Program was established.[57] This program allowed an official designation of nationally significant status to be conferred on important natural areas held by state and local governments, conservation organizations, and private owners. These lands were entered on a National Registry of Natural Areas, and although the registry designation conferred no formal protection on the natural areas, it was hoped that owner pride and the high public visibility that registration prompted would make more specific and expensive measures unnecessary.[58]

By the 1970s, it was becoming clear that such nonacquisition methods would have to be expanded since direct purchase was becoming practical in fewer and fewer cases. The economy had soured, and the vast surplus of public funds anticipated in the 1960s had not materialized. Moreover, the rapid increase of land prices had cut sharply into the purchasing power of those dollars which could be earmarked for land acquisition. As a result, in recent years there came to be a feeling in the Park Service and among the agency's supporters that heavy reliance on direct acquisition was a luxury neither the Park System nor the nation could afford any longer. As one agency administrator expressed it, "The old philosophy was to draw a circle and say everything inside is mine and everything outside is yours. That's gone now . . . there isn't the money to draw a satisfactory line." Those in the Service who planned with an awareness of these limits in mind were considered the realists. One top administrator said that the good planners "are the ones who look at the economy and realize we can't afford to buy the world." Former director Whalen expressed a similar thought when he said that "planners who have grappled with the question of protection and who are realists come up with less-than-fee solutions." Internal doubts about the economic practicality of a continued strong reliance on fee simple acquisition to defend natural and historical areas were matched by external doubts. In 1979 the General Accounting Office criticized the Park Service for what it saw as its history of excessive reliance on outright land purchase when "other protection alternatives, such as easements, zoning and federal controls . . . would have done the job better and at a more reasonable cost.[59]

Beyond the economic reasons for lessened reliance on outright acquisition as a means of attaining traditional goals were political ones. It was as much due to politics as to monetary costs that the Park Service turned to less-than-fee acquisition and to the idea of new parks which included large permanent inholdings of private land under public regulation. In the late 1950s the Park Service proposed the establishment of the Cape Cod National Seashore. Unlike previous parks where the intention was the eventual elimination of inholdings, at Cape Cod the developed areas, including entire towns, were accepted as permanent parts of the park (as long as they abided by agreed-upon zoning and development standards). Although it was argued, and probably with some justification, that the presence of towns made up of quaint clapboard and cedar-shingled buildings enhanced the national seashore, it was also true that eliminating the towns would have involved displacement and condemnation on a scale not previously seen in connection with the establishment of national parks. The political costs of this kind of mass displacement would have been enormous.[60] So if the national seashore was going to be established, and agency leadership was strongly committed to it, the Service would have to reconcile itself to large permanent inholdings. The idea of establishing national parks with permanent inholdings subject to restrictions on commercial development, housing styles, and so forth took hold within Congress and the Park Service, and, appropriately enough, it became known as the "Cape Cod formula."[61] It was considered an especially appropriate arrangement for the areas worthy of preservation yet where there was already considerable settlement.

In recent years the Park Service has been asked with increasing frequency to protect areas considered nationally significant and therefore worthy of preservation yet where local opposition to a fully public, traditional park made the political costs of such an approach prohibitively high. One recent example was a proposal to give federal protection to the Big Sur region of California. The *National Journal* reported that, "one standard approach—making the Big Sur a national park or seashore—[had] little· support."[62] The area's 1,720 residents consisted of ranchers, farmers, artists, writers, and urban escapees "who had managed to keep it to themselves. The last thing they wanted was to turn the Big Sur into another Yellowstone." Reflecting local sentiment, Representative Panetta asserted, "I specifically didn't want a traditional federal approach, like a national park or scenic area."[63] In such a case only some formula which involved less than total acquisition would do. Agency Director Whalen expressed what was probably the prevailing view among Park Service leadership in the 1970s when he told Congress that in the future, new parks would be different from those added in the past in that a wide variety of land ownership patterns would prevail within them. Some lands would still be acquired outright, others would be protected by easements, still others would be zoned by a local agency.[64]

Agency involvement with the less-than-fee acquisition or regional planning should not be wholly attributed to economic and political problems with full fee, however. Nor should they be seen as the result of pure political opportunism. As mentioned above, views within the Park Service on what the agency should be preserving changed and expanded in the 1960s and 1970s. When agency leadership started thinking of cultural as well as historical and natural preservation, and when whole living landscapes came under its consideration, there evolved a new preservation synthesis which allowed the agency's mission to be viewed in entirely new terms. As one Park Service administrator told the author, "the Service's mission is [to] preserve nationally significant areas, be this through its own management in full-fee, less-than-fee, statutory planning authority, or no permanent role at all." This put the National Park System into a new perspective. The administrator continued, "The traditional national parks are just nationally significant areas in which a certain preservation technique has been used, probably one which will be used less frequently in the future." Given this expanded sense of preservation responsibility, it logically followed that concentration on outright acquisition as an instrument of preservation could cause more damage than good in the long run. An administrator in the Bureau of Outdoor Recreation expressed a sentiment which the author's interviews showed had considerable support within the Park Service:

> Buying to prevent despoilment wins the battle but loses the war. Look at California. You concentrate on buying Point Reyes but in the meantime you've lost the rest of the coast. The same with rivers. You buy up a few wild and scenic rivers and the rest go to hell. [If the Park Service] keeps paying out and buying, it establishes precedents that undermine good planning everywhere.

Thus a number of ideological, strategic, political, and economic factors drove the Park Service away from that part of the Mather and Albright formula which emphasized full acquisition of unique spectacular places and toward a concern for the whole environment and a role in guiding the future of landscapes on which mixed patterns of public and private ownership prevailed. We now examine the specific nature of the agency's involvement in these areas.

GREENLINE PARKS AND LESS-THAN-FEE

The Park Service's involvement with land beyond that which it fully owned and managed took two principal forms. First, the agency acquired less-than-fee rights (i.e., less than complete ownership) to land as a means of protecting the parks from incompatible land uses on their borders, as a way of making its hard-pressed acquisition funds stretch farther, and as a way of reducing the disruptive effects of private inholdings within the parks themselves. Second, it promoted a new type of park, the greenline park, as a means of involving itself in regional planning and as a way of establishing new parks where more traditional ones seemed inappropriate or were impossible to establish.

Less-than-Fee

Less-than-fee purchases did not involve discrete breaks from earlier practices. However, the agency, continually nettled by the General Accounting Office, did expand its purchases of easements and development rights when direct public access to the land was not needed. In 1968 the Park Service had less-than-fee rights to 10,000 acres of land. By 1974 that number had risen to above 25,000 acres, and by 1981 it stood at slightly above 63,000 acres.[65] At Piscataway Park, Maryland, the Park Service acquired scenic easements to 2,750 acres of largely upper-class residential land to protect the views from Mount Vernon, just across the Potomac in Virginia. The Service continued to hold easements to acreage along the scenic parkways it managed. The agency made use of the less-than-fee rights to protect the sections of the Appalachian Trail from incompatible development. Along the national scenic rivers, it held more than 19,000 acres, and at the national seashores it held more than 14,000 acres.

At some park units, the agency's less-than-fee holdings were more extensive than its fully owned lands. For example, at Fire Island, the Park Service held 2,936 acres outright and had less-than-fee rights to another 3,151. At the Grant–Kohrs Ranch National Historic Site, the Park Service had 216 acres of fully owned land and almost five times that amount in less-than-fee. In other cases, easements were used much more selectively, for example, to defend a unit against land use change on small, isolated inholdings, to protect a key right of way, or to preserve an important but small element of a scenic vista. On Cumberland Island, Georgia, the Service had but one small less-than-fee tract of 1.63 acres. At Gettysburg, 12.95 acres were under restrictions. At Valley Forge there were 7.93 acres. At the Vicksburg Battlefield, it held restrictive rights to 5.78 acres.

Although the Park Service's policy manual formally committed the agency to reliance, where appropriate, on less-than-fee acquisition techniques, and the agency has increased its reliance on them considerably in recent years, agency rank and file was of two minds on the issue. There was resistance within the agency to any lessened reliance on full acquisition. This resistance and its consequences were at the root of the seeming contradiction between the sincere desire on the part of top agency leadership to break free of what one agency administrator called a "full-fee mentality" and the frequent criticism of the agency by both outsiders and insiders for its excessive use of direct, complete acquisition.[66]

Advantages

As an idea, less-than-fee had great appeal at first sight. It appeared to be a way of reducing the expenses associated with land acquisition; it was a way of getting around the problems caused by outright condemnation of private property; it even appeared as a way of letting private owners maintain land which otherwise would have to be managed at government expense. These first-sight virtues made less-than-fee popular with those outside the agency who had a tendency to be critical of the agency and who did not need to delve below the surface of the idea to its management consequences. This characteristic also made it popular with those in the agency, especially planners and those in the ranks of agency leadership, who were attuned to either external political demands or to the strategic opportunities associated with less-than-fee acquisition.

Those below the top ranks of agency leadership, or not in politically sensitive positions, however, responded to other things. One was undoubtedly pure inertia. It is typical of bureaucracies that some procedures have a life force of their own and can resist changes in policy which might otherwise be dictated by a realistic assessment of goals and policies.[67] Within the Park Service, acquisition had been the standard procedure by and large. Until recently, less-than-fee had been used infrequently and only in unique circumstances. Even today, acreage in which the Park Service has less-than-fee rights comes to slightly less than 1 percent of that which it owns outright. Such reluctance to leave the beaten path is by no means always irrational. Bureaus are no different than people in that they learn to perform given tasks better with experience, and the deepest and most successful experience of the Park Service was in park management.[68] If the agency could define the problem as one of park management, the solutions were familiar, practiced ones.[69] (One is reminded of Maslow's observation that when you have a hammer in your hand, everything looks like a nail.[70]) The reasons for these attitudes went beyond inertia, and reluctance to try new ideas. As an instrument of land use control, less-than-fee acquisition has disenchanted many park managers who have relied on it.

Costs

The acquisition of less-than-fee rights was frequently touted as an inexpensive alternative to full acquisition of land, an alternative which had the added benefit of creating good will since it did not involve the displacement of people. Experience showed that it was seldom inexpensive, or even productive of good will for that matter. As Boyce, Kohlhase, and Blaut demonstrated in their study of the components of property value in New Jersey, where the threat of development was strong, the value (and therefore the cost to the government) of less-than-fee rights to preclude development could approach the cost of full fee.[71] Thus, money spent on the purchase of less-than-fee rights where development pressure was heavy, as it frequently was near a national park, would not go much further than money spent on complete acquisition. On the other hand, where development pressure was slight, less-than-fee rights could be bought at a small fraction of full fee costs. But where there was little pressure, there was little reason for government to want less-than-fee in the first place.

Even in areas of little development pressure, less-than-fee rights could be expensive. The purchase of scenic easements along the Natchez Trace Parkway in rural northern Mississippi in the 1950s cost 20 to 30 percent of full land value while those purchased along the Blue Ridge Parkway had ranged from 40 to 75 percent. More recently, in Piscataway Park, the cost of the easements the agency acquired was approximately 50 percent of the total value, and there the land was developed into a prestigious residential area and was already covered by strict zoning restrictions on further development. Elsewhere costs were even higher, running well above 50 percent of full fee in parks such as Gettysburg, Vicksburg, and Valley Forge, places where strong development pressures were present.[72]

The ambiguity of less-than-fee purchase, that is to say, the difficulty in determining exactly what had been bought or sold, led to frequent disputes with landowners and this meant expenses for the agency. One Park Service administrator with much experience with less-than-fee told the author that when the agency acquires less-than-fee, "it winds up paying lawyers more than it would have cost to buy the land in the first place." Whether this was the case or not, the ongoing costs associated with less-than-fee rights were considerable. An internal agency document, assessing the virtues and drawbacks of scenic easements, asserted that the money the agency had saved by purchasing less-than-fee rather than full-fee rights was soon spent on administering and enforcing its rights.[73] And even with proper policing, enforcement was sometimes difficult. Although a billboard could be removed, there was little the agency could do after a woodlot had been clearcut or a building had been put up without prior agency approval.[74] In fact, by the 1970s the forty-year accumulation of headaches caused by the agency's less-than-fee holding along the Blue Ridge and Natchez Trace parkways prompted it to start trading its scenic easements for small strips of land owned outright along the parkways.

It was suggested that, although expensive, less-than-fee acquisition was worth its cost in good will, since people were not evicted and, in essence, were given money for not doing anything.[75] This good will was by no means a sure thing. Many park neighbors living under restrictive easements found them a constant source of annoyance and friction with the Park Service. Since maintaining easements required a high level of policing, those who were the object of this vigilance often came to resent it and to feel they had been singled out for special harassment by the Park Service. Resentment also grew out of the fact that development rights were not tangible and therefore it was often difficult to clearly establish exactly what had been sold. For example, along the Blue Ridge Parkway, the terms of the scenic easements sold to the Park Service stipulated that no new buildings would be constructed without agency permission and that no trees were to be cut without prior agency approval. Immediately there began to arise disputes, some of which reached the courts, over what distinguished a building from a mere shed and over where to draw the line between tree cutting and brush clearing.[76]

There were also cases where those who sold development rights later felt they had been cheated by the government, or at least that they had sold out too cheaply and were owed further compensation. An example of the latter occurred on the high plains where, in the 1950s, many property owners sold conservation easements for wildlife refuges. Under the terms of the easements, the land could not be farmed. When the development rights were sold, the land was not irrigable under existing irrigation techniques so the easements had little opportunity cost to the farmers. Therefore the farmers were willing to sell the easements at a relatively low price. Subsequent improvements in technology, however, made irrigated farming practicable on much of the land, increasing its agricultural value and the discontent of the landowners who had signed away their farming rights. Here the fact that what had been sold was not tangible and discrete militated against the sense of finality, for better or for worse, that accompanies a fee-simple sale and ends disputes over it once and for all.

Another problem for the Park Service arose when land under easement was sold from one private owner to another. Although most purchasers were aware of the fact that the land they were acquiring was under easement, the new owners often did not understand the full extent of the restrictions involved. When this was the case, they often felt cheated by the person who sold them the land, and by the holder of the easements, the Park Service.[77] As a result of all these drawbacks, less-than-fee came to be seen by many in the agency who had used it as an impractical, clumsy method of obtaining resource protection which could be provided simply and less expensively through outright acquisition.

This particular quality of the less-than-fee acquisition, i.e., its surface appeal as a concept and its serious drawbacks when transformed into an instrument of land management, meant that the decision to adopt it visited benefits and costs on different parts of the organization. Adoption was a boon to those who had to formulate agency strategy and maintain its external relations. On the other hand, the decision to adopt meant headaches for the managers in the field who were ordered to rely on less-than-fee.

As matters now stand, less-than-fee seems firmly entrenched. The manager's side of the story is little articulated, and one hears much more about its virtues than its drawbacks from agency leadership and official agency pronouncements. Agency leadership's willingness to accept the management consequences of less-than-fee does not mean that the issue is settled, however. It is not. Such an imperfect instrument is more likely to raise policy questions again and again than to settle them once and for all.

The Greenline Park

The second form of agency involvement in land it did not own outright centered on the promotion of the greenline park or, as it was sometimes known, the area of national concern. Briefly, the greenline park concept involved a specially designated region under a mixed pattern of public and private land ownership. Planning for the region was to be comprehensive and aimed at preserving its nationally significant resources.[78] A greenline park, like any other national park, would be designated by Congress. When a region was so designated, federal planning authority would be conferred on a federal agency which might, but need not, directly manage a portion of the land within the area. The National Urban Recreation Study, jointly authored by the Park Service and the Heritage Conservation and Recreation Service, presented an idealized plan for greenline parks which included land use restrictions and management goals for the designated region (figure 7-2). The plan shown was for an imaginary area of considerable scenic value with a mix of commercial forestry and recreation on private land. The plan called for the continuation of these activities in privately held development zones and conservation areas. The public would acquire only selected areas to allow public access to water, to establish a wildlife refuge, and to permit the construction of trails, picnic sites, and campgrounds. Elsewhere within the region, the managing public agency would acquire scenic or conservation easements. Still elsewhere the federal agency would enforce land use regulations promulgated by state and local authorities.

The greenline park idea emphasized intergovernmental planning cooperation, with federal, state, and local levels each having their own distinct functional and spatial responsibilities, yet with each cooperating with the others in the overall planning.[79] Such a multilevel approach assumed and allowed for the fact that the various levels of government might have distinctly different interests within a particular area. For example, the state might have an interest in the health of the timber and recreation industries in the region while the foremost concerns within the local towns might be the preservation of jobs, tranquility, and the tax base. While state and local governments could be counted on to look after the interests of their constituents, the federal presence was also needed because, in the words of the National Urban Recreation Study, "local and state officials operate within local and state constituencies and priorities, they couldn't be expected to put national interests first." [80]

Such a national interest could take several forms. For example, there might be a national interest in preserving the unique scenic splendors of the area. There might also be a national interest in making recreation facilities widely accessible. Ideally, the mix of planning responsibilities in the greenline park would ensure that all the interests in the region would be protected.

This emphasis on multilevel government participation within the greenline park was not only conceptually neat, it also accommodated constitutional reality. The constitution, through its delegation of residual

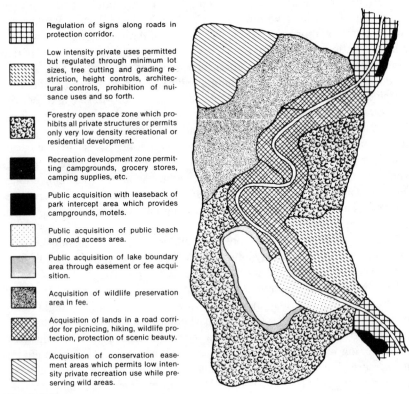

Regulation of signs along roads in protection corridor.

Low intensity private uses permitted but regulated through minimum lot sizes, tree cutting and grading restriction, height controls, architectural controls, prohibition of nuisance uses and so forth.

Forestry open space zone which prohibits all private structures or permits only very low density recreational or residential development.

Recreation development zone permitting campgrounds, grocery stores, camping supplies, etc.

Public acquisition with leaseback of park intercept area which provides campgrounds, motels.

Public acquisition of public beach and road access area.

Public acquisition of lake boundary area through easement or fee acquisition.

Acquisition of wildlife preservation area in fee.

Acquisition of lands in a road corridor for picnicing, hiking, wildlife protection, protection of scenic beauty.

Acquisition of conservation easement areas which permits low intensity private recreation use while preserving wild areas.

Figure 7-2.
Idealized greenline park. (From Heritage Conservation and Recreation Service/ National Park Service, *National Urban Recreation Study*, Vol. 1, p.120.)

powers to the states, made the states the ultimate repository of power to regulate land use.[81] Therefore, when states identified areas of special concern and adopted special regulations and restrictions for them, they were on firm legal footing. Since the federal government had no such legal base from which to attempt similar regulation of land use in special areas, it would have to involve the states and rely on states' constitutional powers in doing so. As a Park Service planner expressed it, "If the federal government wants into the planning business, it will have to be through multi-level government planning bodies. It's the only way to bring the real source of regulatory power, the state governments, into the game." Greenline parks were a way of bringing the federal government, through the Park Service, into the game.

Models

There were several models for the greenline park, and perhaps the first was the English national park.[82] According to Goddard, these English parks were really nothing more than "delimited areas within which a certain degree of control is exercised over economic development in an attempt to preserve the region's existing character."[83] Within them, lands were under several types of ownership, both private and public. Villages and even towns existed within their bounds. Shortly after the Second World War, the first of the English parks was created. The parks were established on populated, working landscapes; they were not carved out of the wilderness, *ex nihilo,* as were the great American parks. The English parks were adjustments to the reality of a small, highly populated countryside, but they included many of the most beautiful regions in England and Wales: a large part of the Lake District, the mountains of northwest Wales, the Pembrokeshire Coast, the North York Moors and the Yorkshire Dales. Also, unlike the American parks, there was very little recreation development in these English national parks, although a few camping sites were established in all of them. The primary management aim of the parks was to preserve the rural landscape and its scenic amenities and, according to Goddard, to fight "a rear-guard action against the steady encroachments of commerce and industry."[84]

Although the American national park tradition was very different from the English one, the selective applicability of the latter to the United States was apparent. In many areas of the United States, especially in the East and along parts of the West Coast, the population density more closely approximated Western Europe than the western interior of the United States. Moreover, the scenic value of these settled landscapes depended more on the character of their cultural imprint than their inherent natural features. Finally, in many of these areas, entrenched local interests were strong enough to ensure that the preservation of scenic amenities, with its minimal disturbances of the status quo, would have to take precedence over the provision of active recreation in any regional management plan.

In 1973, a model was established closer to home when New York adopted a comprehensive plan for its vast, underpopulated Adirondack region. The plan aimed at perpetuating the unique characteristics of the region with a pattern of mixed public and private land ownership and with state-level guidance of development.[85] A state agency, the Adirondack Park Commission, was established and charged with developing a regional plan. The agency was given a great voice in the administration of public land and great regulatory power over the use of private land within its area of responsibility. The regulations it developed were far reaching, establishing many areas of very low density zoning and severely curtailing the development potential of the towns under its jurisdiction.

Support for Greenline Parks

Even before the Adirondack area was designated a special planning region, the idea of federal greenline parks had gained currency and acquired a network of supporters in the federal government and among public interest groups. The Park Service was an important part of this network. According to a long-time observer of the agency:

> In the early 1970s there was a faction of good, smart people in the Park Service who were interested in [the greenline park] concept. They felt these parks were the coming thing, and that if the National Park Service did not get involved in them, it would be left behind. The states would run them themselves or perhaps the Bureau of Outdoor Recreation would move in.

During these years, support for federal involvement in land use planning coalesced around the Jackson bill, which would give federal support to a wide range of state-level planning efforts. Support for Jackson's bill was not unanimous, however.[86] Some thought it was too ambiguous about what should be protected and what the proper federal role in land planning should be. They feared that with such a flaw it would be ineffective and therefore ultimately damaging to the cause of good land use planning. The greenline park, on the other hand, would be a clearly defined region of federal interest and legislation, for its establishment could be explicit about the extent and limits of federal power.

When the Jackson bill was defeated in 1974, the greenline approach looked even more attractive. The fear that the Jackson bill's vague yet seemingly ambitious language aroused had contributed to its defeat.[87] While the vagueness meant ineffectiveness to some supporters of a federal role in land use planning, for those landowners who were not sure of whether the bill would hurt them, the same vagueness was threatening. After the Jackson bill was defeated, the greenline parks concept appeared to the defeated bill's supporters like a way of getting around these fears, even if the concept was a less satisfactory means of involving the federal government in land use planning. For those supporters of a strong federal land policy who had disliked Jackson's bill in the first place, its defeat cleared the way to try what they saw as an inherently superior approach.

In 1976, Congressman Florio (D-N.J.) introduced a proposal to establish the Pinelands National Reserve in New Jersey. The proposal, the product of a working group that included personnel from the Park Service and the Bureau of Outdoor Recreation, was strongly shaped by the greenline concept, and it showed.[88] The reserve would contain a small core of publicly owned land, while the rest of the area would remain in the hands of private owners. The private land, however, would be subject to use restrictions promulgated by the Pinelands Commission, a public agency to be modeled after New York's Adirondack Commission. The Pinelands Commission would contain representatives of the department of interior, the state of New Jersey, and the municipalities within the reserve's boundaries. As was the case with its counterpart in the Adirondacks, the commission was given overall planning authority and was directed to draw up the master plan for the area.

Florio's comments in introducing the bill on the floor of Congress reflected the thinking of those who developed the greenline idea. "It is becoming obvious that [a way] must be found to meet the burgeoning demand for the preservation of outstanding landscapes [yet at the same time] provide a humane living environment."[89] He added that the alternative to the establishment of a national park ought to be something other than "the relegation of outstanding landscapes to indiscriminate development." Florio touched on the need to involve all levels of government in a cooperative planning effort. He stressed that a national park would not be the solution to this resource preservation problem, and he advanced a proposal to fill the gap between complete federal control and none. The rationale Florio advanced for federal involvement in the pine barrens was the presence of resources of national significance, and he noted that it was the same rationale as that for the establishment of national parks.

In the bill, Florio asserted that the federal government's role should be primarily one of providing assistance to New Jersey in preserving the area. Florio's bill was not reported out of committee in 1976, but two years later a similar one was. It became law, and a national reserve was established in New Jersey's pine barrens.[90] Although the Florio plan did not designate the Park Service as lead federal agency in the pinelands, Director Whalen fought hard (and unsuccessfully) to get this designation for his agency because he believed greenline parks would be an important type of park in the future.[91]

In 1977, New Jersey's Senator Case introduced a bill "to establish a national system of reserves for the protection of outstanding ecological, scenic, historic, cultural, and recreational landscape."[92] The bill would have set up a National Reserve Council under the chairmanship of the secretary of interior and would have established unified guidelines for reserve management. It also would have established formal procedures for evaluating candidate areas for the new reserve system. The bill had been quickly drawn up and introduced with little advance work having been done in the Senate, but it did generate immediate interest. While the interest was not sufficient to gain its passage, it did indicate that greenline parks might be the best federal approach to land use planning.

Although a separate, formal category of greenline parks was never added to the National Park System, the concept was to have a strong influence on those who drew up new park legislation. In the late 1970s, many units added to the Park System, including most of the major new ones, incorporated elements of the greenline park. Their authorizing legislation made provisions for large permanent areas of private land within park boundaries. The new park's bills incorporated or allowed for mechanisms for coordinated planning of public and private lands. They made it clear that the maintenance of present landscapes and land uses was to be a planning goal equal to, or in some cases superior to, the provision of recreation opportunities.

Largely as a result of the recent addition of parks influenced by the greenline concept, by the end of 1981 the National Park System had come to include 4,422,000 acres of private lands within park boundaries.[93] While even some of the oldest parks included some private land (Yellowstone included 216 acres of private inholdings, the Great Smoky Mountains National Park had 2,609 privately held acres, Olympic National Park had 4,312 acres of private land), most, perhaps 80 percent, of total inholdings in the system in 1981 had been added since the mid-1970s. Much of the new private land was in the Alaskan additions: Aniakchak, Bering Land Bridge and Cape Krusenstern national monuments; and Gates of the Arctic National Park. Here the private land was to allow indigenous cultures or already established towns to flourish without the inhibiting restrictions of public land ownership. Much private land also came into the System through the establishment of national wild and scenic rivers. For example, the New River Gorge unit in West Virginia included over 54,000 acres of private land, only a small fraction of which would be transferred to public ownership. Park planning called for the rest to remain under essentially unchanged uses. Many of the new parks established near urban areas included large tracts of private land. Cuyahoga Valley, with 12,000 acres of federally owned land, included more than 14,000 acres of private land. Some of this private land was to be acquired but much would remain under current patterns of private use. Jean Lafitte National Historic Park and Preserve contained slightly under 15,000 acres of federal land and over 13,000 acres of private land. Santa Monica Mountains National Recreation Area had slightly under 8,000 acres of federal land yet included

107,000 acres of private land within its boundaries. Here plans called for much of this private land to remain private, with planning for public and private lands coordinated toward the best interests of both.[94]

Thus, by the early 1980s the Park Service had become involved in a form of regional planning, not through a major policy decision, but incrementally through the accelerating accumulation of national parks with greenline characteristics.

ACCUMULATING DOUBTS

Management Difficulties

In practice, Park Service involvement in the fate of land it does not own yet for which it has planning responsibilities has been a source of considerable problems and dilemmas for the agency. Neither widespread recognition of agency interest in regional planning nor agency commitment to such planning has cleared away the obstacles to an effective role for the agency in this area. These obstacles are many and serious.

Problems of Local Control

Congress has historically given the Park Service little power to control land beyond its boundaries.[95] Much of the new park legislation was modeled on the greenline park concept in the mix of public and private land it called for, the provisions it made for less-than-fee acquisition, and the utterances about federal, state, and local planning cooperation it usually contained. The real power to regulate land, however, was invariably left in the hands of local authorities. The Park Service was usually given power to cajole, beg, and appeal to noble instincts. For example, at Cuyahoga Valley National Recreation Area, where one of the purposes of the park was to stay urban expansion, development of the private land surrounding the park was, in the words of the authorizing legislation, "left up to local governments to regulate or not as they see fit."[96] The Chattahoochee River National Recreation Area in Georgia clearly traced its ancestry to the greenline concept. Fourteen separate sites along the river totaling 5,000 acres would be acquired outright and managed by the National Park Service. However, 20,000 acres within the National Recreation Area would remain in private hands, and would be subject only to local regulation, if, in fact, the localities saw fit to regulate. As for sanctions to protect any national interests in the private land in the recreation area, there were none.[97]

To a large extent, Congress's unwillingness to work out some means of giving the Park Service regulatory authority, something necessary if the agency's public statements about protecting national interests in the greenline-inspired parks were to become more than pieties, was due to its sensitivity to local sentiment. While from Washington, intergovernmental cooperative planning might look like a rational way of protecting all interests from local to national, from a local or state perspective, it can look like the usurpation of a traditional right. If anything, the National Urban Recreation Study understated this point when it said that federal involvement in the fate of local lands "may be viewed by some local residents as an intrusion, even where an area is seen clearly by the Congress as one of national importance."[98]

Local governments had long been used to making land-use decisions according to their own calculus of benefits, and they were loath to give up this power. Moreover, they were encouraged to fight to keep what power they enjoyed by the popular notion that the local community ought to be the place where policy is made.[99] This desire to keep decision making local was further reinforced by distrust of bureaucratic power so great that the possibility of bureaucrats coming to dominate all government decisions was a spectre that has haunted American politics.[100] Thus a multigovernment cooperative venture could be viewed by local citizens not as a proper reapportioning of power to make land use decisions but as the first step toward complete loss of local power. In fact, many of the residents of the proposed Big Sur National Scenic Area, where the idea of cooperative planning was bruited about, feared just such a complete loss. Speaking for them, Senator Hayakawa characterized the federal planning effort there as "the golden tipped foot in the door," the first step toward total federal control.[101]

Even without this fear of an usurpation of power, there was a good reason for local citizens to be reluctant to willingly share power with the federal government. Land use decisions in this country have always been characterized by conflict, and that conflict has its roots in the fact that when land use changes, someone usually gains at someone else's expense.[102] In many cases, national interest and local interest in how a region's land was used could be viewed not as complementary overlays of concern but as directly antagonistic interests. What was preservation of a nationally unique landscape to a federal agency's planner mindful of some national interest might have been an intolerable infringement on ownership rights to a local property owner or to a local politician mindful of constituent feelings or the power of the local development interests. Obviously such a conflict was open to a compromise solution, but if the local citizens were holding all the high cards already, should they allow themselves to be put in a position where compromise was necessary?

Given this reluctance of local citizens to voluntarily share regulatory power and Congress's disinclination to mandate any sharing in the face of this reluctance, the Park Service was forced to rely on substitutes for direct regulation to achieve its ends on land it did not own, including private land within the greenline type parks. Principally, the agency encouraged local governments to develop their own comprehensive land use plans by giving them seed money to start a planning process.[103] Although encouraging and supporting local planning was expensive, the agency figured that if it did work to protect the parks, it would be a good investment. It hoped that once local planning started, the federal interests would be considered in the process. There was no guarantee of this outcome, however. Where true differences between local and broader interests were at the heart of land use change in areas of national concern, merely establishing a local planning process was not going to alter the reality of this conflict. If anything, it would heighten conflict by articulating it and making it obvious to all participants. Thus, the mere establishment of local regulation by itself was not likely to strengthen the hand of those who spoke for the broader interests. The National Urban Recreation study observed that "local initiative is required for regulation to work effectively in critical environmental and cultural areas. If the political will is not present, regulation will fail."[104]

When Controls Fail

Local planning worked to protect wider interests only infrequently, and where it failed, the agency was forced to fall back on methods that were clumsy and expensive. For example, at Gettysburg the National Park Service tried to induce the town to prevent the development of a commercial strip at the main entrance of the park. When that failed, it was forced to devise a 2.4 million dollar plan to beautify the tacky four-block strip of motels and fast-food franchises with plantings, benches, and fancy streetlights.

At the national seashores, the Park Service's principal fallback technique was what has been called the "sword of Damocles" provision of the authorizing legislation, that is to say, the Park Service's right to condemn property which was not in compliance with local zoning ordinances.[105] On Fire Island, where the local towns frequently turned a blind eye to violations of the zoning code, the Park Service was often forced to exercise its right of condemnation to enforce the zoning laws of the towns within the parks. Condemnation was an unsatisfactory substitute for zoning enforcement, however. If a landowner opted to build in violation of zoning regulations, the Park Service's chronic shortage of condemnation money meant that his chances of being condemned were slim.[106] If, by some chance, the agency did choose to condemn (and could afford to do so), the courts usually saw to it that the property owner got at least fair market value for his property. Hence there was not likely to be a financial penalty for a zoning violation. Because of the lack of money for land purchases, the agency had to use condemnation sparingly. The superintendent of Fire Island said of his condemnation policy, "It is a selective process, we're careful about where we strike. We can't get them all, so we get the most outrageous ones." [107] He was not optimistic, however, about his chance of controlling the development in his park with only the tool of condemnation available to him. "We're losing the battle. Much of the growth that is taking place now is based on the fact that we can't stop it." Even after the severity of the problem was recognized, efforts to get Congress to allow the National Park Service to directly enforce the zoning law were defeated. [108]

The need to involve many government agencies and different levels of government in the ongoing planning efforts at greenline parks also gave rise to considerable coordination problems. On the upper Delaware, maintaining a sustained, coordinated planning effort proved very difficult, and the whole project came close to collapsing. According to an agency planner involved in it:

> In the beginning the [Park Service regional office] didn't know how difficult [the project] would be and was overwhelmed by it. People got stretched too thin and the Service couldn't keep its commitment. The states [of New York and Pennsylvania] were upset and so were the local planners. There was no progress, no guidelines, and no one knew what the hell was going on.

Gitelson, studying the planning for the Delaware Water Gap National Recreation Area, found that "officials on all planes of government were often totally unaware of what the other actors on other planes of government were thinking, planning or coordinating." [109] He concluded that coordination was an inevitable problem when many officials representing a large number of agencies and constituencies dealt with a problem. The degree to which problems emerged with greenline-style parks can be read between the lines of the National Urban Recreation report, whose authors strongly favored the greenline approach as an alternative to the traditional fully public park, but whose enthusiasm had been tempted by experience: [110]

> Although the application of the greenline concept has been at least a partial success in all areas where it has been attempted, highly intensive regulatory and land-management efforts have proven necessary. To be effective, regulations must be carefully designed and administered, and regulatory and land acquisition efforts carefully coordinated. In short, successful operation of such units requires a new level of sophistication in land and water management.

Park Service personnel and others who had been involved with regulation efforts in the greenline parks were frequently less circumspect, however.

The special assistant to a regional director told the author bluntly, "I don't like [greenline parks] because they're impossible to manage." [111]

Greenline Parks as Inherently Unstable or Inappropriate

The various management and legal problems associated with the greenline concept where it was put into practice led even its supporters within the Service to fear that it represented an inherently unstable middle position between the traditional national park, with complete federal control, and its opposite, a landscape of private property with land use completely under local control. They feared that greenline parks would eventually drift toward the one or the other extreme. The agency's experience on the upper Delaware illustrated just how unstable the middle ground could be.

The Upper Delaware National Scenic and Recreation Area was established to protect, as the name implied, the scenic and recreation values of the upper Delaware River. Prior to agency involvement and the special status designation of the waterway, the canoeists who used the river frequently camped along its largely unposted banks at night. With the national scenic river designation, and the increased canoe use which accompanied it, local property owners became nervous about trespass and vandalism, and posting of the river banks increased. Responding to the owners' fears, the National Park Service increased river patrols, and when canoeists were encountered camping on the newly posted land, they were thrown off. The canoeists had to camp somewhere, however, so the agency was compelled to acquire land for formal campsites along the river. This aroused the resentment of the local residents, who already blamed the agency for the increase in the river's recreation use and felt that formal campsites would encourage even greater use. The canoeists, for their part, blamed the Service for ruining their tradition of casually camping along the river banks and forcing them into crowded official sites. [112] Thus the middle ground, because it aroused the ire of both sides, was unattractive and perhaps untenable.

To a manager faced with such a sequence of events and escalating hostility on both sides, the temptation to move in either direction, toward full fee or abandonment of agency presence, must have been severe. An agency planner involved in developing a management plan for the upper Delaware also saw the middle ground as unstable, but for another reason. He saw the following scenario as a distinct possibility:

> The locals see more litter on the river and think the feds should pick it up and they'll put pressure on us. We will pick it up and little by little we'll take over the rest of the region's services, what the locals should be doing for themselves. Finally, we'll be doing everything so why not buy out the place completely?

Some of those who saw a tendency toward complete buy-outs feared that greenline parks were fiscal time bombs, that their authorization might ultimately lead to enormous acquisition expenditures which were wholly unanticipated at the time of their authorization. They feared that as a result, the Park System might find its future budgets, and therefore its future, shaped not by long-range planning or even congressional oversight but rather by a creep toward full acquisition as frustration over reliance on less-than-fee purchase and regional planning devices mounted. On the other hand, there were those who most feared a drift in the opposite direction, toward decreasing federal involvement. They feared that this drift, born of increasing management frustration, might lead to the loss of the nationally significant resources that the greenline parks were established to protect in the first place. As one administrator told the author, "with [greenline parks] you don't get permanence. Five hundred years from now bought land will still be bought, who knows about [greenline parks]?" A staff member on the House parks subcommittee was making a similar point when he said, "One thing I know about Yellowstone is it's still there!"

The above were, in essence, questions and fears about the efficacy of devices available to the agency for controlling the use of land it did not own. Many, however, questioned the very appropriateness of a large-scale Park Service involvement in land it did not own. There were those who said proper agency responsibility did not extend much beyond traditional park management. One agency official felt that a large commitment to regional planning would be "inappropriate" since it would "lead away from the true mission [of the Park Service] which is based on uniqueness and national significance." The argument went that if the Service became a regional planning agency, ubiquity of concern would undermine its traditional (and proper, according to the official) concern for the special and unique. A former park ranger also felt land use planning was an inappropriate activity for the agency: "I hate the federal zoning aspect of greenline parks. You can't zone from sea to shining sea; it isn't right or proper, and the Park Service will get in trouble if it goes down that path."[113]

Dealing with Pollution

Perhaps it was the Park Service's moves into land use planning to promote environmental quality and to protect the parks from threats from remote sources of pollution that raised the most serious questions of appropriateness. One agency administrator told the author: "If we have problems with air quality we should go to EPA and say 'see what pollution is doing to our national parks, do something about it.' It's something we shouldn't put a lot of energy into. That's EPA's job." This view was expressed in several different forms by agency officials. A regional director said "A government agency has its own mission to pursue and we can't go too far down the trail of regional air quality." Why? "You lose credibility if you overstep, if you carry [your] flag too cavalierly into fights where you don't belong."

Exactly where the agency belonged when it came to fighting air pollution through regional planning efforts was a difficult question for the agency to settle for itself, for the question of appropriateness was occasionally tied up in even thornier ones of practicality. The former superintendent of Death Valley said "I used to get up in the morning and watch the brown haze from the Los Angeles metropolitan region creep toward [the park]. What was I going to do about it? Shut L.A. down?" Another high administrator, referring to the same problem, said "You have to take on dragons you can slay." To her, regional air pollution did not appear to be one the agency could slay, so it was best left to others.

Many of those who feared the practical consequences of a major agency commitment to regional environmental quality through land use control believed that the Park Service should at least be an active participant in the environmental coalition fighting for air and water quality and, in such a role, should be willing to institute litigation and to lend professional support and make data available to public interest groups concerned with the overall environment. Some, however, thought that even this overstepped the line of proper agency concern. One official told the author:

> The air quality fight isn't ours. Perhaps we should instruct our interpretators to say 'yes the air here is polluted, but you American citizens, through the political system, chose to pay the environmental costs for the energy you want to run your homes and cars. . . .' The National Park Service should be concerned [only] with direct impact on the resources, for example it should understand what acid rain does to the Statue of Liberty and take steps to reduce the damage.

This was undoubtedly an extreme view. Most agency personnel interviewed by the author were willing to use a tangible impact on the resources of the park system as a criterion for active Park Service involvement in regional environmental quality matters. Even given this seeming consensus, however, the line of responsibility was not agreed upon, since what constituted an external threat or a tangible impact on a park was not always clear cut. As one official phrased the relevant question, "Is foul air in the park a degradation of the resource? What about the views from the park, does the smoke plume on the horizon constitute degradation?" Some thought yes, some thought no.

Lack of resolution within the agency on this point encouraged decisions to float upward within the executive branch where they naturally became politicized and colored by the general philosophies of particular administrations. Environmental quality was an issue of high-level federal concern through the 1970s, and under President Carter, Assistant Secretary Herbst encouraged the Park Service to, in his words, "aggressively protect the parks from further air pollution damage and to restore the visibility of those parks which are presently degraded."[114] This meant, for example, defining the resources of the southern Utah parks expansively. Views from the parks became as much a resource as the rock formations within them. According to an administrator speaking at the end of the Carter administration, "The last four years have seen an aggressive outreach on the matter of vistas; the Carter administration really pushed it." Litigation to stop threats to vistas such as the coal-powered Kaiporowits electrical generating plant was pressed. Under Carter, Assistant Secretary Herbst's attitude toward his own job also encouraged the Park Service to think more in terms of the total environment. "If there was one Assistant Secretary whose concerns were environmental first and foremost, it was me, and I pushed [the Park Service] along with me, although we were in basic agreement."[115]

As a result of these encouragements, Park Service leadership took increasingly ambitious environmental stances. Director Whalen came to see his role as one of an environmental activist whose responsibility went far beyond managing national parks. He told the author that:[116]

> In any administration there are going to be three chief spokesmen for the environment, the Park Service Director, the Fish and Wildlife Director here in Interior, and the chief of the Environmental Protection Agency—so I used my office as a platform for environmental issues, even when it was technically not my business by a strict definition.

With the arrival of the Reagan administration, the Park Service's view of its responsibility for the health of the general environment, or even to what degree that health had an impact on the parks, was caught up in the general questioning of previous environmental policies. According to an agency administrator responsible for pollution questions, "the Reagan administration isn't sure how far out it wants us to draw the line on pollution threats."

Summary

We might now summarize the agency's recent involvement in the fate of land it does not own. In the past twenty years there have arisen both strategic and defensive reasons for the Park Service to abandon its tradition of reliance on full-fee management and its general neglect of land beyond park boundaries. As the issue of land use planning moved into a high place on the federal agenda, the Park Service looked like an appropriate policy instrument. As the price of land soared and land development near the parks increased, it appeared that full-fee acquisition was becoming a less satisfactory means of protecting established parks or pursuing traditional System goals. Responses to these changed circumstances involved increased reliance on less-than-fee acquisition and the establishment of a number of new parks incorporating greenline features, features which committed the agency to cooperative land use planning ventures.

The agency's experiences in these areas have not been entirely happy ones. First, less-than-fee acquisitions involved many management difficulties and this has pitted managers, who must live with these difficulties, against planners and others in the agency who stand to gain from the idea's first-sight appeal. As Park Service experience with less-than-fee grew, agency leadership increasingly found itself caught between its concern for ease and efficiency of System management on one hand, and public relations and political advantage on the other.

Second, the new parks incorporating greenline features usually incorporated them in a vitiated form. This, plus the inherent legal weakness of the federal government when it comes to land use regulation, made these new parks ineffective instruments of regional land use planning as well as difficult entities to manage. As a result, here too agency leadership was caught between those who had to live with management consequences of an idea and those who stood to gain from its advocacy. In addition, movement beyond simple park management raised objections from those within the agency who felt such a move was simply inappropriate and that it violated the agency's mission. This tended to place the question of appropriateness at higher levels of the federal administration and take them out of the agency's hands. Thus in this policy area, as in so many others, decisions and actions in response to changes in the wider world around it did more to exacerbate than to settle questions about the Park System's place in modern life and on the landscape of late twentieth century America.

8 Conclusion

PUBLIC VALUES AND AGENCY GOALS

In a democracy, any public organization must ultimately base its goals on society's interests and values. Since these are constantly changing, organizations must continually reassess their goals and, if necessary, realign or even replace them. In fact, this process of adjustment is so important that Selznick is led to observe that "organizations are not unlike personalities [in that] the search for stability and meaning, for security, is unremitting."[1]

All public organizations, however, operate in uncertain environments. Neither the demands made on an organization nor its bases of support are likely to be fully understood because demands are often indirect while sources of support and opposition are often obscure.[2] Furthermore, demands for public action can arise seemingly from nowhere and without warning.[3] Finally, even if an organization's environment were entirely knowable, there are practical limits to how much information can be gathered.[4] These factors combine to make an organization's environment, in Holden's words, "a place of confusion and uncertainty, with false signals strewn around like dandelion seeds in an open meadow."[5]

Some prediction is usually possible because some important actors and demands can be identified in advance. Some events are linked in foreseeable ways, allowing an organization to plan for the future.[6] All organizations are not equal in this regard, however; the mix of environmental randomness and predictability varies greatly from one organization to another and from one time to another.[7]

In retrospect, the environment of the Park Service during the Mather and Albright era appears to have been a kind, hospitable, and knowable one for the agency. This does not mean that Mather, Albright, and their

261

fledgling agency faced no uncertainty or change; they certainly did. But there were robust, emerging trends in American life to act as clear guides to decisions affecting park management, System expansion, and other matters of policy. For example, the growing number of automobile owners indicated a trend which could be extrapolated into the future to serve as a policy guide. More summer vacation time for the American worker was another trend that served as a policy guide as the Park System geared up to serve millions of Americans arriving with their families. The era produced ideas as well as material trends which were useful as policy referents. For example, visitor policy took its cue from the progressive ideal of a broadly defined, undifferentiated public as the proper beneficiary of government action.

The National Park Service was also lucky to have had dependable sources of support in the nation's professional and commercial elite. This added an important element of predictability by freeing the agency from reliance on fickle congressional favor. Moreover, it protected the Park Service from becoming a politicized agency where the rewards and retribution which attend changing electoral fortunes would have become a fact of life. This wealth of policy referents and sources of predictable support was not mere happenstance; a whole complex of trends, attitudes, and interests gave Mather's era a marvelously optimistic vision of a rational, benevolent, and materially rich future.

It is easy then to see why the Park Service developed a strong sense of mission in the early days. It had a simple, forceful image of the good society, that is to say, the progressive vision, to give its policies confidence and a moral suasion, and to align the agency with the emerging forces and ideas of modern industrial society. It was the notion of progress above all that guided Mather to his early successes. As his biographer succinctly understated it, the Park Service's first director "was certainly one for going along with progress."[8] Progress meant people and autos in parks. It meant partnership with private enterprise, both to promote the parks and to accommodate visitors in them. It meant state park systems supported by the National Park Service and servicing visitors on their way to the national parks. It meant public places of preservation, relaxation, and contact with the sublime. As a means of turning that vision into reality, the policies Mather set for the agency were as correct as the vision itself was appealing. There was little uncertainty over the agency's role as long as the vision was held by those on whom the agency depended.

Decay of the Progressive Vision

The progressive vision lost its freshness and a good deal of its charm, however, as the century wore on, and the trends of the early twentieth century with which it was associated played themselves out. What had been seen as progress came to look like "growth-o-mania."[9] Cities which were to have been liberated by the auto were choked by it. The expressways which were to put the urban dweller in touch with nature only put it further from his reach by encouraging the growth of suburban rings around the cities. Many, especially the poor, were left behind by progress, left dependent on inadequate public transportation in the automobile era, left in inner cities and decaying, depopulating rural places.

Nor did progress as it was expressing itself in material prosperity prove to be a respecter of history or of nature. In the cities, urban rebuilding and road construction demolished historic sites and even whole districts, while on the periphery of metropolitan areas, nature and stable rural landscapes went under the blade of the bulldozer. The dichotomies of the exploited and the preserved lands, the public and the private lands, no longer seemed satisfactory. Whereas once they seemed like simple, elegant guides to policy, increasingly they seemed simple minded.

Public problems and discontents often accumulate slowly, unnoticed, and then suddenly upwell into the public's consciousness and onto the government's agenda. Such an upwelling over the side effects of economic growth occurred in the early 1960s. Perhaps nowhere were these discontents better summarized than in *The Quiet Crisis*. In his book, Udall touched on all the problems mentioned above, and he made it clear what he thought was the root of the problem; it was the uncritical acceptance of conventional, traditional notions of progress. He also made it clear that the federal government had a responsibility for dealing with these problems. In the Kennedy and then the Johnson administration, confronting the accumulated discontents which had sprouted from the material progress of the twentieth century became a matter of high priority.

This meant two things for the Park Service. First it meant that the old policy guides that progress provided were no longer sufficient. It was Wirth's sense of values and his policies, which were based on what seems in retrospect like an uncritical acceptance of the old vision of progress, that led to his downfall. Second, it meant that the agency would be drawn into the fight against the problems that decades of devotion to progress had caused for the nation's cities, its nature, the material remnants of its history, and even the national parks themselves. The National Park Service found itself continually involved in these matters over the past two decades. It tried to recast its own sense of mission on the basis of these involvements, a sense of mission which would free it from old traditions yet would act as a policy guide the way the notion of progress had in the past.

Mixed Signals

A clear, coherent sense of mission has not emerged from the experiences of the Wirth era. Whereas there was a broad consensus for the policies worked out by Mather and Albright, the policy areas in which the Park Service had hoped to find new relevance and a recharged sense of purpose have turned out to be full of conflict and uncertainty of vision, sometimes rooted in the deepest of contemporary dilemmas.

When the agency came to the cities, it found the forces of localism pitted against those of metropolitanism. It found a vision of the future city as a collection of discrete, small-scale communities pitted against that of a metropolis as a single, integrated social unit with high levels of functional specialization and interaction. Similar material and ideological conflicts prevailed in other policy areas. When the agency came to land use regulation, it found itself between forces with a vested interest in local control and those which favored higher levels of regulation, and each side had a compelling image of the proper landscape of the future. On one side, local inhabitants worked in harmony to decide their own future and did so without outside, and inevitably destructive, interference. On the other, comprehensive rational planning in the broadest public interest prevailed. Even in the oldest parks, policy conflicts reached new heights as the image of the crown jewels as the sublime vacation spot for the American public in all of its variety was pitted against the image of the preserve where the few tread lightly and respectfully.

In short, the landscape of power and ideology on which the National Park System is set has become an uncertain one. It has not been easy to find a way through these conflicting images and forces since the discredited progressive vision was not replaced by a coherent image of the good life, or by any clear rallying point. Who then can be surprised that the agency, having its sense of mission tied up with how the nation feels about its nature, its history, its cities, and its landscape, should itself be uncertain and should have undergone all the reversals and inner conflicts which stem from the wider uncertainty that prevails in these areas. This being the case, what is to be done?

First of all, the Park Service is unlikely to find a unifying vision in the foreseeable future; its responsibilities are too disparate, its policies too surrounded by contention. The popular image of the System and the Service, enshrouded as it is in the mythology of the long ago and the far away, obscures this fact but cannot alter it. The Park Service should accept as a central fact of life that it must look for different referents and philosophies to guide it when it comes to each of its disparate tasks. It must work out a separate set of principles, or at least policy guides, for each of its major areas of responsibility: nature, history, the cities, and the land beyond park boundaries. Let us first look at how it might do this with regard to nature and the natural wing of the Park System.

TRADITIONAL CONCERNS

Nature and Natural Areas

The early leaders of the Park Service and the recent leaders of the environmental movement have operated from very different premises about some basic characteristics of civilization, and these different premises strongly informed the policies they advocated for the natural wing of the System. Mather and Albright shared the essential optimism of the early twentieth century. The emerging modern world of industry and science was a better world than the one it was replacing. The mix of culture and nature on the new landscape would be a good, balanced one, and bringing this landscape into being was part, perhaps the most important part, of the job of public land managers.

Multiple View of Civilization and Nature

This assumption of balance was evident in the early management plans for the great national parks, plans which undoubtedly strike today's readers as too accommodating of the inroads of civilization and almost careless in their disregard for nature. But a civilization which was viewed as basically benign and beneficial was not something to be on guard against. Exactly this attitude and these assumptions about nature and civilization can be seen in Albright's recent defense of the Service against charges that it made serious management mistakes during Mission 66 because it was blind to the pernicious inroads of development in the parks. Albright was willing to concede that the Park Service might have made some management mistakes, but it was really guilty of nothing more than occasional excesses when developing the parks to accommodate visitors. In essence, it was guilty of allowing too much of a good thing.[10]

On the other hand, the environmentalists of the past two decades have taken a far more pessimistic view of civilization's relationship to nature. To them, the relationship was heavily marked with conflict, and the contest was unequal. As Schell wrote, "taken in its entirety, the increase in mankind's strength has brought about a decisive shift in the balance of strength between men and the earth."[11] Civilization would inexorably obliterate its opponent unless it was stopped by the clear-sighted and strong-willed every place it threatened to make a gain at nature's expense, and the national parks were on the front lines. To modern environmental activists, the question was not one of striking a balance but rather one of yardage won or lost, and time looked more like an enemy than a friend. Thus, agency administrators are the recipients of two profoundly different and even antagonistic philosophical messages which bear directly on their management of the natural areas of the System. One comes from the agency's traditions, the other from its current political environment, and each has its adherents within the Service.

There is yet a third view of the relationship of civilization to nature which is subscribed to by some agency personnel and which, like both of the above views, has implications for proper management policy. I have termed it the ephemeralist view, and it stresses neither the benign qualities of modern civilization nor its malevolent incontinence. Rather, it emphasizes (and presumes) civilization's transitory nature. This view takes a geological time frame as its point of departure and stresses that the things of man's making which are here today will soon be gone.[12] In a way, the ephemeralist view is similar to the leveling, transitory view of civilization which was dominant during the dark ages. The world was something which would soon be brought to an end by the Second Coming. And like medieval man's world view, the ephemeralist perspective has a sedating effect: Why get excited about that which will eventually be gone, be its immediate effects good or bad? The superintendent of Grand Canyon National Park spoke from such a view when he dismissed the importance of promptly removing park facilities (motels, maintenance buildings, restaurants, and so forth) from the rim of the canyon—something environmental groups had been pressing for and which the superintendent himself favored: "In the long run none of these facilities will have the slightest impact on the canyon itself. I'm in favor of moving everything back from the rim but I'm not in a hurry because whatever we do, someday there won't even be traces of it left." One is reminded of the adage of the middle European peasant regarding his ultimate imprint on the land he tilled, "The forest always returns to conquer."[13]

The Need to Choose

The presence of these three views points up a real quandary for Park Service leadership. In a very basic sense, the correct selection criteria and management policies for the natural wing of the Park System seem to depend on finding the right view of modern civilization itself in regard to some of its most elemental yet abstruse properties: its limits, its stability, its future, even its basic goodness. It is a quandary which overtook the Park Service only as the era of progressive optimism faded into history and the appeal of its reflexive, unhesitating answers to these questions faded with it.

On closer examination, however, the quandary is not necessarily a simple matter of correctness. There are questions of appropriateness to be addressed as well. For example, few would question the basic correctness of the geological view of our civilization. Clearly, by the time the sediments of the Mississippi Delta have been thrust up into the mountain ranges of the future, the presence or absence of a hotel in the redwood groves of Sequoia National Park in the late twentieth century will not matter very much. But does this kind of correctness matter? This is not the time scale on which we operate. Nor is it one from which we can take our values. If we tried to do so, we would find very little value to take; all would be leveled to an equifinal meaninglessness. "In the long run we're all dead," John Maynard Keynes said in reply to an objection that one of his projections would not hold into the far future. The geological view is simply inappropriate as a policy referent for the natural wing of the System. To incorporate it into policy making is to incorporate pure nihilism.

Civilization as a destructive juggernaut is a more difficult view to deal with in that it can make claims to both correctness and appropriateness as a guide to policy. However, while the claims can be made, they cannot be convincingly substantiated. The present times make the idea less appealing today than it was in the mid-1960s. During the Johnson era, people could look back on two decades of nearly uninterrupted economic expansion and land development. Today the economy is lackluster, the bubonic plague is back, and pirates again haunt the Malaccan Straits. Disorder and anarchy seem like more serious threats to a happy future than the stifling effects of an overordered and developed world.

Then there is the matter of appropriateness. Even if the juggernaut view is correct in general with regard to civilization's expansive, consuming tendencies, does this mean that it is a proper referent when we get down to specifics like the management of the natural wing of the National Park System? I think not. First, national parks are not subject to the same inroads of civilization as private lands, or even other federal lands for that matter. Although it has been fashionable to see park visitors as a serious threat to the parks, the equal of any of civilization's pernicious inroads, this hardly seems like a reasonable view. The impact of a visitor access road, even a paved one, on a tract of several hundred square miles is not on the same order of magnitude as clearcutting the tract. The impact of one thousand pairs of feet on a mountain is not the same as that of a chalet development. To ignore these facts is to throw away an important point of perspective. When Albright objected to including the National Park System in the Wilderness Bill because the great national parks, even with their visitor facilities, were already managed as wilderness areas, he was more right than wrong. The failure to distinguish between development to accommodate the park visitor and other possible forms of land exploitation turns the visitor into an enemy of, or at least a threat to the parks, and it does so on unreasonable grounds.[14]

Second, the juggernaut view leads naturally toward seeing the parks as a biological lifeboat. Save whatever can be saved by throwing it in the National Park System now. We can worry about sorting things out later. Even if this view is correct in its essence, the juggernaut is not so ferocious a beast that questions of opportunity costs are irrelevant. For example, there is still time to make choices as to the best possible sites to add, or to deliberate over whether a unit really belongs in the Park System or not. It is even appropriate to decide if we want to add another unit to the Park System or spend the money on other things. When the juggernaut looked more formidable the economy looked more robust. One could siphon off large amounts of wealth to save whatever land remained available, and the cost would be made up so quickly it would not be noticed. Of course, an all-consuming material culture and an economy as vast as a money machine were two sides of the same coin. One appeared as the problem, the other as the solution. Now that the pace of growth has slowed, the old problem does not look so formidable and the old solution does not seem so appropriate.

Policy Guidelines for Park Selection

The National Park Service above all ought to look to its own traditions for guidelines in managing the natural wing of the System. Although the progressive vision has lost much of its suasion as an all-embracing world view, the past management and selection principles for natural areas that were drawn up in accord with it do have an overriding sensibility, even today. They seem able to stand on their own without the support of a broader progressivism.

First there was the principle of managing a national park as a park, a place to be used and enjoyed by the citizens of a democracy. Both the idea of the national park as a park and the idea of a park in a democracy seem to have gotten lost, or at least obscured, in debates about the National Park System in recent years. A park is anthropocentric; its special quality comes from its appeal to humans. It strikes people as grand or sublime, or it just makes people happy to be there, for whatever reason. This was certainly a principle which Mather followed. He never lost sight of basic human appeal as a selection criterion, and certainly most of the early national parks were well endowed with it. Some persons have tried to deprecate this unique and direct human appeal of the great natural areas of the System by dismissing them as freaks of nature.[15] To do this is to confuse the queer with the sublime. Calling Grand Canyon a freak of nature is like calling one of Rembrandt's self-portraits a freak of art or calling *The Tempest* a freak of literature.

The National Park Service should accept the responsibility for d ciding what is and is not sublime and then factor its decisions into evaluations of new units. The prospects of finding absolute, eternal standards of human appeal are nil, or close to it, but America does have elements of an aesthetic canon when it comes to nature. It can be seen in the paintings of the Hudson River School and in the early photos of the West. It is a canon which is bound up with American history and patriotism and in many ways it has withstood the test of time. Albert Bierstadt's grand paintings of the Rocky Mountains still leave people in awe. Yellowstone's beauties still have the same effect they did a hundred years ago. There is the stuff of policy here.

Granted, sensibilities do change. (I am reminded of the scene in *Pride and Prejudice* where the family discusses how horrid mountains are.) There is no reason the System cannot change with them, however. It is also true that tastes in the sublime can vary widely, even within the same era. One person's taste might run to one kind of scenery while another may prefer an entirely different type. There is certainly room for both charm and majesty in the National Park System, and for the subtle, the austere, and the exuberant, for the great wonders and the intimate ones. Decisions about human appeal need not rule out differences in human taste any more than they need rule out the possibility of change over time.

In recent years the Park Service has avoided the admittedly difficult issue of direct human appeal in its selection criteria by falling back on scientific criteria. It has made ecosystemic, physiographic, and geological characteristics important in evaluation, while making questions of appeal irrelevant. This comes close to abandoning the idea of a park altogether. Perhaps some representative of exposed Silurian rock face should be preserved on a federally owned site (although I cannot see why). There is no reason for such a site to be called a park, however, or for it to be part of the National Park System unless it has more to recommend it than pure representativeness.

The idea of the national park as a park should also be put back into System management. It would give a solid referent to what are now thorny questions, such as those associated with clearing trees for vistas or building roads in the parks. If one looks at the parks from a strictly biocentric point of view, then vista clearing is wrong, as is the construction of roads, comfort stations, and other facilities. Each will alter the environment and ecosystem, never mind how little, and thus each will be a violation of biocentric management principles. The same applies to mowing the sides of roads, either for driving safety or to make a parklike impression. All these things become reasonable, however, if the essentially anthropocentric nature of a national park is recognized, and direct human appeal is given at least some of its former weight in park management. Does this mean that all criteria of ecosystemic integrity should be abandoned? Of course not; there is still a large place for ecosystemic study and for the incorporation of the results into management. But it does mean keeping things in perspective, which in turn means that the Park Service cannot, in the words of one agency planner, "treat every damned bayberry bush as if it were a national treasure." Does it mean that it should worship mass access and run blacktop into the farthest cranny of every national park? Before giving the obvious answer to this question, let us examine it in the context of the larger question of what democracy should mean with regard to the national parks.

By and large, the most vocal advocates of biocentric management, the environmental activists, have been the most contemptuous of the park visitor. The contempt takes various forms. There are those who view wanting to sleep in a bed while in a national park as typical of the worst materialist, consuming aspects of American life. They argue that only those willing to visit the parks in a primitive-ascetic style should be welcomed into the national parks on more than a day's visit.[16] Others should be compelled to seek lodgings and accommodations beyond park boundaries.

Should these park visitors be thrown on the tender mercies of the standardized roadside motels that loiter in gaudy clusters around park entrances the way they do around the less attractive interstate interchanges? Surely the Park Service owes the American public both more respect and a better deal.

Accommodating the Visitor

Here too the Park Service would do well to look to its own traditions. Mather and Albright believed in opening the parks up to as wide a range of personal tastes in accommodations as possible, and that was very wide indeed. The gamut ran from primitive campsites to luxury hotels. As we have seen, this idea is very much out of fashion. It was killed in part by the excesses of Mission 66 and in part by the emerging idea of biocentric park management. It was also killed by fear of "Say's law," i.e., the maxim that says that in certain circumstances, supply creates its own demand.[17] This certainly seemed to be the case with national park accommodations in the late 1950s and early 1960s, when a large increase of park visitor accommodations was immediately followed by a large increase in visitors to be accommodated. It appeared that there was a vicious circle in operation: more accommodations led to more visitors, and this in turn created pressures for still more accommodations. Since there seemed no end to demand lurking out there among the public at large, the only way to break the circle before it got out of control was to limit visitor accommodations. Such a strategy was probably reasonable at the time. Population increases coupled with increasing wealth and mobility meant that demand was likely to keep up with and exceed any increase of facilities. This is probably not the case today.

With a nearly steady population and static or even declining mobility and real income, the potential demand for park accommodations is probably static at best. If this is the case, it is not unreasonable to expect the Park Service to attempt to meet the demand. The Park Service is well aware of these changed socioeconomic conditions since the Mission 66 era, but it uses them as an excuse for exclusion. It argues that in a world of diminished resources, quotas and limited facilities are only reasonable. It tells us that we must learn to live with less, and the national parks can teach us how to do that. Rather than look at these changed circumstances as an excuse for pious moralizing and a pretext for teaching questionable lessons about future lifestyles, it should look at them as an opportunity to return to the more truly democratic policies of a happier era for the national parks. It should gauge present demand and attempt to meet it. For perhaps the first time in its history, it has the chance of doing so.

The Park Service has often touted its demonstration role, and it could certainly fall back on such a role to justify a policy of democratic, pluralistic accommodation. It could provide car camping facilities which are models of aesthetic quality and contact with nature. Grand Canyon's El Tovar Hotel is a relic of past elegance; it has textures and spaces rarely seen these days. This too might serve as a model for a future which shows more respect for such elegance than did the recent past. Next to the El Tovar is the Bright Angel Lodge. It is a model of tasteful, economic accommodations. Grand Canyon National Park is not diminished by the presence of these two structures. It would be diminished, however, if they were replaced by large clutches of Holiday Inns and Waffle Houses at the park entrances.

What about the auto in the parks? There is nothing wrong with accommodating auto tourists, but here too a principle of pluralism and a respect for each in his own taste might be adopted. The parks should be managed to accommodate all styles of visits to the extent that they do not damage that which makes the park special or do not unduly interfere with other types of park enjoyment. Most of the parks are big enough to accommodate different styles if demands are kept reasonable. A road running right along the south rim of Grand Canyon is unreasonable because, while it provides for an auto tour of the rim, it precludes enjoying the vista of the canyon without the presence of autos and the distractions they bring. This does not mean that auto access to the rim in certain spots is bad, however, because such access is not an unreasonable infringement on others. It would be unreasonable to ban autos from the park entirely because the thought of their presence will degrade the experience of the park for the backpacker down in the canyon. It would be reasonable to ask such a backpacker to use his imagination a little and imagine that the cars were not there.

In sum, the great nature-based parks of the System should be thought of as parks above all. They should also be managed as parks in the service of a people with a wide variety of tastes, a variety which can be accommodated with a bit of compromise.

Historic Preservation

When history took on a new importance to the Park Service in the 1960s, the agency's own particular circumstances and traditions raised questions about perspective and approach: First there was the question of whether the agency should mean what it said about history, that is to say, should it really treat history as seriously as formal policy statements said it should? One might smile at such a question but it was a very real one. Such seriousness could have considerable cost. A cynical attitude toward history allowed the agency to use historic sites as bargaining chips, and this gave its administrators flexibility in their dealings with Congress and environmental groups. On the other hand, a real concern and respect for history would have disallowed much of this bargaining and would have reduced agency flexibility accordingly. Thus, when the question of sacrificing historic preservation's covert uses to its elevated formal status came up, it created tension between those who spoke from an interest in the integrity of the historical wing and those with an interest in (but who would not be foolish enough to speak in public for) its debasement.

Increased emphasis on history within the National Park System also raised the question of basic approaches to history; it was forced to choose between selective commemoration and comprehensive preservation. As the 1972 plan illustrated, by formulating a scheme which was comprehensive and preservationist in its outline, the Park Service made a bid for enhanced professional status and the right to control the gate to the System. Things did not work out as intended, however. Because Congress assumed much of the initiative on national parks in the 1970s, it, more than the Park Service, came to enjoy the freedom and flexibility the 1972 plan bestowed. Congress used that freedom to turn national parks, especially historical ones, into distributive goods. This in turn raised the question of proper agency response. Should it object to this? Should it attempt to fight Congress? Should it simply acquiesce to the changed circumstances?

Finally there were the questions of revisionism and of partisanship. Should history be used to legitimize emerging social forces? The 1972 plan was formulated to allow this kind of commemoration beneath the facade of impartial preservation, and agency leadership took advantage of it. The Service's leaders during the late 1960s and through the 1970s tended to be politically liberal, and they operated in what (at least in retrospect) was a liberal political environment. Their notions of historical significance were colored by their sentiments as well as by their sense of strategy. However, liberalism was not so dominant as to place the use of its values to guide historic preservation above contention. For example, if the women's rights movement had been the object of united national opinion, then supporting Women's Rights National Historical Park would have been a good strategic move for the agency. It was not backed by such opinion, however. In fact, as later events proved, it was the focus of a good deal of social conflict. With the establishment of the park, that conflict was funneled right into agency ranks. Some in the Service saw the park's establishment as an act of bravery, morally if not strategically correct. Others saw it as an act of opportunism, the conversion of a site of little intrinsic worth into an altar of a dubious cause. In a sense it was both an act of opportunism and a moral commitment. But in that it was subject to such radically different interpretations within the agency, it did not contribute to a pervasive, unifying sense of mission when it came to historic preservation. If anything it only served to diminish consensus.

Failings of the Criteria

As the reader must have surmised, I feel that the core of the agency's current problems with its mission in relation to history lie in the 1972 plan, in the assumptions which went into this plan, and in its failure to establish a clear, reasonable perspective on the agency's relationship to history. Because of this failure, things which should have been settled when history became important to the agency and whose settlement would have put history policy on a firm foundation, were not.

The first thing that the 1972 plan should have made clear, but did not, was the limit of agency responsibility with regard to history. The National Park Service has never been the keeper of the nation's history; that is the job of scholarly institutions. It is not even responsible for the nation's historical education; that is the responsibility of the nation's schools and universities. The Park Service is properly the keeper of certain physical remnants of the past and certain commemorative sites. It is also reasonable for the agency to consider itself an interpreter of these remnants and sites. The 1972 plan obscures the degree to which the historical wing of the Park System had always been based on the availability of physical artifacts and sites. In fact, the plan appears to reverse things, designating an event or historical period as important, and then implying that the Park System is lacking if it does not include some remnant of the event or period. The difference is not merely semantic play, for the latter approach obscures the fact—and for the historical wing of the Park System, perhaps the central fact—that some important events are bound up with physical objects or key sites while others of equal importance are not, or are associated with physical objects or sites only in the most trivial ways. For example, both the battle of Gettysburg and the development of American pragmatist philosophy are events in our history worth knowing about and understanding. But while standing on the battlefield surrounded by the monuments might give one some understanding of the battle and its significance to later generations, standing in Benjamin Pierce's house is not likely to give one much understanding of pragmatist philosophy or of its importance. Pragmatist philosophy is preserved in the writings of pragmatist philosophers.

Some epochs have left remains that can be readily seen and appreciated, while others have left nothing. Because of climate and levels of material culture, the Indians of the Southwest left spectacular remnants, while those of the eastern woodland cultures did not. It matters little that the two might have been of equal importance or that an understanding of the two might be equally valuable. In only one case are artifacts available to aid the nonexpert in understanding the culture. Yet the agency is formally committed to the preservation and interpretation of remnants of both cultures.

The 1972 plan obscures differences in the relationship between past events and their material remnants. By doing so, it leads the agency to attempt to interpret pragmatist philosophy on wall plaques. The Park Service knows there is something wrong, sometimes comically wrong, with some of its attempts at interpretation, but it does not know the source of the trouble. It may lie in a failure to fully appreciate the fact that only part of the past lends itself to interpretation through physical remains and that it is this part that is the proper realm of the Park Service. Accepting this fact would make consideration of historical resources the full equal of considerations of historical significance.

Policy Modifications

The Park Service should reexamine its history plan, identify the grossly underrepresented categories, and ask itself if perhaps they are underrepresented not because of negligence but because no one in the past could find appropriate sites associated with them. The agency should then ask itself if perhaps the discrepancy between the 1972 plan and the System's holdings reflects poorly on the plan rather than the System.

Adherence to this double criterion of significance and interpretable remnants would go a long way toward settling what is probably the most acrimonious and long-standing split within the agency, that which pits the historians against much of the rest of the Service, especially its managers and interpretive specialists. It would allow the roles of both the historian and the interpretive specialist to be clearly defined. And although clear definition necessarily involves circumscription, it would allow considerable integrity for both. Since historical significance would no longer be the sole criterion for determining the proper shape of the historical wing, this would be a loss of formal authority for the historians, but as we have seen, their formal authority is usually hollow. Furthermore, this combination of considerable formal authority and little actual power is a source of much trouble: The historians are mindful of their weakness and act out of a feeling of impotence, while others, fearing the great formal authority of the professional historian, act out of that fear. With historical significance being only one of two evenly matched criteria, the rest of the Service might be more willing to listen to the historians when they speak. With the interpretation and preservation specialists involved in the most basic decisions about the historical wing, their roles would also have more integrity. To them would fall the questions concerning the appropriateness of preservation and the possibility of meaningful interpretation. They would no longer find themselves asked to preserve sites or artifacts when there is little or nothing left, or trying to use material remnants to teach lessons which cannot be taught through physical artifacts.

A second failing of the 1972 plan is that it did not settle the relationship between preservation and commemoration within the Park System's historic wing. In a sense, the agency tried to deal with the question by overtly ignoring commemoration. This by no means settled things, however; the commemorative impulse will not go away nor will the reminders of commemorative past of the historical wing of the Park System simply disappear. Some, perhaps most, of the System's major historic sites were originally established to commemorate something. This cannot be masked by such ploys as calling the Statue of Liberty a relic of late nineteenth-century immigration.[18] The official dominance of preservation over commemoration also will not mask the System's blatantly commemorative new units, such as the Martin Luther King park. Because of the agency's policies, commemoration remains officially a nonissue, and because of its limbo status, it cannot be addressed squarely. Granted, it is a difficult issue to address, but it must be addressed for two reasons.

Commemoration and Revisionism

First, an overemphasis on preservation presents a real danger of losing touch with those parts of the past that the National Park System now makes accessible. Let us look at Gettysburg again. The more preservation-minded in the agency would have the battlefield restored to the exact conditions at the time of the battle. One agency historian told the author that Gettysburg was a battlefield, not a park, and therefore mowed lawns were not appropriate there. He also wanted the cannons moved out of their ceremonial positions and back to where they were during the battle. He tolerated the monuments, but just barely. I am usually pretty bored in the presence of the past but I was deeply moved by Gettysburg, and especially by the great profusion of monuments to the fallen that it contained. On the snowy late winter day when I visited the park, the impact of the long line of statues and columns along the fence where the Union forces repelled Pickett's charge was overpowering. One can learn about the tactics and conditions of the battle in books; one can learn about the strategic importance of the battle from them as well. However, it would be hard to understand the importance of the battle to the American psyche unless one saw the outpouring of commemoration it prompted. The National Park Service tries to accommodate the fact that so much of the original purpose of the Gettysburg Park was commemorative by referring to it as an outdoor museum of bronze and stone. The agency thus sees itself as a preserver of historic resources, and in so doing assumes the role it finds most comfortable. This view misses the very essence of Gettysburg's value. By itself, most of the statuary is common, even second rate. It is the context that makes it important, and the context has been ignored so long that many in the Park Service can no longer see it. Something whose value cannot be seen is in danger of being lost.

Second, commemoration gives a legitimacy or a heightened sense of value to the event being commemorated. When this legitimacy and value is conferred by the U.S. government, a powerful signal is sent out. Since no judgment about the past is free of implicit judgments about the present, these signals are important in what they say about the government's attitudes and intentions toward the classes, races, and groups with special attachments to the events being commemorated. Thus the matter of commemoration in the National Park System is too important to be handled in an under-the-table manner, no matter how thorny a question it turns out to be if it is faced squarely.

A third issue in need of resolution is that of revisionism, that is, adjusting criteria of historical significance to include historical events, or even whole classes of events, which were ignored, forgotten, or misinterpreted in the past. The problem is similar to that of commemoration because here too decisions will necessarily confer respectability and value. If the Park Service assumes responsibility only for that part of the past that has left important remnants, part of the problem will be solved, if not philosophically, at least in a practical sense. The agency will no longer be responsible for making grand decisions about underrepresentation and misinterpretation of the past. (Which it is probably unqualified to do anyway.) It will only be required to make ad hoc judgments about specific sites. The value of this should not be underestimated, since with such a policy in effect, the Park Service will not be compelled by the logic of its comprehensive classification scheme to go looking for sites which reinterpret history. It will only be called on to make judgments about historical significance in cases where the remains are unquestionably important to interpreting the event. When this occurs, the agency will simply have to do the best it can, seek all the advice it can get, and make the most informed decision possible.

Its own professional historians can be of great help on those occasions when the Service does get drawn into questions of historical revision; they can make sure that revision is just that, and not surreptitious commemoration. The agency's waiting period for evaluating historical significance, discussed earlier, is a good one and can help it here. It is insurance against being swayed by the fads of the moment and it is a bulwark against pressure. The agency should respect it so that it will be unimpeachable when it is really needed.

Pork Barrel Parks

Finally, there is the matter of pork barrel historical parks. First, one should recognize that their money costs are not very high. Most of these parks are small in acreage and have small staffs and budgets. Often they were state historical parks or privately preserved sites which came into the System with little or no acquisition expense to the federal government. Perhaps one could argue that sites commemorating events of little significance trivialize the Park System and by doing so pose a threat to the public's esteem for the System. This seems unlikely, however. As we have seen, visitors seldom, if ever, ask questions of historical worth. Besides, the public does not often think of these parks when they think of the National Park System; it thinks of the great natural parks of the West.

In the best of all possible worlds, a Park System without many of the inconsequential historical parks would be a better one. It would be less expensive to run, and it would be more respectful of the nation's history. But the world is not an ideal place, and perhaps Capitol Hill is the least ideal place of all. The wheels of congressional operations need constant lubrication with pork barrel benefits, and parks are a relatively inexpensive benefit. Taken together, all the pork barrel parks in the National Park System probably cost a small fraction of what the Clinch River breeder reactor or Tennessee-Tombigbee waterway will eventually cost. Historical parks with little historical significance are also a relatively benign form of pork. Compare the potential for harm in preserving the house of an undistinguished signer of the Declaration of Independence with that of an Army Corps of Engineers project that tears up whole landscapes or that of a nuclear reactor which breeds plutonium. Historical pork barrel parks are quite innocuous, and if they substitute for malevolent forms of pork, they are a positive good. The Park Service should view them as such.

This does not mean that the Service has to be active in seeking out such sites, or encourage their addition to the System. The 1972 plan did just that, either by design or more probably inadvertently. The agency should accept them when it has to, and when they do come into the System, it should aim to keep their significance in perspective. The Service does not have to tell the visitor that the park is the result of politics and that nothing significant ever happened there, but it should resist the urge to overinterpret, to put on a big show. Park Service interpreters and graphic experts occasionally get carried away and let flashy displays give an undeserved air of importance to what is being interpreted, but this is the exception rather than the rule. By and large, Park Service interpretive planners do keep matters under control and in perspective and the agency as a whole acquits itself well in dealing with the pork barrel parks. It need do no more.

THE NEW RESPONSIBILITIES

Urban National Parks

When the Park Service came to the cities and made its commitments, it appeared to have a clear policy guide in the Great Society's determination to create a better life for urban dwellers through direct federal intervention in the problems of the cities. Urban National Parks were to be an element of this policy, and with these parks, both the System and the Service were to find a new political, and one might say moral, relevance. Today, however, there is uncertainty over the agency's role with regard to the nation's cities.

In retrospect, we see that the Johnson era was the end of a long period in which there seemed to be little confusion about what was right for the cities, or at least about the direction government should be taking. One has only to think of New York during the decades from the 1920s to the 1960s, the era of Robert Moses. What was right was progress, and that meant bridges and tunnels, highways, massive urban redevelopment projects, and large, accessible parks on suburban peripheries. It meant the encouragement of metropolitan economic expansion and physical growth. Opposition to these things, either to save an urban neighborhood from an arterial expressway or a rural landscape from obliteration through large-scale recreation development, was viewed as local obstructionism.[19] It was considered selfish and unbecoming, and it was publicly treated as such by prominent politicians and the popular press.

With the erosion of optimism about progress and spreading disillusion with what progress was doing to the cities, things became muddied. Antiurbanism, or at least an antimetropolitan sentiment, became respectable and invested opposition to the expansion of the city, or in some cases to its internal redevelopment, with a legitimacy it had not previously enjoyed.[20] One now saw a more equal combat for public approval between those who continued to promote urban expansion in the style of Robert Moses and those who opposed it.

For the Park Service, newly involved with the city in a major way, and charged with forging policy which would determine metropolitan land use and activity patterns, this competition created some fundamental problems. With whom should it cast its lot, which interest should it serve, and from which side should it take its policy guides? Should the parks be made into mass recreation facilities, major and integral parts of the metropolises, or should they be preserves for nature and the few, in short, refuges from the forces of urban growth and change? There was no avoiding these choices because as soon as the urban parks—Gateway, Cuyahoga Valley, Fire Island, Golden Gate, Santa Monica Mountains, and the rest—were established, they became the battlegrounds between metropolitanism and localism within their regions. Choosing sides was made more problematic for the agency by the fact that it could draw on its own traditions to justify either course, and by the fact that factions of support for each policy soon developed within the agency. As a result of these difficulties, the Park Service was unable to form a clear philosophical or moral view of its own role in the nation's metropolitan areas. Was it an agent of social redress through metropolitan expansion or a defender of all that such expansion destroyed? It could not dodge the issue; urban park planning necessarily involved making decisions that would further one side or the other.

The agency's strategy thus far, as we saw in chapter 6, has been one of expedience and bureaucratic *realpolitik*. It has given in to localism on many development issues (certainly it has done so in the two cases examined most closely here—Gateway's Breezy Point unit and the Fire Island National Seashore), while maintaining a more balanced rhetorical stance. Let us look at the circumstances which gave the Park Service the freedom it needed to pursue this strategy.

Generally speaking, two characteristics of an organization's goals will increase its freedom of action. The first is multiplicity, that is, having many goals which it may pursue. The more goals an organization has, the more opportunity it has for shift and substitution as needed.[21] The second is generality or ambiguity. When goals are fuzzy, they can be all things to all people, thus allowing an agency to avoid the conflicts which arise when goals are specific, and winners and losers are obvious. Ambiguous goals also allow latitude in implementation; an agency might resolve general goals into specifics of policy in the manner most advantageous to it at a particular time, and then adopt a different course of resolution under different circumstances.[22] The set of overall goals which the Park Service brought to the urban park planning process was rich in ambiguity and in the potential for substitution.

Promoting access to the nation's great historic sites and to its great natural areas has always been an important agency goal. Preserving the national treasures under its care has been an important goal also, so the use and preservation goals could be mixed in urban park planning.[23] The 1960s and 1970s also saw the rise within the Service of what might be called *engagé* goals, those of promoting environmentalism, social change, and equal access to life's benefits. With the acceptance of these as legitimate agency goals, education and demonstration became increasingly important in park planning throughout the System. The Park Service has undertaken programs to bring the disadvantaged of all stripe to the parks and to use the parks as sites for their education. The agency has also used the parks as environmental labs and public information centers. The roofs of visitor centers have sprouted solar collectors and lawns are cut with mowers fitted up to run on the alcohol products of biomass conversion. The Park Service has become a patron of the arts, and many parks have regular art exhibits. In several areas of the System, artist-in-residence programs have been established to allow artists to live, work, and teach full time in the parks. These *engagé* goals have added to the choice of goals which a park may be designed to pursue, thus conferring the advantage of flexibility on the agency when it is faced with conflicting demands in its urban parks.

The urban parks had a conceptual ambiguity not shared by other parks in the System. They were defined more strongly by notions of what they were not supposed to be than by what they were supposed to be. Although this presented planning problems and put great stress on the agency's creative capacity, it also conferred the advantage of flexibility that ambiguity usually confers. General concepts and principles could be refined into specific policies and actions in the most convenient manner in any set of circumstances. As a result, the agency had the flexibility to find its way through the rocks and whirlpools of urban politics and to accomplish something in the urban parks, even if sometimes, as was the case at Gateway, all that was accomplished was the park's survival (in which former director Whalen took such pride). But how are we to judge accomplishments of this sort? What should be done with these urban parks of such modest success?

Judging the Urban Parks

The entire urban wing of the National Park System was, in essence, an experiment. It was also a gamble; commitments were made in the hope that they would find "sustaining social conditions."[24] For a public agency, there is nothing inherently wrong with experimentation, with shifting resources from one area of social concern to another or from policies in accord with one perspective to those in accord with another. As society evolves, new concerns arise and old ones fade.[25] From a strategic point of view, it would be unwise for a public agency to ignore the opportunities for role change that these shifts present. From a societal perspective, it would be unfortunate if public agencies were unimaginative and ignored opportunities to respond to these changes.

While imagination and creativity in an organization can bring great rewards, they are also fraught with potential for error. Because new ideas have not been tried before, the chances that they are impractical are great.[26] This should not discourage exploration for new ideas or experimentation with new policies, but it should encourage agencies to be mindful of the possibility of failure, to be aware of what failure looks like (this is not as easy as it seems, given the propensity of organizations to accept weak indicators of success), and to be willing to write off failures when the time comes to do so.[27]

The Park Service has shown little willingness to back out of failures. Perhaps because many traditions and aspirations have been brought to the urban national parks, the Park Service can fabricate justifications for any stance it takes, decisions it makes, and choices forced on it. For example, it could rationalize its continued presence on Fire Island as a bulwark against development when it was forced to abandon any hope of making the park accessible to the public at large. Perhaps because its planning skills are so good, the temptation to cover failures with virtuoso planning and call them successes has been too great to resist. One need only think of the way it has used the Gateway development plan to cover its failure to get the necessary political support to establish a park which lived up to its promise.

Rationalizations, however, can get only so convoluted and artful before they collapse of their own weight. The agency must be willing to establish criteria for success and write off ventures which fail to live up to these criteria. Indiana Dunes is probably such a failure. It was doomed before it was even established, when U.S. Steel leveled the high dunes and built a steel mill there.[28] Preserving the site on which early experiments in plant succession were conducted, a justification for the park offered by agency personnel, is not sufficient reason for a national park with a steel mill in the middle of it. Fire Island is also without doubt a failure. It is grasping at straws to say that it is preserving the barrier island from development. The island has no nationally significant characteristics which qualify it for inclusion in the System, not even by the agency's own liberal and elastic standards. It does not contain ecosystem types unrepresented elsewhere in the System. It is not biologically or geologically unique.[29] The argument has been advanced that it is culturally unique, but if so, it is because of the high concentration of the very rich in some of the towns on the island and the high concentration of homosexuals in others. This is hardly the kind of cultural uniqueness Albright or Mather, or anyone else for that matter, had in mind preserving in the National Park System. Fire Island is not accessible to the public at large; furthermore, the development cap imposed on it means it is unlikely to be so in the future. The agency cannot even control private development on the islands. It has neither the money for condemnation nor the power to attain this end through regulation. The Park Service should fold its tent and slip away across Great South Bay, leaving the island residents to preserve their little enclaves by their own efforts, not at national expense.

More troublesome for the policy analyst, however, are those urban parks which cannot be written off as unambiguous failures. Gateway is an example. Clearly, it has fallen far short of its original goals, but it is not obvious that those goals are unattainable. Here the agency is facing a basic problem in uncertainty. The park's ultimate success will depend on circumstances over which the Park Service has no more than minimal control and very little capacity to predict. They include future attitudes toward our cities, the urban policy of the next administration, and even future transportation innovations (or the lack of them). There is no way of knowing now whether future events will continue to make it increasingly difficult for the Park Service to attain its goals in its urban parks or whether the tide will turn. The latter seems unlikely, but it is by no means impossible.

Given this uncertainty, the Park Service should not lose sight of why it is in the urban parks business in the first place; that is, to provide access to the special recreation resources of the region. It would be easy to lose sight of this; all the agency has to do is start believing that protecting the modest natural and historic resources found in the urban national parks justifies its continued presence. It does not. The only thing that justifies its continued presence in the urban parks it currently manages is a reasonable hope that it can someday turn them into what they were meant to be. Only if the Park Service keeps this in mind and at the very center of its urban presence will it be willing to continually assess its urban parks and decide whether the borderline cases are moving toward more complete success or are becoming unredeemable failures. The jury is still out on many of the urban national parks, but when it returns, the agency must be willing to listen to the verdict, understand it, and act accordingly.

Beyond Park Boundaries

As was the case with the urban parks, the agency's recent involvement with land beyond park boundaries drew it into the center of a national controversy and funneled discord into agency ranks. Although the agency's concern occasionally went beyond the boundaries of the national parks during the early days of the Park Service, any efforts which extended beyond the management of fully owned public park land were incidental and tactical and did not need to be accommodated in any overarching sense of mission. The National Park Service was free to be "the best damned park manager in the world."

It was in part the agency's isolation behind the drawn wagons of Mission 66 which led to Wirth's removal and to the strong rebuke of agency leadership that the establishment of the Bureau of Outdoor Recreation represented. It was the same involuted sense of responsibilities that made the agency appear to be an unsuitable policy instrument when the New Conservation made recreation opportunities and accessible nature in everyday life an important goal. Thus it was not surprising that agency leadership in the modern era developed an increased concern for the fate of land beyond the national parks as part of its search for relevance from the mid-1960s onward and eventually came to see in regional planning and private land use regulation an opportunity for profitable new additions to the Park Service's range of responsibilities.

Reading the Times

In this policy area as well as the others, however, the times proved difficult to read, and the federal interest in private land development, which looked like the wave of the future in the late 1960s and early 1970s, became considerably less certain as the 1970s turned into the 1980s. Thus, as was the case with its involvement with urban parks, the Service found itself in the middle of conflicts of interests and values with uncertain outcomes, and as a consequence it was unable to develop an effective, or even clear, role for itself. Today there are unknowns and perhaps unknowables which haunt the agency's efforts to formulate coherent policies for dealing with land beyond that which it owns. First, as was mentioned above, the Jackson land use bill, which was the centerpiece of the efforts to establish a permanent federal presence in land use regulation, was defeated in 1974 and was never resurrected. For the Park Service, the proper interpretation of the cause of this defeat and understanding its long-term implications are of primary importance.

It is possible to view federal interest in private land use regulation as a logical extension of the National Environmental Policy Act and of the federal concern for the environment which NEPA expressed.[30] It is also possible to see it as a logical extension of the progressive tradition, in which the federal government responded to and accommodated the forces of modernism by clearing away the arcane obstructions and allowing rationality to usher in a better future for all. Indeed, Reilly and others seem to view the need for state and federal land use regulations in just such progressive, rationalist terms. They talk about the "new spirit" abroad in the land and about the damage local governments cause with their anti-quated rules for land use.[31]

Within such a view, Park Service involvement in land use planning could be seen as a logical extension of the agency's original mission. It was a response to the needs of late twentieth-century American society in the way the preservation of parts of the public domain was a response to society's needs earlier in the century. Such a perspective not only makes land use control appear to be a logical extension of the agency's original mission, but it also imparts an element of inevitability to it. The defeat of the Jackson land use bill in 1974 could be seen as a parallel to the defeat of the early efforts to establish the Park Service itself; that is, as the neces-sary preliminary gropings before the formula for success was found.

Looked at in this light, agency involvement in land use control is not only proper and logical, it is also fine strategy. If federal land use planning is the wave of the future (and many agency personnel feel it is), then a role for the Park Service in this planning puts it on the side of progress and helps it redefine its mission in a relevant, modern context. Also from this perspective, the defeat of the Jackson bill and the broad-based approach to planning it embodied was a blessing in disguise for the agency, since it cleared the way for the greenline approach as an alternative. It meant that the path to the future ran right past the agency's front door.

Convincing arguments (made even more convincing by the passage of time) have been advanced that the Jackson bill was not a harbinger of things to come or a preface to the inevitable, but rather a high-water mark from which the tide of interest and support ebbed, perhaps never to re-turn.[32] If this is correct, then the prognosis for the success of agency involvement with greenline parks, or any extensive involvement in land use beyond park boundaries for that matter, is very different. Such in-volvement would be far more likely to lead to trouble than to respect and relevance. Promoting greenline parks would be bucking the current of the times rather than running with it.

There is another element of uncertainty the agency faces when it enters land use planning. The success of its ventures in this area undoubtedly will also depend on the pattern of judicial decisions which emerges on the takings issue. One of the most disputed parts of the U.S. Constitution is what is commonly known as the takings clause, the passage in the fifth amendment which reads "nor shall private property be taken for public use without just compensation." Exactly what constitutes taking private prop-erty, or simply "a taking" to use judicial parlance, is not further defined and remains unclear. When the government requires a tract of land for an expressway and moves through eminent domain to acquire it, it is unquestionably a taking for which just compensation must be made. There is also a judicial tradition of interpreting the takings clause to exclude normal town zoning (since such zoning is in theory undertaken in the interest of the property holders rather than the public at large). There is a large gray area between these two clear cases and into it falls most public land use restriction beyond that of simple town zoning. In general, the rule is that if the restrictions government has placed on an owner's options are sufficiently severe, then a taking has occurred and the owner is entitled to compensation. This general rule has not led to any consistent pattern of specific interpretation, however, and the result is uncertainty about how far restriction can go before compensation must be paid.

The viability of regional planning efforts such as the greenline park will depend on the emergence of a clear pattern of judicial interpretation which extends the justificatory logic of zoning to other types of land use regulation. If such a pattern does not emerge, the greenline parks, including those which have already been established, will remain largely unworkable and the agency will be forced further into compensatory forms of protection, such as the purchase of easements and outright acquisition. Only time will tell how this larger issue will resolve itself, but in the meantime the Park Service is stranded in its commitments. Many major units with greenline features have come into the National Park System, but the agency has neither the effective tools at its disposal to make them work, nor the unambiguous support of Congress, to say nothing of the Reagan administration, for its aims of making them work. As federal budget problems increase, the agency is even losing the threat of condemnation as an instrument of land use control.

This present ineffectiveness, coupled with uncertainty over the future, raises, or at least heightens, internal differences over the wisdom of a large-scale commitment to the greenline park. (There is no real dissension over the need to protect parks from deleterious development on their boundaries.)

Agency Responsibility

I would argue that even given these uncertainties, the extension of the Park Service's interest beyond park boundaries is justified by the possibility of overcoming the denaturization and the dehumanization of the landscape on which most of us lead our daily lives. This does not imply a Luddite attitude or a return to a preindustrial past; rather, it is a call for the Park Service to lend a hand in smoothing down the rough edges of the landscape of modern life. Helping bring about a landscape on which nature is woven into everyone's life, where the changes undertaken for profit are publicly controlled and well thought out, and where the scale of development is kept close to human needs is as admirable a goal for the Park Service as any it could pursue.

Admirability is not enough, however. As we have seen above, there are real obstacles to the success of any move into this area of policy, and great uncertainty as to whether these obstacles will be diminished by future events. Given this, the Park Service must adopt a strategy which takes this uncertainty into account and minimizes the impact of unpleasant turns of fate on its commitments. At the same time, it should maintain its ability to capitalize on favorable occurrences and, where possible, help make them happen. Specifically, the Park Service might continue to try to make the greenline parks in the System work. It should abandon its wishful thinking about federal, local, and state cooperation and it should not paper over the failures of these parks to live up to their promise any more than it should oversell its successes with the urban parks. Instead, it should make clear the extent of their shortcomings and make the problems explicit; to do otherwise would only add to the problem. Also, as things now stand, the agency probably should oppose the creation of new parks with greenline characteristics if they do not have effective land use controls. Supporting them only continues the illusion that the approach is working well. It is not. It should be emphasized that saying no to unworkable greenline parks and owning up to failures is not the same as backing out of commitments. If anything, such a strategy will focus attention on the problems of the greenline parks and the approach to rational land planning that they represent. This will increase their chances of eventual success.

Before the Park Service makes any more commitments to land use planning through such concepts as the greenline park, there must be a broader federal commitment on which to base it. The Park Service can help promote that commitment, but it cannot do much without it.

LOOKING AHEAD

In closing, I might suggest (again?) that the agency probably will never find a sense of mission as coherent, with such visionary appeal, or as successful a strategic guide as the one Mather and Albright fashioned for it. The forces of emerging modern society in the early twentieth century and the notion of progress which accompanied it had the power and charm to give the era a uniquely coherent vision of the good life, and it was on this vision that the agency's original sense of mission was founded. We are not likely to see such a coherent and persuasive vision again, but this need not condemn the Park Service to relic status as an agency which has outlived its usefulness. Exactly the contrary is the case. A creative Park Service with equal dedication to experimentation in meeting new social demands, to a tough pragmatism in evaluating those experiments, and to protecting the integrity of the System entrusted to its care would remain an important part of the federal government and would ensure that the national parks remain an important part of American life.

APPENDIX A. BACKGROUND INFORMATION

I. *General*

The National Park Service is part of the U.S. Department of the Interior. As of August 20, 1983 it had 11,297 permanent employees. It had an FY 1983 budget of 909 million dollars. The Park Service is headquartered in Washington, D.C. It has nine regional headquarters, each with first-line responsibility for the parks in its region and each under a regional director. The Park Service has two major service centers which supply planning, design and interpretive services to the entire system. One is in Denver, Colorado, and the other is in Harpers Ferry, West Virginia.

II. *Formal Expression of Agency Purpose*

The agency's manual, U.S. Department of Interior, National Park Service, *Department Manual*, Part 145, p. 1.2, gives a formal expression of agency purpose when it names five "long-range objectives." The first is: "To conserve and manage for their highest purpose the natural, historical, and recreational resources of the National Park System."

The second and third long-range objectives bear directly on the National Park System. One directs the Park Service to make the system accessible and attractive: "To provide for the highest quality of use and enjoyment of the National Park System by increased millions of visitors in the years to come." The other directs it to expand the system: "To develop the National Park System through inclusion of additional areas of scenic, scientific, historical, and recreational value to the nation." The fourth calls on the Park Service to use the System as a base from which to educate the public:

"To communicate the cultural, natural, inspirational and recreational significance of the American heritage as represented to the National Park System." The final one directs the agency: "To cooperate with others to protect and perpetuate natural, cultural, and recreational resources of local, state, regional, and international importance for the benefit of humankind."

III. Types of Units in the National Park System

The National Park Service's *Index, National Park System and Related Areas, 1979*, pp. 7–8, gives the following definitions and explanations of the various types of units in the National Park System:

Generally, a national park covers a large area. It contains a variety of resources and encompasses sufficient land or water to ensure adequate protection of the resources.

A national monument is intended to preserve at least one nationally significant resource. It is usually smaller than a national park and lacks its diversity of attraction.

The national preserve category is established primarily for the protection of certain resources. Activities such as hunting and fishing or the extraction of minerals and fuels may be permitted if they do not jeopardize the natural values.

The national lakeshores and national seashores focus on the preservation of natural values while at the same time providing water-oriented recreation. Although national lakeshores can be established on any natural freshwater lake, the existing four are all located on the Great Lakes. The national seashores are on the Atlantic, Gulf, and Pacific coasts.

National rivers and wild and scenic riverways preserve ribbons of land bordering on free-flowing streams which have not been dammed, channeled, or otherwise altered by man. Besides preserving rivers in their natural state, these areas provide opportunities for outdoor activities such as hiking, canoeing, and hunting.

In recent years, national historic site has been the title most commonly applied by Congress in authorizing the addition of [historical] areas to the National Park System. A wide variety of titles—national military park, national battlefield park, national battlefield site, and national battlefield— have been used for areas associated with American military history. National historical parks are commonly areas of greater physical extent and complexity than national historic sites.

The title national memorial is most often used for areas that are primarily commemorative. But they need not be sites or structures historically associated with their subject. For example, the home of Abraham Lincoln in Springfield, Illinois, is a national historic site, but the Lincoln Memorial in the District of Columbia is a national memorial.

Originally, national recreation areas in the Park System were units surrounding resevoirs impounded by dams built by other federal agencies. The National Park Service manages many of these areas under cooperative agreements. The concept of recreational area has grown to encompass other lands and waters set aside for recreational use by acts of Congress and now includes major areas in urban centers. There are also national recreation areas outside the National Park System that are administered by the Forest Service, U.S. Department of Agriculture.

National parkways encompass ribbons of land flanking roadways and offer an opportunity for leisurely driving through areas of scenic interest. They are not designed for high speed point-to-point travel.

Two areas of the National Park System have been set aside primarily as sites for the performing arts. These are Wolf Trap Farm Park for the Performing Arts in Virginia, and the John F. Kennedy Center for the Performing Arts in Washington, D.C. Two historical areas, Ford's Theatre National Historical Site, Washington, D.C., and Chamizal National Memorial, Texas, also provide facilities for the performing arts.

IV. Composition of the National Park System

As of November 11, 1983, the National Park System consisted of the following:

Type of Area	Number of Areas
National parks	48
National historical parks	26
National monuments	78
National military parks	10
National battlefields	10
National battlefield parks	3
National battlefield sites	1
National historic sites	63
National memorials	24
National seashores	10
Parkways	4

continued

Type of Area	Number of Areas
National lakeshores	4
National rivers	11
National capital parks	1
Parks, other	10
National recreation areas	17
National scenic trail	1
National preserves	12
National Mall	1
White House	1
National Park System, total units	335

APPENDIX B. INTERVIEWS

Albert, Lewis, park manager, Cuyahoga Valley NRA, 3/4/81, Peninsula, Ohio.

Babb, Frederick, planner and landscape architect, NPS, 12/11/80, Denver, Colorado.

Beatty, Laura, cultural resources specialist, NPCA, 9/19/80, Washington, D.C.

Bevinetto, Tony, staff, Senate Committee on Energy and Natural Resources, 11/11/80, Washington, D.C.

Biderman, George, chairman, Fire Island National Seashore Advisory Committee and Fire Island Association, 12/22/80 et passim, New York, New York.

Billings, Linda, former staff member, Sierra Club, Washington office, 9/2/80 et passim, Washington, D.C.

Binneweis, William, superintendent, Morristown NHP, 2/9/81, Morristown, New Jersey.

Bowman, Ann, program analyst, NPS, 11/10/80, Washington, D.C.

Brown, Barbara D., program manager, NPS, 3/16/81 (phone) Washington, D.C.

Butowsky, Harry, historian, NPS, 9/19/80 et passim, Washington, D.C.

Chapman, Howard, western regional director, NPS, 12/3/80, San Francisco, California.

Clawson, Marion, special consultant, Resources for the Future, 9/8/80 et passim, Washington, D.C.

Curry, Richard, special assistant to the director, NPS, 9/3/80 et passim, Washington, D.C.

Davis, Jack, deputy western regional director, NPS, 11/28/80 San Francisco, California.

Dickenson, Russell, director, NPS, 4/3/81, Washington, D.C.

Eck, Arthur, legislative assistant, NPS, 2/3/81 (phone), Washington, D.C.

Edwards, Robert, consultant to the NPS and former chief of staff, Interagency Advisory Council on Outdoor Recreation, DOI, 9/8/80, Washington, D.C.

Egan, Paul, legislative officer, American Legion, 1/30/81 (phone), Washington, D.C.

Ehorn, William, superintendent, Channel Islands NP, 12/5/80 et passim, Ventura, California.

Evison, Boyd, superintendent, Sequoia NP, 11/26/80, Sequoia NP, California.

Fries, Nancy, ecologist and park planner, Santa Monica Mountains NRA, 12/4/80, Woodland Hills, California.

Galvin, Denis, director, NPS Denver Service Center, 11/20/80 et passim, Denver, Colorado.

Gardiner, George, biologist, NPS, 9/18/80, Washington, D.C.

Giambardine, Richard, supervisory landscape architect, NPS, 11/20/80 et passim, Denver, Colorado.

Good, John, former superintendent, Everglades NP, 1/8/81, Homestead, Florida.

Harrison, Thomas, chief of planning and resource management, Gettysburg NMP, 3/6/81, Gettysburg, Pennsylvania.

Hart, Judy, legislative specialist, North Atlantic Regional Office, NPS, 1/27/81 (phone), Boston, Massachusetts.

Hasbury, Jeanne, administrator, HCRS, 9/10/80 (phone), Washington, D.C.

Hauttman, Jack, chief, Division of Natural Resource Systems, HCRS, 9/25/80, Washington, D.C.

Hawkins, Albert, superintendent, Delaware Water Gap NRA, 2/10/81, Delaware Water Gap NRA, Pennsylvania.

Hendrix, Gary, research director, South Florida Research Center NPS, 1/28/81 (phone), Everglades NP, Florida.

Herbst, Robert, former assistant secretary of the interior, 4/7/81, Washington, D.C.

Hill, Michael, chief ranger, Channel Islands NP, 12/6/80, Ventura, California.

Holland, Ross, acting assistant director for cultural resources, NPS, 9/30/80, Washington, D.C.

Hornbeck, Kenneth, chief, NPS Denver Service Center Statistical Unit, 11/21/80 et passim, Denver, Colorado.

Humphrey, Donald, program analyst, NPS, 10/29/80 et passim, Washington, D.C.

Ingram, Jeffrey, Arizona environmental activist, 12/8/80, Tucson, Arizona.

Jones, R. Arthur, legislative assistant, NPS, 9/17/80 et passim, Washington, D.C.

Kaplan, Susan, policy analyst, Office of Policy Analysis, DOI, 9/19/80 et passim, Washington, D.C.

Kriz, Willis, chief, Land Resources Division, NPS, 9/14/82 (phone), Washington, D.C.

Lautz, Henry, staff, Appalachian Trail Conference, 9/10/80 (phone), Harpers Ferry, West Virginia.

Levy, Benjamin, historian, NPS, 10/1/80 et passim, Washington, D.C.

Libengood, Richard, ranger, Grand Canyon NP, 11/24/80, Grand Canyon NP, Arizona.

Lienesch, William, administrative assistant, NPCA, 3/20/81 (phone), Washington, D.C.

Little, Charles, editor, *American Land Forum* and formerly of the Congressional Research Service, 3/16/81 (phone) et passim, Washington, D.C.

McCone, John, staff, Sierra Club, Washington Office, 10/3/80, Washington, D.C.

Marks, Richard, superintendent, Grand Canyon NP, 11/24/80, Grand Canyon NP, Arizona.

McCann, Theodore, park planner, NPS, 3/17/81, Washington, D.C.

Milne, Robert, chief, International Park Affairs Division, NPS, 10/29/80, Washington, D.C.

Mintzmyer, Lorraine, Rocky Mountain regional director, NPS, 11/21/80, Denver, Colorado.

Moody, Joan, assistant editor, *National Parks and Conservation Magazine,* 9/3/80 et passim, Washington, D.C.

Moorehead, Jack, superintendent, Everglades NP, 1/8/81, Everglades NP, Florida.

Mortimer, Ronald, consultant to the NPS and former park planner, 12/2/80, San Francisco, California.

Nelson, Susan, environmental activist and member, Santa Monica Mountains NRA Advisory Commission, 12/4/80, Woodland Hills, California.

Olsen,Russell, chief, Management Consulting Division, NPS, 10/9/80 et passim, Washington, D.C.

Pickelner, Joel, superintendent, Fire Island National Seashore, 12/24/80, Patchogue, New York.

Pinner, Clementine, chief, Training Division, NPS, 9/10/80 (phone), Washington, D.C.

Pinnox, Cleveland, staff, House Interior Subcommittee for Parks and Insular Affairs, 10/10/80 et passim, Washington, D.C.

Prichard, Paul, executive director, NPCA, 10/6/80, Washington, D.C.

Reynolds, John, assistant superintendent, Santa Monica Mountains NRA, 12/3/80 et passim, San Francisco and Thousand Oaks, California.

Ritche, David, Appalachian Trail Conference liaison officer, NPS, 10/10/80, Washington, D.C.

Rittersbacker, David, staff, Council on Environmental Quality, 9/10/80, Washington, D.C.

Ryor, Elton, legislative officer, NPS, 1/27/81 (phone) et passim, Washington, D.C.

Sagan, Mark, director, NPS Harpers Ferry Service Center, 4/7/81, Harpers Ferry, West Virginia.

Sheaffer, C. Bruce, supervisory budget analyst, 9/24/80 (phone), Washington, D.C.

Sherman, David, park ranger, National Capital Region, NPS, 11/6/80, Washington, D.C.

Sherwood, Susan, historian, NPS, 3/20/81 (phone), Washington, D.C.

Siehl, George, analyst and public lands specialist, Congressional Research Service, 2/17/80 et passim, Washington, D.C.

Snape, Dale, budget examiner, Office of Management and Budget, 9/15/80,Washington, D.C.

Spratt, Mike, outdoor recreation planner, NPS, 12/11/80, Denver, Colorado.

Strumpf, Emmanuel, public affairs officer, Gateway NRA, 1/27/81 (phone), New York, New York.

Tipton, Ronald, staff, The Wilderness Society, 9/30/80 et passim, Washington, D.C.

Trombello, Lawrence, interpretive specialist, Allegheny Portage NHS, 3/5/81, Allegheny Portage NHS, Pennsylvania.

Utley, Robert, former chief historian, NPS, 12/10/80, Santa Fe, New Mexico.

Wauer, Roland, park manager, NPS, 2/3/81 (phone), Washington, D.C.

Weeks, Nicholas, landscape architect, NPS, 12/1/80, San Francisco, California.

Williams, Lawrence, staff, Council on Environmental Quality, 9/10/80, Washington, D.C.

Wilson, Betty, administrator, HCRS, 9/10/80 (phone), Washington, D.C.

Wittpen, Richard, park planning specialist and landscape architect, NPS, 11/20/80, Denver, Colorado.

Whalen, William, former Director, NPS, 12/1/80, San Francisco, California.

Wright, David, chief, Office of Park Management and Environmental Quality, 11/12/80 et passim, Washington, D.C. and Denver, Colorado.

Wright, Robert, consultant to the NPS, 1/19/80, Denver, Colorado.

Zeisman, Joe, Fire Island Wilderness Coalition, 1/30/81 (phone), New York, New York.

Notes

Chapter 1. Introduction

1. Mati Tammaru, quoted in Novogrod, Dimock, and Dimock, *Casebook*, p. 101.

2. Cahn, "Park Service," p. 20.

3. Rowntree, Heath, and Voiland, "United States National Park System," p. 121.

4. Holden, "Imperialism in Bureaucracy," p. 64.

5. Coopers and Lybrand, *"Management Improvement Project,"* p. I-40.

6. Ibid.

7. PL 89-665.

Chapter 2. The First Fifty Years

1. Carhart, in ORRRC, Study Report no. 27, p. 108, writes that: "Albright's work may be less known, but he was the one who carried the workload of organization, planning, and installing administrative staffs. Mather and Albright were a team, each contributing exceptional qualities." Both men have been subjects of thorough, solid biographies. For Mather, see Shankland, *Steve Mather*, and for Albright, see Swain, *Wilderness Defender*.

2. Sax, "America's National Parks," p. 67.

3. Yard, "Historical Basis," p. 7.

4. Lee, *Antiquities Act of 1906*, see especially chapter 4, "Vandalism and Commercialization of Antiquities, 1890–1906," pp. 29–38, and chapter 5, "Temporary Protection of Ruins," pp. 39–46.

5. See Ise, *Our National Park Policy*, pp. 136–142, for a discussion of the circumstances surrounding the establishment of Platt, Sullys Hill, and Wind Cave national parks.

6. Sullys Hill National Park was converted into a wildlife refuge in 1931.

7. Nash, "American Invention."

8. Leonard A. Salter, quoted in Peffer, *Closing of the Public Domain*, p. 326.

9. McConnell, "Conservation Movement," p. 463.

10. Nienaber, *Differentials in Agency Power*, p. 42.

11. McConnell, "Conservation Movement," p. 464.

12. Hays, *Conservation*, see especially chapter XIII, "The Conservation Movement and the Progressive Tradition," pp. 261–276.

13. Ibid., p. 265.

14. Peffer, *Closing of the Public Domain*, p. 319.

15. Ibid., p. 103.

16. Ibid., p. 318. Peffer points out that although the reservation of large parts of the public domain in public ownership was a major issue of the progressive era, large-scale alienation of the public domain continued into the 1920s.

17. Schmaltz, "Raphael Zon," p. 88.

18. McConnell, "Conservation Movement," p. 467.

19. Gifford Pinchot, quoted in McConnell, "Conservation Movement," p. 464.

20. Muir, *Our National Parks*, p. 76. This is not to say that Pinchot was incapable of flights of near poetic expression. In his *Breaking New Ground* he wrote that:

> The Yosemite Fall, with its 1,700 feet of drop, was worth crossing the continent to see. As the wind waved the falling water back and forth across the face of the great cliff, it left me nearly stunned with amazement. And stunned with water too, when like the boy I was, I ran under it, and it beat upon me at the end of more than a quarter of a mile of fall. Such a maelstrom of wind and water I was never in before or since. It was a great experience.

He never seemed to let such feelings guide his official actions, however.

21. Ibid., p. 2.

22. Ibid., p. 1.

23. Horace McFarland, "Address," quoted in Foss, ed., *Recreation*, p. 636.

24. Ibid.

25. Schrepfer, "Conflict in Preservation," p. 65.

26. For a detailed treatment of the establishment and management of the Civil War battlefield parks by the Army, see Lee, *National Military Park Idea*.

27. Organic Administration Act of June 4, 1897 (30 Stat. 34–36, 43, 44, 1897). Cited and discussed in Nature Conservancy, *Preserving Our Natural Heritage*, vol. I, p. 123. For a synopsis of early agency attitudes and policies, see Robinson,

Forest Service, chapter 1, especially pp. 1–10. See also Carhart, in ORRRC, Study Report no. 27, p. 109.

28. Nienaber, *Differentials in Agency Power*, p. 55; Lindblom, "Policy Analysis," p. 306; and Wildavsky, *Politics of the Budgetary Process*, pp. 166–167, have all noted the tendency for different values to become institutionalized in their own government unit, making that unit a watchdog for those values within the political process. Undoubtedly Hetch Hetchy convinced those interested in the preservation of the national parks that their values needed such a representative within government. Ise, *National Park Policy*, p. 186, also suggests that an agency responsible for the entire park System might serve as a check against the addition of insignificant parks.

29. President William H. Taft, Special Message to Congress, February 2, 1912, quoted in Foss, ed., *Recreation*, p. 182.

30. Seidman, *Politics*, p. 141. See Peffer, *Closing of the Public Domain*, chapter 9, "The Problem of the Remnants," pp. 169–180, for a discussion of the ambitions of Interior during the period under discussion.

31. National Park System Organic Act, August 25, 1916 (39, Statutes at Large, 535).

32. Long, "Power and Administration," p. 257.

33. Kaufman, *Are Government Organizations Immortal?*, p. 52.

34. "Administrative Politician"—Holden, "Imperialism in Bureaucracy," p. 944; "Managerial Activism"—Rehfuss, *Public Administration*, p. 228.

35. Long, *Polity*, p. 53.

36. See Downs, *Inside Bureaucracy*, chapter 2, "The Life Cycle of Bureaus," pp. 5–10, for a discussion of the advantages a newly created bureau will possess in its struggle to gain a secure place itself.

37. Rourke, *Bureaucracy*, p. 11.

38. For an example of such friction between the Park Service and its supporters, see Stratton and Sirotkin, *Echo Park Controversy*.

39. Holden, "Imperialism in Bureaucracy," p. 944.

40. Selznick, *TVA*, p. 48, writes that while the search for ideology is general among organizations, it will be most intense "among those born of turmoil and set down to fend for themselves in undefined ways among institutions fearful and resistant."

41. Huth, *Nature and the American*, p. 191.

42. ORRRC, Study Report no. 27, p. 107. See Cleveland, "National Forests," for a discussion of recreation in the national forests even before Graves put the Forest Service leadership's seal of approval on such use.

43. Carhart, in ORRRC, Study Report no. 27, p. 109.

44. Kaufman, "Natural History," p. 142; Downs, *Inside Bureaucracy*, p. 10; Rourke, *Bureaucracy*, p. 35.

45. Downs, loc. cit.

46. Graves, "Crisis in National Recreation."

47. Sax, "America's National Parks," pp. 67–68.

48. Shankland, *Steve Mather*, p. 263. Also see photo opposite p. 275. Before his term as president, Hoover had served as the head of the NPA, so he had already been enlisted as a friend of the National Park System. His policies toward the System (see Ise, *Our National Park Policy*) illustrate the degree to which he remained a friend.

49. Shankland, *Steve Mather*, p. 86. Kaufman, *Are Government Organizations Immortal?*, p. 18, writes that "The normal oscillations of power between the parties and among factions within the parties are yet another potential quicksand. It stands to reason that the proud organizational instruments of one administration will loom in the eyes of the victors as tools of, and monuments to, the predecessors just vanquished at the polls." The Park Service was lucky to have escaped this fate during its early years.

50. Merrium, "National Park System," p. 5.

51. Ibid. As will be seen in chapter 4, the NPA did not always see eye to eye with agency leadership, especially when it came to visitor accommodation. In this matter, the NPA tended to be less development oriented than early agency leadership.

52. Yard, "Historical Basis," p. 3.

53. See Shankland, *Steve Mather*, p. 95; and Swain, *Wilderness Defender*, p. 124, for a discussion of these efforts to court business interests.

54. Shankland, op. cit., p. 145.

55. Rourke, *Bureaucracy*, p. 138, suggests that the best way for administrators within the federal bureaucracy to win congressional favor is to gain the support of powerful interests outside government which those in government must respect.

56. Swain, *Wilderness Defender*, p. 189.

57. Ise, *Our National Park Policy*, pp. 321–322.

58. Rourke, *Bureaucracy*, p. 138.

59. Edelman, *Symbolic Uses of Politics*, also writes of these two strategies. He has argued that the two will usually be found in a complementary relationship.

60. Yard, "Historical Basis," p. 6.

61. Caro, *Power Broker*, chapter 24, "Driving," pp. 468–495.

62. Shankland, *Steve Mather*, pp. 196–197.

63. Rehfuss, *Public Administration*, p. 4. Simon, Smithburg, and Thompson, *Public Administration*, p. 385, write of the value of public support that: "Although public opinion is usually not as influential as are pressure groups in affecting an agency's programs, still an enthusiastic general public support could probably not be negated by any pressure group. Likewise a general public hostility would

probably result in abolition or substantial modification of any governmental program."

64. Stratton and Sirotkin, *Echo Park Controversy*, p. 17.

65. Caro, *Power Broker*, p. 143.

66. Lee, *Family Tree*, p. 18.

67. Franklin K. Lane, letter to Mather dated May 13, 1918, reprinted in Foss, ed., *Recreation*, pp. 185–189. Ise, *Our National Park Policy*, p. 195, writes that the letter was "doubtless written in cooperation with Mather, or more likely written by Mather." In any case, it embodied Mather's management principles.

68. *American Forests*, "Former Directors Speak Out," p. 50.

69. Franklin K. Lane, letter to Mather dated May 13, 1918, reprinted in Foss, ed., *Recreation*, pp. 185–189.

70. Shankland, *Steve Mather*, p. 65.

71. Ibid., p. 161.

72. Ise, *Our National Park Policy*, p. 244.

73. Shankland, *Steve Mather*, p. 183.

74. *American Forests*, "Former Directors Speak Out," p. 27.

75. Mather and Albright also shared the progressive era's suspicion of politically appointed government officials. They made strenuous efforts to remove politically appointed personnel from the parks which the new agency inherited and to institute high professional standards for incoming Park Service employees. For more on this point, see Shankland, *Steve Mather*, pp. 243–256.

76. Caro, *Power Broker*, p. 243.

77. Stratton and Sirotkin, *Echo Park Controversy*, p. 17.

78. Ise, *Our National Park Policy*, p. 646.

79. Downs, *Inside Bureaucracy*, p. 9.

80. On this point see Starbuck, "Organizational Growth."

81. ORRRC, Study Report no. 27, p. 115.

82. Robinson, *Forest Service*, pp. 154–158. Also see Nash, *Wilderness*, pp. 200–237.

83. Ise, *Our National Park Policy*, p. 644.

84. ORRRC, Study Report no. 27, p. 113.

85. Ibid., p. 113. The National Park Service was aided in its strategic maneuvering by internal divisions within the Forest Service on its own proper mission. Within the USFS, there remained true disciples of Pinchot, utilitarians who saw recreation as a frivolous diversion from serious agency business, or who saw recreation as incompatible with producing timber and therefore best left to the National Park Service. Moreover, some forest managers felt that recreationists only increased the incidence of fire, so to actually encourage them to enter the

forests was pure foolishness. The Forest Service also had its allies to consider. Undoubtedly, forest product users viewed any shift from commodities production with the same uneasiness as did the National Park Service. Thus there were limits to how far the Forest Service could go in managing its lands for preservation and recreation before the costs in internal dissension and constituency dissatisfaction became great. Nevertheless, these restraints must have been little comfort to Mather and Albright. Since the Forest Service had many times more land under its jurisdiction than the Park Service, and was a much bigger agency, it did not have to move very far toward preservation and recreation to become a serious threat to the Park Service. It could consign most lands and agency resources to "no-nonsense" production of forest products and still have enough men, money, and land left to stake a claim to being the National Park Service's equal in recreation and preservation.

86. Holden, "Imperialism in Bureaucracy," p. 945, writes that "in the nature of government, there is an almost inevitable mixture of missions between agencies." This suggests that the mission purity, and hence the functional distance, which Mather managed to keep between the two agencies was unusual and likely to be transitory. Clark and Wilson, "Incentive Systems," p. 157, write that "there is almost no way for an organization to preserve itself by simply seeking ends for which there are no other advocates." Organizational ends are usually scarce resources and, like other scarce resources, they will be contested.

87. Ise, *Our National Park Policy*, p. 643. As an example of the ambitions of Park Service leadership, it included Grand Canyon, which was not a national park but a Forest Service administered area, in its Portfolio of National Parks. Identifying it in the public mind with the National Park System was part of a successful campaign to bring the canyon under Park Service jurisdiction.

88. Franklin K. Lane, letter to Mather dated May 13, 1918, reprinted in Foss, ed., *Recreation*, pp. 185–189.

89. See Elson, *Guardians of Tradition*, pp. 15–40.

90. National Resources Planning Board, *Report on Land Planning*, part 11, p. 202. According to Huth, *Nature and the American*, pp. 189–190, the early twentieth-century supporters of the National Park System thought of the parks in similar spiritual and patriotic terms.

91. The NPA's displeasure over the continued inclusion of Platt in the system can be read between the lines of its humorously prosaic description of the park's distinctive characteristics in its *National Parks Bulletin*, August, 1923, p. 7. While Zion National Park is: " 'The Rainbow of the Desert.' A gorge cut 2,500 feet down through the White Cliff and the Vermilion Cliff of the colorful Plateau Country of Utah. Magnificently carved by erosion—carries the Story of Creation from the rim of Grand Canyon up through millions of years." Platt is dryly described as: "Conserving mineral springs—serves city of Sulphur as a city park."

92. Ise, *Our National Policy*, pp. 186–187.

93. Ibid.

94. Shankland, *Steve Mather*, p. 184.

95. Nienaber and Wildavsky, *Budgeting and Evaluation*, p. 33.

96. Rourke, *Bureaucracy*, p. 85.

97. In fact, Rourke, *Bureaucracy*, p. 65, cites Interior as a department with a long-standing interest in expanding the geography of its operations and its constituency.

98. Ise, *Our National Policy*, p. 248. Also see Lee, *Family Tree*, p. 62.

99. Prophet, "Recreational Resources," p. 541.

100. Albright, *Origins*, p. 23.

101. Ise, *Our National Park Policy*, pp. 344–345; Swain, *Wilderness Defender*, p. 197.

102. Lee, *National Military Park Idea*, pp. 38–52.

103. Albright, *Origins*, pp. 15–16.

104. Shankland, *Steve Mather*, p. 185.

105. Quoted in Papageorgiou, "Architectural Schemata," p. 1164. Ironically, Mather's espousal of state parks along such lines aroused resentment similar to that which his efforts to attract a mass clientele to the national parks themselves caused. After the speech to a state park conference which contained the passage quoted in the text, a state park official rose to angrily denounce this view of what a state park should be: "Our parks and preserves are not merely picnicking places. They are rich storehouses of memories and reveries. . . . A state park cannot be planned until it is found. . . . Speaking for myself I would not be at all interested if the function of parks and recreation would merely be to provide shallow amusement for bored and boring people."

106. Swain, *Wilderness Defender*, p. 24. Downs, *Inside Bureaucracy*, p. 95, observes that "the selection of functional areas in which to expand will be influenced just as much by power considerations as by any logical linkage with [present roles]." Thus we should not assume that the addition of such disparate functions is an unusual event in the world of bureaucratic politics. The degree to which the spirit of aggressive bureaucratic entrepreneurship in search of new responsibilities characterized the leadership style of Mather and Albright is illustrated by an incident recounted by Albright's biographer (Swain, *Wilderness Defender*, pp. 202–205). One weekend President Hoover went on a retreat to a fishing camp on the upper Rapidan River in the Blue Ridge Mountains to the west of Washington. Since the president wanted to get some work done, he brought with him several officials, including Albright, with whom he needed to confer on various matters. Albright knew that Hoover would get up very early Sunday morning for a horseback ride so he made a point of being up and around very early himself—and therefore available for an invitation to go along on the ride. During the morning's ride through the mountains with Albright, Hoover decided that the Park Service should look into the possibility of constructing a parkway running along the crest of the Blue Ridge. The agency did look into the possibility. It also built and managed the parkway. Albright later prided himself on the way he grasped the opportunity and turned it to such an advantage for his agency.

107. Executive Orders of June 10, 1933, and July 28, 1933.

108. Albright, *Origins*, p. 23.

109. Downs, *Inside Bureaucracy*, p. 213.

110. Stratton and Sirotkin, *Echo Park Controversy*, p. 17.

111. Arrow, *Limits of Organization*, p. 57.

112. Lassen Volcano was one of these exceptions. Mather opposed its inclusion in the Park System, but to no avail. According to Shankland, *Steve Mather*, p. 172, Mather later changed his mind and came to view the Lassen Volcanic National Park favorably.

113. Ise, *Our National Park Policy*, p. 355.

114. Lipset, *Agrarian Socialism*, p. 274.

115. For a more detailed discussion of the role of the Park Service in administering the CCC program, see Lee, *Family Tree*, pp. 52–53. Also see Wirth's autobiography, *Parks, Politics and the People*, pp. 94–157.

116. Lee, *Family Tree*, pp. 52–56; Ise, *Our National Park Policy*, pp. 367–369.

117. Park, Parkway and Recreational Area Study Act of 1936 (49 Statutes at Large, 1894), reprinted in Foss, ed., *Recreation*, pp. 239–240.

118. Recreation's potential for leading to bigger things is clear in the National Resources Planning Board's *Report on Land Planning*, Part 11. On p. 241 the report reads that:

> The recreational use of the highway system of the United States plays a very large share in our national recreation scheme. Traffic over this system is now more than one-half recreational traffic.
>
> Since the automobile plays such an important part in the recreational scheme of the Nation, it is recommended . . . that the National Park Service be authorized to advise and assist other governmental agencies in their studies of the recreational use and design of parkways and highways.

119. Peffer, *Closing of the Public Domain*, chapter 14, "Revival of the Jurisdictional Feud," pp. 232–246.

120. Marion Clawson, personal communication, September 25, 1980.

121. Holden, "Imperialism in Bureaucracy," pp. 943–951.

122. Ise, *Our National Parks Policy*, p. 365.

123. Ibid., p. 437.

124. Ibid., p. 438.

125. Schrepfer, "Conflict in Preservation," p. 64.

126. Ise, *Our National Park Policy*, pp. 443–444.

127. Schrepfer, "Conflict in Preservation," p. 64.

128. *American Forests*, "Past Directors Speak Out," p. 28.

129. Newton B. Drury, speech to the Inter-American Conference on Conservation of Renewable Resources, September, 1948, reprinted in Foss, ed., *Recreation*, pp. 190–194.

130. Ise, *Our National Park Policy*, pp. 447–453. Agency appropriations were cut from 21 million dollars in 1940 to 5 million in 1943. That same period saw the agency's full-time staff reduced from 3,510 to 1,974. In one year, 1941 to 1942, the number of visits to the National Park System decreased from 21 million to 6 million. Source: Lee, *Family Tree*, p. 37.

131. Stratton and Sirotkin, *Echo Park Controversy*, is the definitive study of the controversy. Also, see Richardson, *Dams, Parks and Politics*, for a more recent treatment of Echo Park and the entire question of dams in parks during the Truman and Eisenhower administrations.

Stratton and Sirotkin, *Echo Park Controversy*, suggest that Drury assumed that a compromise with the Bureau of Reclamation would have brought the issue to a satisfactory conclusion without conflict—which he felt would have hurt both bureaus. In any case, a compromise certainly would have been in keeping with Drury's modus operandi.

132. Kaufman, *Are Government Organizations Immortal?*, p. 13.

133. Stratton and Sirotkin, *Echo Park Controversy*, p. 50.

134. As a point of fact, Arthur E. Demary served as agency director for several months between Drury and Wirth. However, little of consequence occurred during this period and Ise, in his history of the Park Service, *Our National Park Policy*, merely mentions Demary in passing.

135. Stratton and Sirotkin, *Echo Park Controversy*, p. 50.

136. DeVoto, "Let's Close the National Parks," p. 51.

137. Stratton and Sirotkin, *Echo Park Controversy*, p. 66.

138. Ise, *Our National Park Policy*, p. 648.

139. Such a dependence would restrict the agency's freedom of action, and here there was a lesson in Echo Park. Stratton and Sirotkin, *Echo Park Controversy*, write that although the Park Service and the Bureau of Reclamation might have come to an amicable understanding if they had had the freedom to do so, the two bureaus were fixed in their antagonism by their allies, who once engaged in the conflict were not disposed toward a peaceful solution.

140. McConnell, "Conservation Movement," p. 463.

141. Nienaber, Ingram, and McCool, *The Rich Get Richer*, p. 30.

142. Lee, *Family Tree*, p. 39.

143. Wirth, *Parks, Politics and the People*, p. 335.

144. Ise, *Our National Park Policy*, pp. 546–550.

145. Stratton and Sirotkin, *Echo Park Controversy*, p. 95.

146. Wirth, *Parks, Politics and the People*. See chapter 9, "Mission 66 and the Road to the Future," pp. 237–284 for a detailed discussion of Mission 66.

147. Ise, *Our National Park Policy*, p. 567.

148. Wirth, *Parks, Politics and the People*, p. 358, writes that:

Some organizations are always looking for a fight. They have got to have a cause for raising money. In some conservationist publications I've seen photographs that make it look as though the Park Service were taking a whole mountain down to build a park drive. The organizations may even be in agreement with a project but write up their campaign in their books in such a way as to suggest they fought for a long time, finally forcing the Park Service to take action. They end up taking full credit for the accomplishment. When I retired I wrote a letter to one organization stating that I'd been a member for over thirty years, that I had read its booklets and pamphlets, and that although they had championed the national parks they had never said a kind word about the service or given it credit for anything it did.

149. Shankland, *Steve Mather*, p. 315.

Chapter 3. The Modern Era

1. Marshall, "Problem of Wilderness," p. 122.

2. Schrepfer, "Perspectives on Conservation."

3. The Sierra Club would later change its position and fight the ski resort proposal for Mineral King.

4. Schrepfer, "Conflict in Preservation," p. 66.

5. Stegner, "Wilderness Idea," p. 97, wrote that:

Something will have gone out of us as a people if we ever let the remaining wilderness be destroyed; if we permit the last virgin forests to be turned into comic books and plastic cigarette cases; if we drive the few remaining members of the wild species into zoos or to extinction; if we pollute the last clear air and dirty the last clean streams and push paved roads through the last of the silence, so that never again will Americans be free in their own country from the noise, the exhausts, the stinks of human and automotive waste.

6. Schrepfer, "Conflict in Preservation," p. 67.

7. Jeff Ingram, interview with the author, December 8, 1980.

8. Green, *Recreation*, p. 39.

9. Brower, "New Decade," p. 3. This new predominant view is, in a sense, a grotesque inversion of the vision of human progress and evolution set forth earlier in the century by Teilhard de Chardin, *Phenomenon of Man*. In Teihard's view, man was the instrument of a teleological evolution in which conscious purpose would come to dominate and order the material universe—for the good. Although environmentalists shared the view that mankind was increasing his impress on the natural world, to them, such an impress meant neither order, good, or progress as it should be understood.

10. See Wildavsky, "Analysis of Issue Contexts," for a discussion of how bringing different frames of reference to a political problem can practically guarantee conflict between actors.

11. Lee, *Family Tree*, p. 38. Also see Foresta, *Open Space Policy*, pp. 37–40.

12. Clawson and Knetsch, *Economics of Outdoor Recreation*, p. 192, show that in the decade before 1960, the use of the nation's state parks went from a little more than 100 million visits annually to over 250 million visits. However, during that decade, state park acreage increased only about 20 percent, from slightly less than 5 million acres to about 6 million.

13. Foresta, *Open Space Policy*, pp. 67–73.

14. Ise, *Our National Park Policy*, p. 546, notes that agency leadership was "not enthusiastic" about the ORRRC study.

15. Fitch and Shanklin, *Bureau of Outdoor Recreation*, p. 52, suggest this reason for the neglect.

16. ORRRC, Study Report no. 13, p. 1.

17. ORRRC, *Outdoor Recreation for America*, p. 127.

18. Ibid., p. 49.

19. Ibid., pp. 9–10. The report had three other important recommendations: (1) That a national outdoor recreation policy be established. (2) That guidelines for the management of all outdoor recreation resources be developed. (3) That present recreation programs be expanded, modified, and intensified to meet increasing needs.

Each of these recommendations could be read as an indictment of the Park Service, since in the 1930s the agency had been given explicit responsibility for these things.

20. The first thought among ORRRC's members was to make the new bureau an executive office and link it organizationally to the Bureau of the Budget. The rationale was that such an arrangement would best enable it to use budget review to bring order to the many federal recreation programs. Udall, on the other hand, wanted to place it in Interior. Although many feared that such a location would expose the new bureau to capture by the Park Service, whose antipathy toward the whole idea was well known, Udall argued that the new bureau's tasks would be functionally related to those already entrusted to his department. It so happened that the departments of the Interior and Agriculture were in one of their rare periods of cooperation and Secretary of Agriculture Orville Freeman went along with the idea of placing the new bureau in Interior. This provided sufficient support for the bureau's creation—and for its location within Interior. Rourke, *Bureaucracy*, p. 65, writes that:

> with the establishment of the Bureau of Outdoor Recreation under its jurisdiction, [Interior] has for the first time gained an administrative foothold in the metropolitan areas in the Northeast where both the population and the demand for outdoor recreation have enormously expanded. This new constituency gives the department an opportunity to

serve substantial population groups in a part of the country in which it has not previously played a significant role.

21. Fitch and Shanklin, *Bureau of Outdoor Recreation*, p. 92. Wirth, *Parks, Politics and the People*, p. 282, writes that, "we felt that a new bureau was not necessary; it was our responsibility . . . to plan and finance a national recreation program."

22. Edward C. Crafts, speech to Izaak Walton League's 40th Annual Convention and Conservation Conference, June 21, 1962.

23. On the cost of "free" federal money to states and municipalities, Sharkansky, *Public Administration*, pp. 275–276, writes:

> The 'purpose' nature of federal grants and the requirements that come along with the money are frequent sources of conflict between federal, state, and local administrators. It is alleged that the federal carrot leads recipients to undertake activities that are not in their own best interest and that requirements are frequently inconsistent with their social or economic problems. Federal money is not 'free.' If the aided program is not uniformly popular in a state or locality, the recipient agency encounters some political costs as well as benefits.

24. Smith, "Gateways," p. 216. In 1976, the Bureau of Recreation was given historic preservation responsibilities and its name was changed to the History, Conservation and Recreation Service. It was abolished under President Reagan, who considered the agency a waste of tax money.

25. Fitch and Shanklin, *Bureau of Outdoor Recreation*, p. 81, refer to Udall as "one of the most conservation-minded Secretaries of the Interior."

26. Udall, *Quiet Crisis*, p. 163.

27. Ibid., p. 159. He wrote that

> In a great surge toward 'progress,' our congestion increasingly has befouled water and air and growth has created new problems on every hand. Schools, housing, and roads are inadequate and ill-planned; noise and confusion have mounted with the rising tempo technology; and as our cities have sprawled outward, new forms of abundance and new forms of blight have oftentimes marched hand in hand. Once-inviting countryside has been obliterated in a frenzy of development that has too often ignored essential human needs. (Ibid., p. 159)

28. Ibid., p. 176.

29. Udall's conceptual link between conservation of natural resources and human resources was not a new one. According to McConnell's "Conservation Movement," progressive conservationists proclaimed that the ultimate aim of their movement was the conservation of man himself. Also, Ise, *Our National Park Policy*, p. 360, writes of another prior connection, "The word conservation in civilian conservation corps had a double meaning here, for the agency was designed not for the resources but the character and morale of unemployed young men by providing productive work. . . ." Huntington, "Conservation of Man," also saw a close connection between the conservation of natural and human resources. What Udall did was, in essence, to update the conceptual link, and apply the conservation theme to the Great Society/New Frontier agenda.

30. Udall, *Quiet Crisis*, p. 184.

31. Ibid., p. 176.

32. President Lyndon Johnson, in his message of February 8, 1965 to Congress on natural beauty (*Public Papers of U.S. Presidents*, Lyndon B. Johnson, 1965, Book 1, p. 155), wrote that conservation meant "correcting the side effects of our careless technology." He also wrote that "it is true that we have often been careless with our natural bounty. At times we have paid a heavy price for this neglect. But once our people were aroused to the danger, we have acted to preserve our resources. . . ." In this, too, Johnson seemed to subscribe to Udall's view.

33. Udall, *Quiet Crisis*, p. 161.

34. Udall, in turn, was not well liked by Park Service leadership. "Udall had a vision but he had no support among the old line [of the Park Service]. He was just a young punk intellectual to them," is how one agency administrator remembers the Service's attitudes toward the secretary.

35. Wirth, *Parks, Politics and the People*, pp. 297–298.

36. Rourke, *Bureaucracy*, p. 129.

37. Wirth, *Parks, Politics and the People*, p. 309. One might naturally suspect that Wirth had a hatchet to grind in the matter. However, Fitch and Shanklin, *Bureau of Outdoor Recreation*, p. 81, who were clearly not on the side of the Park Service in the dispute, wrote that:

> John Carver, Jr., former congressional aide to Senator Frank Church of Idaho, who had been appointed Interior's assistant secretary for Public Land Management, . . . gave firm content to the education of National Park Service leaders in the outdoor recreation facts of life.

38. Wirth, *Parks, Politics and the People*, pp. 312–313.

39. Nienaber, Ingram, and McCool, *Rich Get Richer*, p. 6.

40. Schrepfer, "Conflict in Preservation," p. 73.

41. Mazmanian and Nienaber, *Can Organizations Change?*, p. 24.

42. For example see Rourke, *Bureaucracy*, p. 141.

43. Vogel, "Public-Interest Movement."

44. Ibid., pp. 610–611.

45. For their part, some Park Service administrators resented the self-righteous and pugnacious attitude of the environmentalists, which they felt was sometimes disingenuous. One administrator said: "It's easy to throw rocks at the Service and the environmental groups love to fight so we get it from them. But some of it's only show, they have to appear militant to their members." Others resented what they saw as the extreme egotism of the environmental leaders. A ranger at the Channel Islands National Park spoke of a nationally prominent environmentalist-scholar who taught at a nearby university, "He's a real potshotter at park management here. You know why? When he went out to San Miguel [Island, a part of the park], we kept him on the trail like everyone else. He's still pissed off about it."

46. Ise, *Our National Park Policy*, pp. 520, 522–523.

47. Robert Utley, interview with the author, December 10, 1980.

48. In 1980, Wirth, *Parks, Politics and the People*, p. 165, would write of the board's diminished status:

> I believe Congress still relies on the board's advice, but, I am informed, to a lesser degree than formerly. In my opinion the reason is that people have been appointed who do not have the qualifications to make sound professional recommendations in accordance with the intent of the law. In fact, there have been several appointed for political reasons or for their contributions to political parties.

In 1972, a park study group (Conservation Foundation, *National Parks for the Future*, p. 69) criticized the board for being of too narrow a social composition:

> We do not believe that the Advisory Board on National Parks represents the needs and interests of the full sociocultural range of American society. The present membership of the Board does not reflect the broad range of social and economic classes that the National Park Service is presumably set up to serve. It is therefore recommended that the board should be reorganized, and expanded if necessary, to make it representative of the broad range of social and economic groups that comprise American society.

Although in keeping with the vogue of social pluralism, such a recommendation was naive. The board's source of power had been its elite, even clubby membership and the respect the notables on it received from Congress. The dilution of this elite aura would undercut its effectiveness even if it did create a climate of legitimacy by broadening representation. While undoubtedly well intentioned, the NPFF report played into the hands of those who would consciously undercut the board's power or who would use its seats as political favors.

49. Former Director Whalen made this point during an interview with the author, December 1, 1980.

50. For an example of how Yates's House Appropriations Subcommittee handled the importuning of the National Council of the Arts, see House Appropriations Committee hearings on the 1981 budget, vol. XI, pp. 1054–1057. The reader might like to compare this with the tone of the Park Service's budget hearing before the same subcommittee (House Appropriations Committee Hearings on the 1981 budget, vol. XI).

51. Rourke, *Bureaucracy*, pp. 24–25.

52. Simon, Smithburg, and Thompson, *Public Administration*, p. 522.

53. Long, "Public Policy," p. 22. Herbert Finer was perhaps the foremost spokesman for total legislative supremacy. He summarized his position thus: "The elected representatives of the public, . . . are to determine the course of action of the public servants to the most minute degree that is technically feasible." Quoted in Long, "Public Policy."

54. Simon, Smithburg, and Thompson, *Public Administration*, p. 523. They write that:

> As a result of the deficiencies in the tools that the legislature may use to hold administration accountable, administrators exercise a much more positive role in policy formation,

and legislatures are forced to delegate in much broader terms to administrators than is contemplated in the traditional democratic theory which asserts that legislatures 'establish the will of the state' while administrators 'execute the will of the state.'

55. Ise, *Our National Park Policy*, p. 375.

56. Long, "Public Policy," p. 22.

57. Schick, "Congress and the 'Details'," p. 518, writes that Congress has "sought new methods for holding administrators to account and has applied old controls more extensively."

58. Lee, *Family Tree*, p. 49. The big exception to this disuse of the antiquities act was in connection with the Alaska Lands Settlement, when President Carter invoked his powers under the act to declare vast tracts of the public domain as national monuments. In this case, however, the use of the monument status was only a temporary measure to allow more time to arrange for a more widely acceptable settlement. There was no thought of assigning all this land permanent monument status.

59. See McPhee, "Profiles—George Hartzog."

60. Ibid., p. 68.

61. Vogler, *Politics of Congress*, pp. 185–186.

62. Woll, *Public Policy*, pp. 188–189.

63. This was a theme which emerged repeatedly in the author's discussions with Park Service administrators on the subject of the agency's relationship to Congress.

64. Vogler, *Politics of Congress*, p. 265.

65. Ibid., see chapter 5, "Rules and Norms," pp. 205–238.

66. Domenici, "Can Congress Control Spending?" p. 52.

67. Fitch and Shanklin, *Bureau of Outdoor Recreation*, p. 3.

68. Nienaber and Wildavsky, *Budgeting and Evaluation*, p. 83.

69. Natchez Trace Parkway funding had the very persuasive personal testimony of Congressman Whitten, Chairman of the House Appropriations Committee, through whose district the parkway ran. On March 28, 1979 (1980 Budget Testimony, vol. XI, p. 1063), Congressman Whitten started his testimony before the appropriations subcommittee for Interior with, "All of you on the subcommittee are aware of my deep interest in the Natchez Trace." In view of that deep interest, neither Congress nor the Park Service was likely to raise questions about the intrinsic merits of the parkway or about the propriety of spending a large part of the agency's 1980 capital improvements budget on it.

70. Simon, Smithburg and Thompson, *Public Administration*, pp. 402–409. For Congress, only a growing National Park System presents opportunities for log rolling and pork barrel benefits. Balancing new parks by decommissioning others would be fought fiercely by congressmen since it would mean a loss of benefits to local constituents and, probably worse, a loss of face. In fact, when Park Service

Director Walker, newly installed and responding to demands for economy from the Nixon White House, asked his subordinates to draw up a list of fifty units of marginal quality which might be decommissioned, this confirmed to agency administrators what they suspected about their new director, i.e., he was hopelessly naive. They knew from long experience that such decommissions would not be approved by Congress.

71. Rourke, *Bureaucracy*, p. 65.

72. Vogler, *Politics of Congress*, p. 185.

73. Phillip Burton, interview in *National Parks and Conservation Magazine*, May 1979, p. 25.

74. Ibid.

75. However, it should not be thought that the subcommittee was completely indiscriminate in what it supported for inclusion in the Park System. By the late 1970s, the volume of congressionally initiated proposals had increased to such an extent that, in the words of a subcommittee staff member, "Burton has to resist the majority [of the proposals]. A lot of polite deflecting goes on." Nevertheless, in spite of this deflecting, system expansion quickened under Burton's subcommittee chairmanship.

76. PL 94-458 (General Authorities Act of 1976, Sec. 8).

77. Burton's view of what should be saved was very broad indeed. For years the Park Service tried to get Mar-a-Lago, a garish Florida mansion donated to the federal government in the early 1970s, out of the Park System. The mansion could not be opened to the public without a large investment in facilities to accommodate visitors and just maintaining it was costing the agency nearly half a million dollars a year. Burton recognized the mansion's shortcomings but he was reluctant to allow the agency to abandon it for fear of what would happen to it afterward. Only after four years of pleading was he swayed, and with his support the mansion was quietly removed from the System.

78. For a discussion of the use of omnibus legislation in distributing pork barrel benefits, see Shepsle and Weingast, "Political Preferences."

79. PL 95-625, 92 Stat. 3467 et seq.

80. Fenno, *Power of the Purse*, pp. 7-8.

81. Schick, *Congress and Money*, p. 428.

82. Fenno, cited in Vogler, *Politics of Congress*, p. 235, fn. 47.

83. Lieberman, *Fiscal Policy Formulation*, p. 180.

84. Activity in Congress takes place in the House. As Fenno, *Power of the Purse*, p. 503, writes, "The House Committee . . . dominated appropriations politics in Congress." For corroborative findings, see Horn, *Unused Power*. Although the existence of the Land and Water Conservation Fund has reduced the appropriation committee's control of acquisition expenditures (disbursements from the fund do not require approval of the appropriations committee), it still has control over other national park expenditures, such as those for construction and

management. And here, according to McCool, *National Park Service*, it has not been inclined to give the agency what it asks for.

85. Swain, *Wilderness Defender*, p. 189.

86. Kaufman, *Are Government Organizations Immortal?*, pp. 8–9. Kaufman writes that: "cabinet and subcabinet officers stay in office for comparatively short periods, and even the presidency turns over frequently from the viewpoint of an agency and its allies in Congress and outside the government. Time is therefore on the agencies' side; their superiors are gone before many changes can be formulated or implemented."

87. Rourke, *Bureaucracy*, p. 118. Rourke writes that: "It is asserted that administrative policy decisions tend to reflect too narrow a set of interests. Because authority for designing programs is widely dispersed among subordinate units of the executive branch, policy responds not to the needs of broad segments of the public, but more commonly to the pressures of small client groups that hold individual bureaus in captivity."

88. Ibid.

89. Ronald Walker became director during the era of Nixon's "imperial presidency," when loyalty to the president seemed to be a more highly valued trait than administrative competence. The man Walker replaced, George Hartzog, was acknowledged by everyone to have plenty of the competence but was suspected by Nixon's inner circle of lacking loyalty to the administration.

90. Wirth, *Parks, Politics and the People*, p. 318, writes that, "It is customary in the National Park Service for the director and the assistant director for administration to defend the service's request for appropriations and legislation before the committees of Congress."

91. By the terms of the Wilderness Act, the Park Service, as well as other federal land-managing agencies, had to review their holdings for areas suitable for addition to the National Wilderness System. NEPA required careful assessment of planned park development and management changes for their potential impact on the environment. For a more thorough discussion of these points, see Nature Conservancy, *Preserving Our Natural Heritage*, vol. I, especially chapter II, part B and chapter XIII.

92. Robert Herbst, interview with the author, April 7, 1981.

93. For example, Herbst showed a special interest in air and water pollution problems in the parks, and he pushed the agency into taking more of an interest in these problems than it had previously shown. He also insisted on the establishment of an Office of Technology and Environmental Education within the agency. Service leadership probably would not have created this office if the choice had been entirely in its hands.

94. Robert Herbst, interview with the author, April 7, 1981. Herbst recalled that the environmental groups were very important to him. "I consulted them a lot and I worked with them," he said. "They came up with some of the best ideas on what the National Park System ought to be, and what it ought to be doing."

95. The former assistant secretary told the author:

I played a larger role than the director because so much was going on at a high level. We were adding to the System. We were making important decisions on Alaska, which involved Congress and the White House. I was in a better position to deal with them than the director.

96. Yamada, "Improving Management," p. 765, suggests that OMB became Nixon's "principal arm" in exercising his management functions.

97. *Washington Post*, August 23, 1974, quoted in Seidman, *Politics*, p. 122. In some cases, OMB's budgetary concerns and political sensitivity appeared to reduce its freedom and effectiveness rather than enhance its power. For example, under Carter, OMB sometimes found itself unable to oppose natural park wilderness proposals, even those it thought lacked merit. One official told the author that "Wilderness is cheap to manage, so even when the arguments for wilderness are purely specious, we don't have much leverage because we can't argue bucks."

98. See Selznick, *TVA*, p. 146. Downs, *Inside Bureaucracy*, p. 70, writes that "Each official in a bureau has a set of specific goals connected with his own self-interest. Therefore, the goals of every bureau member are different to at least some degree from those of every other member."

99. Simon, *Administrative Behavior*, p. 198, writes that:

to a large extent the values gradually become 'internalized' and are incorporated into the psychology and attitudes of the individual participant. He acquires an attachment or loyalty to the organization that automatically—i.e., without the necessity for external stimuli—guarantees that his decision will be consistent with the organization's objectives.

100. Wamsley and Zald, "Political Economy," p. 68, discuss the importance of outside support as a power base for internal factions.

101. Any large public agency is perforce linked to what Heclo (cited in Pressman and Wildavsky, *Implementation*, p. 165) calls issue networks, networks of individuals or groups concerned with a particular public question and in contact with each other concerning it. These networks tend to be amorphous and informal. They can cut across organization boundaries, thereby linking public employees and private activists, public agencies and private groups.

102. Simon, Smithburg, and Thompson, *Public Administration*, p. 456.

103. One Park Service administrator with a background of environmental activism outside the agency said, "Luckily in my job I've never had to choose between what's good for the environmental movement and what's good [for the agency]. I don't know what I'd do if I had to." This attitude was expressed by others with similar backgrounds.

104. Downs, *Inside Bureaucracy*, pp. 75–78 suggests that "bureaucracies have few places for administrators who are loyal to society as a whole." This may be the case, but in the Park Service those who act in conformity with widely held social values show great ability to hold their own against agency leadership. Perhaps it is more accurate to say that strong bureaucracies have few places for those loyal to society as a whole—weak ones have little to say about the loyalties of their members.

105. See Moore, "Public Administrator," for a discussion of the concept of organizational democracy. Also see Ostrom, *Crisis in Public Administration*. Ostrom makes a strenuous argument for spreading decision-making authority widely within formal organizations and in doing so he is typical of the advocates of organizational democracy.

106. Conservation Foundation, *National Parks for the Future*, p. 54. In August 1971, the National Parks Centennial Commission and the Conservation Foundation held a symposium on the parks and invited over two hundred participants, most of whom were experts on some aspects of the National Park System. The reports of the task forces and the symposium proceedings were published as *National Parks for the Future* in late 1972. It quickly became one of the most important and widely quoted documents on the National Park System.

107. Ibid., p. 68.

108. William Whalen, interview with the author, December 1, 1980.

109. Deutsch, *Nerves of Government*, pp. 221–222.

Chapter 4. Nature Policy

1. Russell Dickenson, interview with the author, April 3, 1981.

2. Cahn, "Park Service," p. 22.

3. Smith, "Gateways," p. 215, reflected a similar view when he wrote that "the National Park Service . . . since its inception has had as its primary mission the preservation and protection of the natural scenic splendors of the nation."

4. Ise, *Our National Park Policy*, p. 647.

5. Robert Herbst, interview with the author, April 7, 1981.

6. Mati Tammau, quoted in Novgorod, Dimock, and Dimock, *Casebook*, p. 110.

7. Schiff, "Innovation," p. 14.

8. M. F. Burrill, personal communication with the author, March 30, 1981.

9. Stratton and Sirotkin, *Echo Park Controversy*, p. 38.

10. Ise, *Our National Park Policy*, pp. 645–649, Irland, "Citizen Participation," p. 263, writes that "we have inherited from a simpler past a belief that the public lands can be managed to 'harmonize' conflicting use." And this included preservation and popular access.

11. PL 88-577, 16 U.S.C. 1131–1136.

12. Ibid., section 2(a).

13. Schiff, "Innovation," p. 15.

14. Simmons, "National Parks," p. 405. Simmons writes that:

National parks inevitably form one of the larger reserves of biota . . . parks are thus outdoor biological gardens where bird, beast or flower may be observed in either its natural habitat or a little altered version of [it]. Completely wild areas are important in

constituting a pool of genetic diversity in their biota. The necessity for such a reservoir
. . . is one of the objectively scientific reasons for national parks.

15. Schrepfer, "Conflict in Preservation," p. 68.

16. Herbst, "Threats to the National Parks," p. 5.

17. Paul Pritchard, interview with the author, October 6, 1980.

18. Conservation Foundation, *National Parks for the Future*, p. 93.

19. Shankland, *Steve Mather*, p. 270.

20. Clements, *Plant Physiology*. Leopold, *Sand County Almanac*. On this
point, Ise, *Our National Park Policy*, p. 320, writes that Mather did not understand the balance-of-nature principle of wildlife; but neither did many others of his
era.

21. Conservation Foundation, *National Parks for the Future*, p. 105.

22. Smith, in *National Parks for the Future*, Conservation Foundation, p. 121,
advanced what might be called a cultural lifeboat argument for the national parks,
one in which the parks, if managed properly, might help bring into being a utopian
future in which the current social maladies and environmental discontents of
civilization are no more:

> The significance [of the great national parks] is much greater than that of miniatures of
> isolated and imperiled ecosystems, they [are] the models of the future, by which the
> architects of future civilization must guide their work in the centuries ahead. A high
> civilization, built upon a stabilized and reduced population, presupposing a differential
> stabilization of commodity production, assuming emphasis upon education and the prac-
> tice of the arts, will surround itself and immerse itself in scenery of the kind we find
> surviving now only in the natural areas of the national park system.

The vision here is not the progressive one, nor is its landscape that of progressive conservation.

23. Gardon, "Time to Save," p. 4.

24. McCool, *National Park Service*, p. 25.

25. Selznick, "Foundations," p. 272.

26. Irland, "Citizen Participation," p. 264. See also Fitch and Shanklin, *Bureau of Outdoor Recreation*, p. 31, for a similar view of the Park Service mission
with regard to the great natural areas of the System.

27. Rowntree, Heath, and Voiland, "United States National Park System," p.
107. Coopers and Lybrand, in a study of Park Service management systems,
conclude that the resolution of the use vs. preservation dilemma is one of the
agency's major challenges.

28. Wirth, *Parks, Politics and the People*, p. 359. Shrepfer, "Conflict in Preservation," pp. 66–67, makes clear that from the 1940s on, an active faction of the
preservation community considered roads in parks a faustian bargain. They might
allow popular access and thereby increase the security of the parks from resource
exploitation—everyone agreed that they did this—but that the cost in overdevel-

oped parks was just too high a price to pay. In the late 1940s the antiroad faction gained ascendency in the Sierra Club and thereafter opposition to road building in and to the parks became progressively more intense.

29. See, for example, Simmons, "National Parks," p. 399; Darling and Eichorn, *Man and Nature*, p. 33. For a discussion of the impact of park visitors, see Siehl, "Visitor Pressures." See also Sharpe, "National Parks," p. 208.

30. Sax, "America's National Parks," p. 83.

31. Udall, *Quiet Crisis*, p. 154.

32. Merrium, "National Park System," p. 4, referenced questionable history when he wrote that "clearly the Park Service is in the recreation business by default rather than design." When Fitch and Shanklin, *Bureau of Outdoor Recreation*, p. 31, writing to justify the creation of the recreation-oriented Bureau of Outdoor Recreation said, "when people and scenery clashed, Mather was rarely on the side of the people," they skirted the fact that in Mather's early twentieth century world-view people and scenery mostly complemented each other; thus their assertion was a moot one by and large.

33. Distortion of history also seeps into Rowntree, Heath, and Voiland, "United States National Park System," p. 96, when they write that "the original mission of the National Park Service rearticulated in 1972 was preservation and interpretation of natural landscapes and ecosystems." Projecting ecosystem protection as an agency goal back to 1916 is not rearticulation, for the term conveys the notion of reexpressing or rephrasing original ideas—and ecosystem as a concept simply was not ambient in the early twentieth century. Revision or perhaps historical reconstruction would be more accurate terms, but of course they are terms which bestow less legitimacy on present policy recommendations. It is hard to see these interpretations of the past as anything other than justifications for demands that the Park Service limit park visits, and beyond that, for demands that the agency completely change the way it viewed its user-clients.

34. Conservation Foundation, *National Parks for the Future*, p. 22.

35. Ibid., p. 34.

36. Darling and Eichorn, *Man and Nature*, p. 33.

37. Sax, *Mountains Without Handrails*, p. 10.

38. Ibid., p. 106.

39. Conservation Foundation, *National Parks for the Future*, p. 32. They wrote: "The choice is ours, whether the parks shall remain the 'crown jewels' of our outdoor heritage to be cherished, protected, preserved and worthy of our rigorous self-imposed restraints, or permitted to degenerate into the commonplace. It is a difficult choice, but it must be made."

40. Sax, "America's National Parks," p. 76. As Sax expanded on this point:

Olmsted was not just a builder of parks; he was the author of a distinctive theory about the role parks ought to play in a democratic society. Nothing was further from his view than the now widely held idea that in a democracy the sole acceptable park policy is to facilitate access for the greatest number of people that can be accommodated and then to establish whatever activities the popular sentiments of the hour appear to demand. Instead he held to what might elaborately be called an intertemporal theory of democratic legitimacy: that the justification for the use of the parks must be sought in the long-term judgement of the people and that there was a legitimate role for leadership in a democratic society.

41. Conservation Foundation, *National Parks for the Future*, p. 102.

42. Sax, *Mountains Without Handrails*, p. 83.

43. Paul Pritchard, interview with the author, October 6, 1980.

44. Conservation Foundation, *National Parks for the Future*, p. 21.

45. National Parks and Conservation Association, *Preserving Wilderness*.

46. Ickes, "Keep It a Wilderness," p. 9.

47. McCool, *National Park Service*, p. 21. McCool surveyed a group of sixty-seven park service personnel taking courses at the agency's Albright Training Center at Grand Canyon National Park. The group was a heterogeneous one, with length of service ranging from four months to twelve years. They ranged in rank from GS 4 to GS 13 and in education from some high school to Ph.D.

48. David Ochsner, quoted in Novgorod, Dimock, and Dimock, *Casebook*, pp. 104–105.

49. William Whalen, interview with the author, December 1, 1980. McCool, *National Park Service*, p. 55, found that respect for, and appreciation of, the conservation groups was widespread among agency rank and file. Seventy percent of those he surveyed agreed with the statement, "Conservation groups help the Park Service a great deal in implementing Park Service goals." Seventeen percent were not sure and fourteen percent disagreed.

50. David Ochsner, quoted in Novgorod, Dimock and Dimock, *Casebook*, p. 103.

51. Ise, *Our National Park Policy*, p. 370.

52. *American Forests*, "Former Directors Speak Out," p. 50.

53. Ibid.

54. Memorandum from Secretary of the Interior Udall to National Park Service Director Hartzog, July 10, 1964.

55. Nienaber, *Differentials in Agency Power*, p. 61.

56. Ibid., p. 61, fn. 38.

57. McPhee, "Profiles—George Hartzog," p. 62.

58. Ibid.

59. Siehl, "Visitor Pressures," p. 7.

60. U.S. Department of the Interior, National Park Service, *Management Policies*, p. III-1.

61. Ibid., p. III-8.

62. Boyd Evison, interview with the author, November 26, 1980.

While pleasing the environmentalists, such a shift was not likely to be viewed kindly by the concessionaires, those businessmen who provided lodging and eating facilities in the national parks under special leasing arrangements with the Park Service. To the concessionaires, such a shift was viewed as a direct threat to their business. It was especially disliked by those concessionaires (the managers of the lodges in Sequoia National Park are a good example) who felt that the only way to continue to make a profit was to increase the size of their operations and thus realize economies of scale. It is not clear how much influence on park management the concessionaires really have had, however. Stories have circulated in the Park Service about covert influence being used at all political levels, even up to the Oval Office, to get the agency to abandon, or at least pursue with less vigor, plans to reduce facilities in the parks. Such stories are difficult to substantiate or refute, since they deal with a very murky area of political power and influence.

I suspect, however, that the power of the concessionaires was not great, or perhaps more accurately put, while it might have been able to influence an occasional decision, this power was not important in forming system-wide policy. It must be remembered that there was a natural balance of forces in operation, and while an agency decision to close a facility in a park might have meant a foreclosed opportunity for the park's concessionaire, it meant a new opportunity to provide equivalent facilities outside the park.

63. U.S. Department of the Interior, National Park Service, *State of the Parks-1980*, p. 1.

64. Ibid., p. IX.

65. The International Biosphere Program is part of UNESCO's program on man and the biosphere, established by UNESCO's 16th General Conference, 1970. According to the Nature Conservancy, *Preserving Our National Heritage*, vol. I, p. 3, the biosphere reserves are "protected natural areas for research and conservation of genetic materials. . . . The areas are to be baselines against which change can be measured and the performance of other ecosystems judged."

66. U.S. Department of the Interior, National Park Service, *Management Policies*, p. I-1.

67. 39 Statutes at Large, 535, August 25, 1916.

68. Lee, *Family Tree*, p. 43.

69. Everhart, *National Park Service*, p. 182. Also see Robinson, *Forest Service*, p. 285, fn. 30.

70. U.S. Department of the Interior, National Park Service, *Part Two of the National Park System Plan*, lists the adequacy of natural region representation as follows:

100% Virgin Islands	29% New England—Adirondacks
94% Cascade Range	24% South Pacific Border
90% Chihuahuan Desert—Mexican Highland	17% Gulf Coastal Plain
	16% Central Lowlands
89% Sierra Nevada	12% Columbia Plateau

86% Northern Rocky Mountains
86% Mohave-Sonoran Desert
84% Middle Rocky Mountains
78% Colorado Plateau
72% Interior Low Plateaus
71% Southern Rocky Mountains
71% Superior Upland
70% North Pacific Border
63% Florida Peninsula
56% Pacific Mountain System,
  Alaska
54% Appalachian Ranges
54% Island of Hawaii
53% Maui
45% Interior Highlands
34% Great Plains

10% Great Basin
0% Wyoming Basin
0% Appalachian Plateaus
0% Piedmont
0% Interior and Western Alaska
0% 0% Brooks Range
0% Arctic Lowland
0% Oahu
0% Kauai, Niihau
0% Leeward Islands
0% Puerto Rico
0% Mariana Islands
0% Caroline Islands
0% Marshall Islands
0% Guam
0% Samoa

71. Ibid., lists the adequacy of natural history themes as follows:

Landforms	Land Ecosystems
75% Hot water phenomena	80% Pacific forest
59% Works of volcanism	67% Chaparral
56% Caves and springs	60% Boreal forest
50% Works of glaciers	56% Desert
47% Sculpture of the land	44% Dry coniferous forest
44% Mountain systems	31% Tropical ecosystems
43% Eolian landforms	30% Grassland
43% River systems and lakes	27% Eastern deciduous forest
42% Seashores, lakeshores, islands	
30% Cuestas and hogbacks	
25% Plains, plateaus, mesas	
18% Coral islands, reefs, atolls	

Geologic History	Aquatic Ecosystems
78% Precambrian	39% Lakes and ponds
44% Mississippian–Triassic	37% Streams
34% Permian–Cretaceous	24% Marine environments
33% Paleocene–Eocene	24% Estuaries
21% Cambrian–Early Silurian	20% Underground ecosystems
15% Oligocene–Recent	
0% Late Silurian-Devonian	

72. U.S. Department of the Interior, National Park Service, *Management Policies*, p. I-8.

73. In 1975, Platt National Park, always a bit of an embarrassment to the agency, was done away with as an administrative entity. It was linked with a recreation area on a man-made lake and the combined unit was named the Chickasaw National Recreation Area.

74. U.S. Department of the Interior, National Park Service, *Part Two of the National Park System Plan*.

75. U.S. Department of the Interior, National Park Service, *Index, 1979*, p. 49.

76. Ibid., p. 47.

77. Cahn, "Park Service," p. 22.

78. Alaska Regional Director Cook, quoted in *Courier: The Newsletter of the National Park Service*, January 1981, p. 5. The agency rank and file apparently felt the same about the value of the Alaskan additions, and there were usually many more applications than job openings in the Alaskan parks. According to an agency administrator in Washington, Alaska has "lots of sex appeal for John Q. Ranger, especially the younger ones. The mystique is there."

79. Robert Herbst, interview with the author, April 7, 1981.

80. Everhart, *National Park Service*, p. 182.

81. Robinson, *Forest Service*, p. 258.

82. Everhart, *National Park Service*, p. 182.

83. Robinson, *Forest Service*, p. 258.

84. Ibid., p. 34.

85. Moch, *Economic Analysis*, U.S. Forest Service, sent under a covering memo dated October 25, 1961 to Dr. John Zivnuska, Acting Director, Wildland Research Center.

86. Nienaber, Ingram, and McCool, *Rich Get Richer*, p. 11.

87. Robinson, *Forest Service*, p. 159.

88. Ibid., p. 159.

89. Wild and Scenic Rivers Act (1968), 16 U.S.C. 1271–1287 (1968); National Trails System Act (1968), 16 U.S.C. 1241–1249.

90. Seidman, *Politics*, chapter 4, "Nixon's New American Revolution," pp. 110–124.

91. Nienaber, *Differentials in Agency Power*, pp. 38, 71.

92. Ise, *Our National Park Policy*, p. 650. It must be added that this situation changed with the Alaska land settlement. Much of the new national parkland in Alaska was designated wilderness and this has given the Park Service a much larger relative role in the management of the national wilderness system.

93. Rourke, *Bureaucracy*, p. 67.

94. Selznick, *TVA*, p. 64.

95. Sax, *Mountains Without Handrails*, p. 106.

96. Helen Ingram, personal communication with the author, December 8, 1980. An agency administrator with long experience working with environmental groups explained their tendency to criticize the Park Service in public thus: "Environmental groups are conflict oriented. If the Park Service takes a stand they agree with, they'll ignore the issue and put their energy elsewhere. It's just the way they operate."

97. Rourke, *Bureaucracy*, p. 67.

98. Nienaber, *Differentials in Agency Power*, p. 69, takes a similar view, "The general public tends to demand a great deal from the service yet offers very little real constituent support."

99. PL 88-587, September 11, 1964.

100. Joe Zeismann, interview with the author, February 3, 1981.

101. In the 93rd Congress, Senator Aiken, one of the System's supporters, said on the floor of the Senate that off-road vehicles were "inimical to the wilderness concept." Senator Nelson agreed and added that banning them was the unanimous intent of Congress when it passed the original wilderness bill. (*Congressional Record* vol. 120, May 31, 1974, S 9393).

102. *Fire Island Wilderness Coalition Newsletter*, "Victory," January 1981, p. 1.

103. Joel Pickelner, interview with the author, December 23, 1980.

104. Wilson, *Political Organizations*; Browne, "Organizational Maintenance."

105. PL 93-662 (1975).

Chapter 5. History Policy

1. Lee, *Military Park Idea*, p. 7.

2. Ibid., p. 12.

3. In a sense the Park Service had taken on a sacerdotal role when it assumed management of the great natural areas of the West since they were so linked to American patriotic mythology. (See Elson, *Guardians*, pp. 14–50, for a discussion of just how tight the links were.) With the assumption of the management of historic sites, however, this role was deepened and broadened.

4. Lee, *Antiquities Act*.

5. Lee, *Military Park Idea*, p. 54. Lee writes of the circumstances leading up to Wissler's statement:

> Dr. John C. Merriam agreed to serve as chairman [of the Educational Advisory Committee]. Its five members, all of them eminent scholars, included Dr. Clark Wissler, Curator

of Anthropology of the American Museum of Natural History. To Dr. Wissler was assigned the task of envisioning the future of the National Park System in the field of human history. Dr. Wissler gave much thought to this subject and in addition spent the summer of 1929 in field investigations in the Southwest, visiting, among other places, Chaco Canyon, and Mesa Verde, as well as Sante Fe and the surrounding district. At a meeting of the committee held November 26–27, 1929, in Washington, D.C., Dr. Wissler presented his conclusions in the form of a report, which received extended discussion. The committee then adopted the [above quoted] statement by Dr. Wissler.

6. Lee, *Family Tree*, p. 48.

7. Ise, *Our National Park Policy*, p. 344.

8. Ross Holland, interview with the author, September 30, 1980.

9. Laura Beatty, interview with the author, September 19, 1980.

10. 16 U.S.C., 470–470t.

11. See Udall, *Quiet Crisis*, p. 169. Lee, *Family Tree*, p. 65, writes that, "Historic Preservation became part of the New Conservation with the enactment of [the 1966 act]." George Hartzog took his cues on historic preservation from Johnson and Udall and what Hartzog wrote about historic preservation's importance as a remedy for the less attractive side effects of modern life was very much in the spirit of the New Conservation. For example, Hartzog, reprinted in Foss, ed., *Recreation*, p. 200, wrote:

> The American people are now far removed from the old frontiers when everyone lived close to the earth on farms, ranches, and in small communities. In those days, there was unlimited space; the out-of-doors was always within easy reach. But today when most people have cut their roots and moved off the land, they find less and less opportunity to enjoy what once was taken for granted.
>
> Spiritually, our people need reminders of their past. Without them, they lose the sense of being part of the old America. This is a major need and it is our opportunity to meet it by providing the areas necessary and protecting them so that future generations may always gain strength and stability from them.

12. Everhart, *National Park Service*, p. 38.

13. Robert Utley, interview with the author, December 10, 1980.

14. Ibid.

15. Ibid. Today the Kosciuszko House is administered as part of Independence National Historic Park, but it is little promoted or visited and is used largely for local Polish-American events.

16. Dennis Galvin, interview with the author, November 20, 1980.

17. U.S. Department of the Interior, National Park Service, *Part One of the National Park System Plan*.

18. Lee, *Family Tree*, p. 75.

19. U.S. Department of the Interior, National Park Service, *Part One of the National Park System Plan: History*, pp. vii–viii.

20. Ibid., pp. vii–viii, gives the following schema of American history:

1. The Original Inhabitants
   a. The Earliest Americans
   b. Native Villages and
      Communities
   c. Indian Meets European
   d. Living Remnant
   e. Native Cultures of the
      Pacific
   f. Aboriginal Technology
2. European Exploration and
   Settlement
   a. Spanish
   b. French
   c. English
   d. Other
3. Development of the
   English Colonies,
   1700–1775
   (no subthemes)
4. Major American Wars
   a. The American Revolu-
      tion
   b. The War of 1812
   c. The Mexican War
   d. The Civil War
   e. The Spanish-American
      War
   f. World War I
   g. World War II
5. Political and Military
   Affairs
   a. 1783–1830
   b. 1830–1860
   c. 1865–1914
   d. After 1914
   e. The American Presi-
      dency

6. Westward Expansion,
   1763–1898
   a. Great Explorers of the
      West
   b. The Fur Trade
   c. Military–Indian Con-
      flicts
   d. Western Trails and
      Travelers
   e. The Mining Frontier
   f. The Farmer's Frontier
   g. The Cattlemen's
      Empire
7. America at Work
   a. Agriculture
   b. Commerce and Indus-
      try
   c. Science and Invention
   d. Transportation and
      Communication
   e. Architecture
   f. Engineering
8. The Contemplative Society
   a. Literature, Drama, and
      Music
   b. Painting and Sculpture
   c. Education
   d. Intellectual Currents
9. Society and Social Conscience
   a. American Ways of Life
   b. Social and Humanitar-
      ian Movements
   c. Environmental Conser-
      vation
   d. Recreation in the
      United States

21. Ibid., pp. 54–55.

22. Ibid., p. 76.

23. Max Weber (Gerth and Mills, *From Max Weber*, p. 232), writes that "under normal conditions, the power position of a fully developed bureaucracy is always overtowering. The 'political master' finds himself in a position of the 'dilettante' who stands opposite the 'expert'."

24. Simon, Smithburg, and Thompson, *Public Administration*, p. 524; Woll, *Public Policy*, pp. 186–188; Lipset, *Agrarian Socialism*, p. 75.

25. Paterson, *Glasgow Limited*, Parsons's concept of "the authority of confidence," *Essays in Sociological Theory*, p. 189, is similar.

26. Hays, *Conservation*, p. 3, writes:

Who should decide the course of resource development? Who should determine the goals and methods of federal resource programs? The correct answer to these questions lay at the heart of the conservation idea. (Since resource matters were basically technical in nature, conservationists argued, technicians, rather than legislators, should deal with them.) Foresters should determine the desirable annual timber cut; hydraulic engineers should establish the feasible extent of multiple-purpose river development and the specific location of reservoirs; agronomists should decide which forage areas could remain open for grazing without undue damage to water supplies.

From its very beginnings, the U.S. Forest Service took pains to put forth an image of professionalism. According to Robinson, *Forest Service*, p. 263:

From the early efforts of Pinchot to develop a core of professional foresters to replace the amateurs charged with forest conservation, the service has actively cultivated an image of an institution of professionals with a grand mission and a distinctive competence to perform it.

The U.S. Forest Service has been successful in keeping its image a professional one and this has accounted for much of the autonomy from congressional whims and pressures that it has enjoyed over the years. Fortunately for the U.S. Forest Service, forest management has lent itself to professionalism; it is the object of a scientific discipline with numerous technically oriented journals and degree-granting programs in universities.

27. Brown, *Politics*, p. 90, holds this view when he writes that:

Some programs, by the nature of the problems they are to solve, deal with more complex, technical questions than others. This, of course, often gives a bureau an advantage over others which must base their programs on more subjective information.

28. Helen Ingram, personal communication with the author, December 8, 1980.

29. Long, "Public Policy," pp. 22–24, writes that:

The power that control of the accepted version of the facts gives to any group must be recognized, and that facts are seldom presented in a fashion that is neutral to all interested parties.

30. Sayre and Kaufman, *Governing New York City*, p. 404.

31. Browne, *Politics*, p. 90.

32. Rourke, *Bureaucracy*, p. 111, observes that the ideal professional standards are those which allow the agency to defend its decisions on the basis of objective criteria yet which will "leave room for the exercise of discretion." The agency's 1972 plan was aimed at doing both. Lee, *Family Tree*, writes that the 1972 plan would "undergird" the future growth of the System.

33. Schiff, "Outdoor Recreation," pp. 546–547. Wamsley and Zald, "Political Economy," p. 65, write that:

> Where goals are clearly defined and subject to surveillance, an agency . . . may be left little room for choice or maneuver in goals, program objectives, and perhaps even means of task accomplishment. But if goals are ambiguous or multiple, an organization's elite may press for one definition or another and, within the bounds of political feasibility, allocate resources internally in pursuit of this choice.

Warner and Havens, "Goal Displacement," p. 543, make a similar point:

> The intangibility of goals makes it possible for the organization to accommodate diverse and even inconsistent subgoals. The program can be 'all things to all people' at an abstract level, so that different people can feel satisfied that their interests are being served . . . the intangibility of goals allows flexibility in the interpretation of the claimed goals.

34. See Johnson, *Whitehead's Theory*, pp. 150–151.

35. Rourke, *Bureaucracy*, p. 127.

36. U.S. Department of the Interior, National Park Service, *Part One of the National Park System Plan*, p. ix.

37. See Wildavsky, "The Agency," p. 69. The long-standing refusal of the New York Port Authority to move into the problem-laden field of mass transit, even though it would greatly increase its size and jurisdiction, is a classic case in point.

38. Ise, *Our National Park Policy*, p. 186, shows that this fear, i.e., that the accumulation of inconsequential parks would lessen the agency's capacity to manage the great parks, went back to the first days of the Park Service.

39. Downs, *Inside Bureaucracy*, p. 13, writes of the "decelerator effect," and the problems that slow growth—or no growth at all—can bring an agency.

40. Swain, *Wilderness Defender*, p. 200.

41. William Whalen, interview with the author, December 1, 1980.

42. Ibid.

43. Conservation Foundation, *National Parks for the Future*, p. 65.

44. Ibid., p. 73.

45. William Whalen, interview with the author, December 1, 1980.

46. *Courier: The National Park Service Newsletter*, "Park Briefs," December, 1980, p. 8.

47. U.S. Department of the Interior, *Management Policies*, p. I-7.

48. U.S. Department of the Interior, National Park Service, *Index, 1979*, p. 36.

49. Such urban preservation districts were not unknown before the 1972 plan. For example, Independence National Historical Park, established in 1948, covered 36 acres of downtown Philadelphia. The Jefferson National Expansion Monument, authorized in 1954, included 90 acres of the historic waterfront. Nevertheless, Lowell National Historical Park, covering 134 acres in the center of a city, was unprecedented.

50. Although the Park Service was rather circumspect in print about Lowell's urban renewal and economy-stimulating function, the joint NPS–HCRS *National Urban Recreation Study*, vol. I, p. 180, was straightforward about it:

> *Purposes*: Preserve and interpret sites in City of Lowell, a 19th century mill town, which symbolize the industrial revolution in the U.S.—and revitalize the local economy.

51. There were, however, some groups whose interests were more than purely idiosyncratic, especially the Civil War Roundtable and the Council on Abandoned Military Posts. But even the interest of these groups extended to only a fraction of the agency's historical holdings (and a much smaller fraction of what it hoped to eventually hold). Although these groups showed signs of taking a wider interest in the System's historical holdings in the early 1970s, relations between them and National Park Service leadership became strained to the breaking point when one of the agency's directors expressed aloud, and in terms they took exception to, the opinion that there were already too many forts in the System, and that the Park System did not need a monument to every Civil War general's horse. When this occurred, the chance to use these groups to support the entire historical wing of the System was lost.

52. Conservation Foundation, *National Parks for the Future*, p. 138.

53. Even historic sites in large cities can receive few visitors. The following sites, all in large cities or metropolitan areas, are among the lowest units in the System in visitors per annum. It should also be noted that most of the visits to these sites are local in origin.

Unit	Location	FY 1979 Annual Visits
William Howard Taft NHS	Cincinnati, Ohio	5,000
Theodore Roosevelt NHS	New York City, N.Y.	14,000
Springfield Armory NHS	Springfield, Mass.	16,000
Fort McHenry NM	Baltimore, Md.	16,000
Longfellow NHS	Cambridge, Mass.	18,000

There are exceptions to this "unvisited is beautiful" local viewpoint. For example, in depressed areas such as central Pennsylvania, where the local tourist industry is one of the few things that keeps the economy afloat, the Park Service is encouraged by local chambers of commerce and tourist boards to promote out-of-state visits to the Allegheny Portage and Johnstown Flood units.

In 1975, 25,826,745 acres of the National Park System were in natural areas while only 501,376 were in historical areas. Since there are many more historical

areas than natural areas in the system, the former are, on the average, more than
an order of magnitude larger than the latter. Even most of the larger and better
known historical parks have relatively small staffs; for example:

Park	1981 Authorized Full-Time Personnel
Harpers Ferry	23
Chickamauga—Chattanooga	24
Gettysburg	44
Valley Forge	44

However, the more typical historical parks are much smaller:

Longfellow	5
Tuskegee Institute	6
Hubbell Trading Post	6
Kings Mountain	7
Allegheny Portage Railroad	8

This can be compared to some of the larger natural and recreation areas in the
System:

Everglades	91
Grand Canyon	94
Great Smoky Mountains	103
Golden Gate	124
Gateway	129

54. Lindblom, *Policy-Making Process*, p. 86.

55. Judy Hart, interview with the author, January 27, 1981.

56. See U.S. Department of the Interior, National Park Service, *Management
Policies*, pp. I-1–I-7.

57. Story related to the author by Robert Utley in an interview, December 10,
1980.

58. Ibid. At times even such creative reaction is not possible and the agency
must concentrate on simply saving face. An example of such a face-saving effort
occurred when the agency maintained the National Register of Historic Places. As
related to the author by Utley, Congressman Kirwan (D., Ohio), a senior member
of the House appropriations committee, told the agency, "either the McGuffy
House gets put on the register or there isn't going to be a register next year." The
agency director realized the congressman meant what he said and there was no way
he could keep the boyhood home of the author of McGuffy's Eclectic Reader, a
nineteenth century grammar, off the register. Yet agency historians, who were
responsible for evaluating proposed landmarks, were aghast. They insisted that
they could never support the proposal on professional grounds. The director
finessed the problem by sending an assistant director rather than a mere agency
historian to evaluate the site, ostensibly because of its great importance. Since the
assistant director was a naturalist rather than a historian, he could (and did)
recommend the inclusion of the site in the System without violating his professional principles.

59. *Congressional Record*, vol. 126, no. 174, December 10, 1980, p. 9755.

60. Heckscher, *Open Spaces*, p. 218.

61. Coopers and Lybrand, *"Management Improvement Project,"* p. I-9.

62. William Whalen, agency director in the mid-1970s, had spent much of his previous career in urban parks rather than in the natural areas of the System. As a consequence, he was considered unsuitable for the directorship by much of the agency's rank and file. See Cahn, "Park Service," on this point.

63. Simon, Smithburg, and Thompson, *Public Administration*, p. 54, discuss the role of such vectors of new ideas in public bureaucracies.

64. Utley, "Toward a New Preservation Ethic," p. 7.

65. Downs, *Inside Bureaucracy*, p. 53.

66. U.S. Department of the Interior, National Park Service, *Historic Camden: Study of Alternatives*.

67. PL 95-625, 1978.

68. As Rourke, *Bureaucracy*, p. 79, writes: "If reciprocity is the unwritten, if not altogether inviolable law of political life, an executive can expect that administrative investments (i.e., helping a congressman maintain the support of his constituency) will bring a high rate of political return."

69. Selznick, "Institutional Integrity," p. 218.

70. The National Park Service reserves a job series, ES 170, for its historians, and the number of incumbents in this series stood at approximately 95 during late 1983. Roughly a third of these were assigned to agency headquarters in Washington, D.C., with the balance assigned to agency service centers, regional headquarters, and the parks themselves. The total number of professionally trained historians in the agency is undoubtedly higher, however, since some are working under other job classifications—park ranger and resource preservation specialist, for example.

71. Sharkansky, *Public Administration*, p. 45; Simon, *Administrative Behavior*, pp. 349–353; Downs, *Inside Bureaucracy*, p. 50; Meltsner, "Political Feasibility," p. 862.

72. Sherman, *It All Depends*, p. 135.

73. Redford, *Ideal and Practice*, p. 71.

74. Agency leadership is aware of the fact that its claims to being one of the foremost preservation agencies of the federal government are contingent on respect for its professionalism by other professionals. The National Park Service encourages its historians to remain active within their wider profession and it has tried to involve professional historians outside the service in its policymaking in an advisory capacity.

75. Selznick, *TVA*, p. 57.

76. Hall, "Some Organizational Considerations," p. 463, writes that, "It is commonly assumed that professional norms and organizational or bureaucratic norms inherently conflict, and this conflict is disadvantageous for the professional

in the realization of his goals." On this point also see Blau and Scott, *Formal Organizations*, pp. 244–247; Glazer, *Organizational Scientist*; Etzioni, *Modern Organizations*, p. 76.

77. Rourke, *Bureaucracy*, p. 22.

78. Robert Utley, interview with the author, December 10, 1980.

79. Rourke, *Bureaucracy*, p. 98.

80. Rourke, ibid., pp. 110–112, observes that administrators are wary of allowing professionals too much decision-making power, lest options which are not correct from a professional's viewpoint but which are nevertheless attractive to the administrator be foreclosed.

81. Ise, *Our National Park Policy*, p. 669.

82. Section 106 of the National Historic Preservation Act of 1966 reads that:

The head of any Federal agency having direct or indirect jurisdiction over a proposed Federal or federally assisted undertaking in any State and the head of any Federal department or independent agency having authority to license any undertaking shall, prior to the approval of the expenditure of any Federal funds on the undertaking or prior to the issuance of any license, as the case may be, take into account the effect of the undertaking on any district, site, building, structure, or object that is included in or eligible for inclusion in the National Register. The head of any such Federal agency shall afford the Advisory Council on Historic Preservation established under title II of this Act a reasonable opportunity to comment with regard to such undertaking.

Historians within the agency are responsible for ensuring that Park Service projects are in conformity with Section 106.

83. Utley, "Preservation Ideal," p. 40. The passage continues, giving the historians' view of the proper role of living history at historic sites in the system:

Park interpretation should assist visitors in gaining certain perceptions, understandings and appreciations of the park's resources. Living history programs that sharpen such perceptions are appropriate; those that blur them are inappropriate. Inappropriate living history, moreover, is not merely harmless diversion. The more 'living' it is, the more likely it is to give visitors the strongest impression, and memory, of their park experience. Thus a program that is not usually supportive of key interpretive objectives may be unusually detractive. It is of urgent importance that park officials critically examine the appropriateness of their living history programs.

84. Utley, "Living History," p. 2.

85. Robert Utley, interview with the author, December 10, 1980.

86. Within the Park Service, Allegheny Portage Railroad NHS is considered of so little historic significance that mention of it will sometimes evoke a story explaining its presence within the System, one which merits retelling here. The story has it that the Pennsylvania park was established as a thank-you to Representative John Saylor, a longtime friend of the agency who, nearing the end of his congressional career, intimated to Director Hartzog that he would like to see a national park in his own district. Hartzog had his agency scour the district for something suitable. It was a difficult task since it seemed that nothing much of

historical interest had ever occurred in the congressman's district. The agency, however, finally came up with the remains of a cog railroad and canal system which briefly carried freight over the east-facing slope of the Allegheny Front in the nineteenth century.

87. Twenty-eight percent of 1980 visitors to Allegheny Portage Railroad were local residents. Forty percent were regional residents, i.e., not from towns in the immediate vicinity but whose homes were not more than two or three hours away. (This category extended west to include Pittsburgh and east to include Harrisburg.) Thirty-two percent were from beyond the region, either from the remaining parts of Pennsylvania or from out of state. For most of those in the last category, the park was a secondary destination. Source: Allegheny Portage Railroad NHS visitor records.

88. U.S. Department of the Interior, National Park Service, *Hopewell Village Interpretive Prospectus*, p. 9.

89. Ibid.

90. *Courier: The Newsletter of the National Park Service*, October 1980, "100 VIP's Serve at Turkey Run Farm," p. 7.

91. Ibid.

92. According to the agency's newsletter, "Staff members recognize the many benefits of a strong volunteer program and support it enthusiastically. In turn, the community members have a personal interest in the [park]."

93. Interviews by the author with various park personnel.

Chapter 6. National Urban Parks

1. Lee, *Family Tree*, see pp. 37–60.

2. In 1936 Congress authorized the Park Service to study coastal areas with an eye toward eventually bringing some of the prime seashore into federal ownership (49 Statutes at Large 1982).

3. Ise, *Our National Parks Policy*, p. 254.

4. Russell Olsen, interview with the author, October 9, 1980.

5. U.S. Department of the Interior, National Park Service, *Our Vanishing Shoreline*, 1955. The survey was followed up by three others in the 1950s, *A Report on the Seashore Recreation Survey of the Atlantic and Gulf Coasts*, 1955; *Our Fourth Shore, Great Lakes Shoreline Recreation Area Survey*, 1959; *Pacific Coast Recreation Area Survey*, 1959.

6. ORRRC was authorized by Congress and the enabling legislation was signed by President Eisenhower on June 28, 1958 (72 Statutes at Large 238). The commission consisted of four members appointed from each branch of Congress plus seven members appointed by the president (Laurance Rockefeller was appointed chairman). In addition there was a twenty-five member advisory council consist-

ing of representatives from the various federal departments with a role in recreation plus representatives of state conservation agencies and citizen interest groups.

7. Foss, ed., *Recreation*, p. 579.

8. Ibid.

9. Udall, *Quiet Crisis*, p. 164.

10. Foss, ed., *Recreation*, p. 715.

11. Smith, "Gateways," p. 218.

12. 75 Statutes at Large 149. Foss, ed., *Recreation*, p. 333.

13. 75 Statutes at Large 149, Title VII, Sec. 701(b).

14. 79 Statutes at Large 897.

15. Ibid. Title I, Sec. 6(a).

16. The value of urban open space as bureaucratic turf was evidenced by the way Robert Weaver, head of the Housing and Home Finance Agency, an agency with a rather weak claim to this policy area, bowed out of competition for it with great ceremony and graciously left it to other agencies. See Schiff, "Outdoor Recreation Values" for a discussion of this and other elements of the struggle for the urban recreation turf in the early 1960s.

17. In fact, ORRRC's original legislative mandate ordered it not to take up the question of open space in the nation's large cities, so within city limits ORRRC's recommendations were of little help in deciding the matter.

18. The degree to which this was recognized in Washington during Hartzog's directorship is one of the central points of McPhee's long portrait, "Profiles— George Hartzog."

19. On increased federal spending during this period, see Elizar, "Fiscal Questions."

20. Fitch and Shanklin, *Bureau of Outdoor Recreation*, p. 70. Smith, "Gateways," pp. 215–216, would later express the argument this way:

> For a very modest fee [the National Park Service] provides area, roads, facilities, campgrounds, maintenance, interpretative programs, and personal protection for the visitor who has the money and mobility to get there. There is nothing basically wrong with this federal subsidization of America's relatively affluent in their recreational pursuits; but, in a democracy, we do presume that those opportunities that the government provides to one class of society will also be made available—one way or another—to the rest of its citizenry.

21. Morris, *Cost of Good Intentions*, especially pp. 107–108.

22. Smith, "Gateways."

23. An illustration of the differences which could arise before custom became established and experience accumulated occurred at Fire Island. Udall threw his support behind the idea of a unit of the National Park System on Fire Island, in part because it fit into his ideas of the National Park Service's urban mission. At its easternmost point, Fire Island was only 20 miles from New York City and the

Mellon report saw it as a natural extension of the metropolitan public beaches which were just to the east of it. Hartzog, however, disagreed and thought it was an inappropriate site for a truly metropolitan facility. In the end Hartzog went along with Udall—he had little choice but to do so—but thereafter he referred to Fire Island National Seashore as "the Secretary's park."

24. Seidman, *Politics*, p. 158.

25. Ibid., p. 122. As Seidman pointed out, "program transplants which are alien to the institutional culture . . . are threatened with rejection."

26. Rourke, *Bureaucracy*, p. 130.

27. Fitch and Shanklin, *Bureau of Outdoor Recreation*, p. 83. Elizar, "Citizenship," p. 1, complained that the 1960s had seen government redefine citizens as clients of services it could provide. The competition between the Bureau of Outdoor Recreation and National Park Service to directly deliver personal recreation services seemed to support this notion; public service delivery was the field on which the battle for power was being fought.

28. It was precisely this type of charge that Hartzog was countering when, in one of his first appearances before Congress, he was asked to make clear his agency's priorities and he responded that "Primarily the National Park Service is a people-serving agency. In addition to that, it is also a resource-managing agency." (Congressional Research Service, Library of Congress, *Review of National Park Service Policies*, p. 23.)

29. The Bureau of Outdoor Recreation also continued to conduct studies of federal recreation resources. But this was viewed as a fairly innocuous activity by other federal agencies since the bureau depended on them for data and it tended to be very restrained in making specific recommendations.

30. When the Land and Water Conservation Fund was originally set up, a sixty-forty ratio of direct federal acquisition expenditure to local grants was envisioned. However, federal agencies were more active in drawing up projects for fund sponsorship than were state and local governments, and the ratio of federal to other expenditures eventually settled at approximately fifty–fifty. This ratio became a convention. The Bureau of Outdoor Recreation would not accept it, however, and according to Nienaber and Wildavsky, *Budgeting and Evaluation*, p. 78, it "kept pushing for the original 60-40 split because it strongly felt that 'the need was in the cities' and that the state assistance program could better meet the need than could the federal portion."

One place where lax procedures were the rule was with regard to Statewide Comprehensive Outdoor Receation Plans (SCORP). These reports, detailing a state's recreation needs and plans to meet them had to be prepared by a state and approved by the Bureau of Outdoor Recreation before it became eligible for the Land and Water Conservation Fund money. Ostensibly, the purpose of the Statewide Comprehensive Outdoor Recreation Plan was twofold. It was to (1) encourage statewide recreation planning and (2) permit nationwide data on outdoor recreation resources and needs to be built up from a base of state-level data. However, the scheduling and, to a large extent, the format of the SCORP reports were left up to the states themselves. As a result, their submission dates and content (as well as quality) varied widely from state to state. Although this made the reports useless as a base for any national planning, the Bureau of Outdoor Recreation hesitated to insist on uniformity for fear of offending the states.

31. Heckscher, *Open Spaces*, p. 9.

32. In addition to these benefits, direct federal open space management could overcome the coordination problems caused by metropolitan fragmentation better than could federal aid. Rowntree, Heath, and Voiland, "United States National Park System," p. 111, recognized this when they wrote that:

> A powerful argument can be made in support of a direct federal role. Open space for urban recreational sites is being usurped at a dramatic rate by other types of land developement on the periphery of U.S. cities. Multiple metropolitan jurisdictions and the continuing urban financial crisis prevent most cities from taking the rapid, forceful actions necessary to develop these areas into larger parks. Hence an urgent 'need' appears to exist which cannot be met at the state or municipal level and which therefore seems to require direct federal action.

33. For a detailed treatment of this expansion, see Lee, *Family Tree*, pp. 61–86.

34. Sharpe, "National Parks," p. 200.

35. Russell Dickenson, interview with the author, April 3, 1981.

36. Sharpe, "National Parks," p. 200, saw the connection between the D.C. parks program and an urban commitment in much the same light: "The successes of the Summer in the Parks program have given rise to an expanded notion of Park Service responsibility to urban dwellers, called Parks to the People."

37. McPhee, "Profiles—George Hartzog," p. 46.

38. Ibid.

39. Morris, *Cost of Good Intentions*, p. 32.

40. U.S. Department of the Interior, Bureau of Outdoor Recreation, *Recreation Imperative*, p. 9.

41. For example, Nixon, "Environmental Priorities," in Nash, ed., *American Environment*, p. 251, sounded very much like President Johnson when he said that:

> Our challenge is to find ways to promote the amenities of life in the midst of urban development: in short, to make urban life fulfilling rather than frustrating. Along with the essentials of jobs and housing, we must also provide open spaces and outdoor recreation opportunities, maintain acceptable levels of air and water quality, reduce noise and litter, and develop cityscapes that delight the eye and uplift the spirit.

42. Nixon seemed to prefer the indirect approach of grants to states and municipalities. Apparently, he also feared that federal recreation expenditures were in danger of getting out of hand. In 1970, under Hickel, the Bureau of Outdoor Recreation completed the nationwide outdoor recreation plan which Congress had ordered it to draw up seven years previously (PL 88-29). The plan, entitled *The Recreation Imperative*, was a document of great ambition and fiscal commitment

which, among other things called for "a concentrated effort to provide recreation in the major metropolitan areas." The report met with administrative disfavor and was never released, in spite of Hickel's strong backing. Finally, in 1974, Senator Jackson made the report available as a committee print of the Senate Committee on Interior and Insular Affairs, of which he was chairman. In his letter of transmittal (June 19, 1974) then Bureau of Outdoor Recreation director James Watt wrote that "it needs to be restated that none of the drafts which we are submitting to you have ever received final approval."

43. Smith, "Gateways," p. 220.

44. Ibid., pp. 221–236, provides an excellent, detailed account of the campaigns which led to the establishment of the gateway parks.

45. Nathan, "New Federalism," p. 111, writes that: "Although the criteria were not systematically arrayed, a reasonably coherent strategy emerged in the Nixon years for selecting functions to be decentralized and those for which central government action was deemed to be necessary and appropriate."

46. Siehl, "Alternative Strategies," p. 24.

47. Smith, "Gateways," p. 219. Also see *American Forests*, "Former Directors Speak Out."

48. PL 88-587, September 11, 1964.

49. Foss, ed., *Recreation*, p. 661.

50. Ibid.

51. See Tannenbaum, *Preservation of Open Space*, pp. 190–211, for a description of Udall's actions regarding Fire Island and an interpretation of the motives behind these actions.

52. Lee, *Family Tree*, p. 77, wrote of the national seashores that they were to provide "important outdoor recreational opportunities in a natural environment while holding back from offering facilities for mass recreation in the Jones Beach style of highly intensive use."

53. Selznick, "Foundations," p. 272.

54. As to the problems left behind by the sheer ambition of the Great Society program, Nakamura, *Politics of Policy Implementation*, p. 11, writes:

When Johnson succeeded John F. Kennedy as president, he could draw upon his superb background as a legislative tactician, but he lacked extensive administrative experience. His strength, which he used with consummate skill, was in influencing Congress to pass his legislation. Within a short time Congress responded by approving a series of diffuse policies designed to alleviate major social problems such as poverty, juvenile delinquency, unemployment, urban decay, racial and sexual discrimination, and a host of other social concerns.

It was not long before disillusionment began to set in as it became apparent that it might be easier to 'legitimize' social policy by passing ambiguous legislation than to carry out such policy by means of effective program implementation.

55. Walter Hickel quoted in "Urban and Urgent," *New York Times*, editorial, May 5, 1972, p. 42.

56. See Runte, *National Parks*, and Sax, "America's National Parks" for discussions of the campaigns which led to the establishment of the first great national parks.

57. Of course there were some local tourist interests which were in favor of the establishment of the parks but the fact remains that the overwhelming support for the establishment of the great national parks was nonlocal. It was only after the parks had developed a mass clientele that a politically potent tourist industry based on them developed.

58. Hays, *Conservation*, pp. 261–276.

59. Tannenbaum, *Preserving Open Space*, pp. 191–211.

60. Lewis Albert, interview with the author, March 4, 1981.

61. Platt, *Open Space Decision Process*, p. 155.

62. A major exception was the Delaware Water Gap National Recreation Area, which was planned in conjunction with a dam and water storage project to be undertaken by the Army Corps of Engineers. There the initiative came from New York City, which wanted the drinking water, plus the federal government and a host of regional interests which stood to gain from it in a number of ways. The residents of the local municipalities organized to fight the project and were aided by national environmental groups. However, this was a metropolitan project per se; it was unambiguously intended to serve the entire metropolitan system, both as a source of water and of recreation opportunities.

63. George Biderman, interview with the author, December 22, 1980.

64. Caro, *Power Broker*, pp. 226–240, 499–509

65. Simon, "Prospect for Parks."

66. Morris, *Cost of Good Intentions*, p. 57.

67. Caro, *Power Broker*, p. 514.

68. Altshuler, "Politics of Transportation Innovation."

69. U.S. Department of the Interior, National Park Service, *Fire Island General Management Plan*, p. 13.

70. Ibid., p. 15. On p. 23 the plan states that "the general plan recognizes that Fire Island National Seashore serves a definable population of known and potential visitors. Basically Fire Island now serves, and will continue to serve, the recreational needs of Suffolk and Nassau counties and to a lesser degree the needs of New York City."

71. The officials of cities and towns with large minority populations—like Newark—would have liked to have seen public transit to Gateway instituted and they said so in public. However, they had neither the political nor the economic means to get the transportation they wanted.

72. In the early 1970s Zimmerman, "Neighborhoods," p. 201, wrote that:

Professionalism and the tenets of the municipal reform movement are under attack in large American cities by populists who contend that 'institutional arrangements have maximized the wrong values by placing too much emphasis upon centralization of power in city hall, professionalism, economy, and efficiency, and paying too little attention to responsiveness as a criterion of democratic government.'

73. U.S. Department of the Interior, National Park Service, *Management Policies*, pp. II-2–II-9. Both the Park Service and local citizens who opposed expansion of facilities used the park advisory boards for their own ends. For example, at Santa Monica the agency hoped to use the board to gain some freedom of action for its planning process by encouraging its citizen members to become actively involved in agency planning and thus identify with agency plans, once formulated. At Fire Island, local citizens on the board sometimes went along with small efforts on the part of the National Park Service to increase general park access, fearing that if the agency were completely thwarted it might want to pull up stakes and leave. From the local point of view, this would have been as bad as a development plan which allowed for mass access, so they were willing to make tactical concessions when they felt it necessary.

74. U.S. Department of the Interior, National Park Service, *Gateway National Recreation Area General Management Plan*, 1979. The few exceptions were from groups which spoke for the city's minority residents.

75. Smith, "Gateways," p. 224, writes that "the citizens' committee appears to be a well-motivated and fairly effective single-issue citizens' organization." Among its members (as of August 1980) were Rene Dubos, a distinguished scientist; Robert F. Wagner Jr., former mayor of New York City; Thomas Kean, future governor of New Jersey.

76. Smith, "Gateways."

77. Ibid., p. 214.

78. Ibid., p. 221, writes that New York and San Francisco seemed "to look on the projects as a way in which to unload some of their own park responsibilities on the federal government and, in so doing, alleviate the strain on their park and recreation department budgets."

79. Shankland, *Steve Mather*, p. 182.

80. Morris, *Cost of Good Intentions*, pp. 147–170.

81. Smith, "Gateways," p. 232.

82. *New York Times*, "Gateway Park Bill Gains Favor of House Members," January 21, 1971, p. 41.

83. Morris, *Cost of Good Intentions*, pp. 56–82.

84. This was something Mayor Lindsay learned the hard way when he allowed himself to be perceived as being too biased in favor of the demands of minority groups for housing, public services, and the attention of the city government, naturally at the expense of the whites and their neighborhoods.

85. The fact that the city administration had little interest in a park which lived up to its rhetoric does not mean that city hall's involvement was purely cynical. It looked at the project primarily as a conduit of aid to be made as wide as possible

and they wanted the aid to be delivered in the most useful form, certainly reasonable goals. Unfortunately, from the author's interviews with National Park Service officials involved in Gateway's initial planning, it appears that the agency was not aware of the possibility that the city government's goals for the urban park might be other than those expressed in the rhetoric which accompanied their establishment. As a result, the Park Service expected the city government to be a staunch supporter of its plans to establish a truly metropolitan facility, and it never ceased to be surprised when it acted more like an opponent, or like an arbitrator between the National Park Service and the local citizens who opposed the agency's ambitious initial plans.

86. FY 1981 Budget Testimony, vol. XI, p. 379.

87. Sax, *Mountains Without Handrails*, p. 84.

88. Ibid.

89. William Whalen, interview with the author, December 1, 1980. Lewis Albert, interview with the author, March 4, 1981.

90. Lewis Albert, interview with the author, March 4, 1981. See Clawson and Knetsch, *Economics of Outdoor Recreation*, pp. 36–40, for a typology and hierarchy of open spaces.

91. Heckscher, *Open Spaces*, pp. 232–233. Turkey Run Farm is on the George Washington Parkway in Virginia. It is now called the Claude Moore Colonial Farm and is administered by the Friends of Turkey Run.

92. Sax, "America's National Parks," p. 74.

93. American Public Works Association, *History of Public Works*, p. 66.

94. See Chadwick, *Park and the Town*, especially chapter 9, pp. 163–220, for a discussion of Olmsted's work within the context of the North American and European traditions of park design.

95. Gifford Pinchot, quoted in McConnell, "Conservation Movement," p. 470.

96. This view that the parks should be preserved as natural landscapes for a national constituency was undoubtedly encouraged by the raw and ramshackle nature of the cultural elements in the open spaces of the West, elements which created according to Lowenthal, "American Scene," p. 62, "the contrast between landscapes fit for the gods and their mean and petty human inhabitants."

97. Ise, *Our National Park Policy*, pp. 262–263.

98. The strong preference of those holding a metropolitan view for straightforward full-fee acquisition is evidenced by the name given them by an agency administrator (who was himself identified with the localist view); he called them "buy-the-bastards-out planners."

99. At the Delaware National Recreation Area, where there was tremendous local resistance to the Tocks Island Dam, the same planner argued strongly for the dam. He took a metropolitan perspective to justify it, "I supported the [dam] because the whole [New York] region is going to need the water someday."

100. This localist view is expressed clearly in the agency's draft environmental impact statement for Golden Gate National Recreation Area:

> But it is only a matter of time before higher density proposals are implemented. Roads will have to be widened and new water and sewage facilities will be required. Wildlife habitats will be destroyed or disturbed by the increased density of population and the adverse effects of more cars on the highway and more people in the once remote areas. It would appear that a substantial and irreversible change in the environment would occur throughout the open lands and adjoining communities if the proposed national recreation area is not established. (Quoted in Smith, "Gateways," p. 234.)

101. Elazar, "Citizenship," p. 8.

102. In the past decade, agency leadership has usually been somewhere between the metropolitan and localists' viewpoints, trying to appease both sides and, in general, steering the middle course ideologically. However, one director was clearly a partisan of the localists. William Whalen, himself very young, said of the Fire Island National Seashore, "I love that park, I love it because we beat Robert Moses there, we kept him from ruining it." Interview with the author, December 1, 1980.

103. Selznick, *TVA*, p. 48, fn. 1, writes that "In the TVA, the group of younger men whose careers were bound up with the fate of the Authority was far more ready to adapt itself to new policies framed in terms of 'realistic' adjustment than was the earlier leadership, the reputations of whose members were already established."

104. Max Weber, in Gerth and Mills, *From Max Weber*, pp. 62–63, 284–285, discusses the characteristics of this attraction and calls it "elective affinity."

105. Department of the Interior, National Park Service, *Gateway: A Proposal*, unpaged. Recreation Advisory Council to the U.S. Department of the Interior, *Policy Circular No. 1*, March 26, 1963, reads that, "National recreation areas should be located and designed to achieve comparatively heavy recreation use and should usually be located where they can contribute significantly to the recreation needs of urban populations."

106. U.S. Department of the Interior, National Park Service, *Gateway General Management Plan*, 1979.

107. Ibid., p. 43. When and if a ferry service is established, this will bring an additional 10,000 visitors to Breezy Point. The Department of the Interior's 1978 report, *National Urban Recreation Study*, p. 104, concluded that whatever their original intentions, national urban recreation areas, such as Gateway National Recreation Area in the New York City metropolitan area, primarily benefit local residents.

108. U.S. Department of the Interior, National Park Service, *Gateway General Management Plan*, 1979. At Fire Island, where local antidevelopment pressures were also intense, agency planning showed the same respect for the quality recreation experience. *Fire Island General Management Plan*, p. 23, reads that "Increased use of federal recreational areas will be minimal, with emphasis placed on the quality of facilities and visitor experiences and not on quantitative increases."

109. The Bureau of Outdoor Recreation, in its *Water-Oriented Outdoor Recreation*, recommends a minimum of 75 square feet of beach per person. The Army Corps of Engineers, in its *Delaware River Basin Report*, recommends a minimum of 50 square feet. The Wisconsin Conservation Department, *Comprehensive Plan*, recommends a minimum of 100 square feet in urban areas and 200 square feet in rural areas. U.S. Soil Conservation Service, *Recreation Memorandum no. 3*, suggests a minimum of between 50 and 100 square feet. Louisiana Parks and Recreation Commission, *SCORP, Supplement 1*, recommends a minimum of 110 square feet.

110. The historical designations were dismissed as unwarranted by New York's chief historic preservation officer (letter from Orin Lehman to agency director Whalen, July 14, 1978). The military structures the agency inherited at Breezy Point were described by an agency historian who had worked on the Gateway development plan as "just so much junk and surplus property."

111. U.S. Department of the Interior, National Park Service, *Gateway General Management Plan*, p. 34.

112. Ibid., p. 5. The urban national park has been seen as an especially valuable site for making dramatic and educating points about the proper ordering of man's relationship with his environment. For example, Sax, *Mountains Without Handrails*, p. 84, writes that: "The urban-regional park provides an ideal opportunity to show city dwellers that the psychology of the spoiled child is not the only choice open to us; that we can draw satisfaction by accommodating to natural forces as well as by harnessing them."

113. Almost all the written protests engendered by the 1979 General Management Plan were from the park's neighbors and most complained about the level of park access envisioned in the plan, never mind that peak day access had been scaled down over 70 percent from original plans. See U.S. Department of the Interior, National Park Service, *Gateway General Management Plan*, pp. 177–244 for the letters.

114. Ibid., pp. 228–244.

115. Ibid., p. 5.

116. To a large extent, such a retreat from stated goals was covered by a notion abroad within the public administration profession that "conflict resolution" or "conflict management" were goals in and of themselves. Nakamura, *Politics of Policy Implementation*, p. 38, writes approvingly that "for the policy maker the job of getting people to agree on a policy can become a goal in itself," while Irland, "Citizen Participation," p. 264, writes that "We may not be able to create a plan that delights everyone, but we can hope to construct a process of planning that leaves participants feeling that their needs have been fairly considered." Of course, as Warner and Havens, "Goal Displacement," point out, such positions simply reflect goal displacement, that is to say, the neglect of initially claimed goals and elevation of means to the status of ends in themselves. For the Park Service, the currency of such ideas, elevated to the status of chic public administration theory, justified bowing to local pressure.

117. U.S. Department of the Interior, National Park Service, *Cuyahoga Valley National Recreation Area, General Management Plan*, p. 17. Both Rourke, *Bureaucracy*, p. 57, and Altshuler, "Politics of Urban Transit," have noted the tendency within American public administration to smother political conflict by, in Rourke's words, "transforming political issues into technical problems." Of course, the very definition of a problem as a technical one can, as it does in this case, represent a victory of one side over the other.

118. U.S. Department of the Interior, National Park Service, *Cuyahoga Valley NRA General Management Plan*, p. 3, summarizes its "planning concepts" thus:

In seeking to identify Cuyahoga's 'essence' as a national recreation area—the type of park it can and should be—it is critical to consider this regional setting. The Cuyahoga Valley lies near rather than within the Cleveland/Akron urban sphere, and it resides in sharp contrast to its surroundings. It preserves a landscape reminiscent of simpler times, a place where recreation can be a gradual process of perceiving and appreciating the roots of our contemporary existence. Urbanization and growth are facts of life in this region, facts manifested daily to city residents. And as the metropolitan areas continue to spread and urbanize, the Cuyahoga Valley becomes more distinct, more significant, more valuable, and the importance of preserving it as a place to seek alternatives to urban lifestyles magnifies.

The summary continues:

In recognition of Cuyahoga's value within the regional milieu, planning for the park is based on the idea of open-space preservation rather than facility construction, of recreational settings rather than formalized developments. The necessity for contemporary facilities to serve the varying recreational interests of our cities' residents is obvious and pressing—and the Cleveland/Akron area is no exception. But in a gradually deteriorating environment where fewer and fewer places allow us time and space to rediscover the beauty of nature, the peace of the countryside, or the substance of our past, the need to protect landscapes that refresh the spirit and restore our perceptions has become one of the most basic requirements of recreational planning.

119. U.S. Department of the Interior, National Park Service, *Santa Monica Mountains NRA General Management Plan*, pp. 23–27.

120. Smith, "Gateways," pp. 228–229.

121. U.S. Department of the Interior, National Park Service, *Golden Gate NRA*, unpaged interpretive brochure.

122. When the National Park Service proposed a plank bicycle path running behind the dunes, the Fire Island Homeowner Association waved the bloody shirt left over from the campaign against Moses's beach road and dubbed the proposed bicycle path "the mini expressway." (George Biderman, interview with the author, December 22, 1980.)

123. U.S. Department of the Interior, National Park Service, *Fire Island National Seashore, Final General Management Plan*.

124. U.S. Department of the Interior, National Park Service, *Fire Island National Seashore, Draft Master Plan*, p. 64.

125. *Village Voice*, "Fire Island data insert," November 18–25, 1980, p. 50.

126. U.S. Department of the Interior, National Park Service, *Fire Island National Seashore Park Development Plan*, p. 63.

127. Browne, *Politics*, p. 164. Ingram and Ullery, "Policy Innovation," p. 674, suggest another reason for organizations to overestimate the success of past policies when they write that:

> Large, irretrievable, sunk costs distort an organization's perception of the marginal benefits and costs of change. The future marginal benefits of innovation come to be discounted in light of the past commitments of resources which would have to be reversed. This is especially painful when some innovation could be taken as an admission of past error or would highlight some poor decisions made in the past.

128. Cahn, "Park Service," p. 22. In a general treatment of organizational change and the internal resistance which greets it, Kaufman, "Limits of Organizational Change," pp. 12–13, writes that "changes in organizational structure or behavior may be opposed on the grounds that they would impair the quality of goods or services rendered." This has certainly been a prominent internal argument against urban national parks.

129. Coopers and Lybrand, "Management Improvement Report," p. I-39.

130. In 1976 Horace Albright, quoted in *American Forests*, "Former Directors Speak Out," said that "No one can foresee the end of additions to the urban park areas as city after city receives the same or similar consideration that has been accorded to New York, San Francisco and Cleveland-Akron." Assistant Secretary of Interior Reed, summarizing the arguments against urban national parks, quoted in Conservation Foundation, *Letter*, July/August 1975, p. 3, raised a similar point:

> A direct federal urban role would politicize congressional consideration of park proposals far more than is now the case. New parks may be created not because of national merit but because of local congressional clout—a process akin to the pork-barrel approach by which many public works projects are federally financed.

131. *Congressional Record* vol. 120, December 9, 1974, H11427-8.

132. See Conservation Foundation, *Letter*, July/August 1975, pp. 2–6; Cahn, "Park Service"; Rowntree, Heath, and Voiland, "National Park System," pp. 110–114.

133. U.S. Department of the Interior, National Park Service, "Budget Justifications, F.Y. 1981," pp. 134–135.

134. Ibid., pp. 75–92.

Chapter 7. Beyond Park Boundaries

1. Letter from Franklin K. Lane to Steven Mather, May 13, 1918, reprinted in Foss, ed., *Recreation*, pp. 158–189.

2. Shankland, *Steve Mather*, pp. 86–87.

3. Quoted in Lee, *Military Park Idea*, p. 40.

4. Today, only 1,298 of the park's 3,300 acres are in federal ownership.

5. U.S. Department of the Interior, Heritage, Conservation and Recreation Service/National Park Service, *National Urban Recreation Study*, vol. I, p. 229.

6. Lowenthal, "American Scene," p. 81.

7. Ibid., p. 84.

8. Lynch, *Managing*, p. 87.

9. Udall, *Quiet Crisis*, p. 157.

10. This discovery is reflected in the way Udall, *Quiet Crisis*, p. 177, characterizes litter as a universal problem: "It was a sad fact, also, that the men, women, and children of America the Beautiful became the litter champions of the world. . . . Our landscape litter has reached such proportions that in another generation a trash pile or piece of junk will be within a stone's throw of any person standing anywhere on the American land mass." With this concern for the total landscape, the developers and the preservationists necessarily parted ways. A preservation movement which concentrated on saving a few unique places was perhaps a threat to the few who had a financial stake in developing them. A movement which concerned itself with the aesthetic and ecological integrity of the entire landscape, however, was another matter entirely, since there was no limit to whom it might affect through restrictions and reduced opportunities for profit. Foresta, *Open Space Policy*, found that land developers in New Jersey were far more willing to accept programs which aimed at preserving isolated parcels of land through state acquisition than programs which involved no state acquisition but which aimed at regulating entire regions, like coastal or mountain areas.

11. Nixon, "Environmental Priorities," in Nash, ed., *American Environment*, p. 251.

12. Babcock, *Zoning Game*, p. 11. In the 1920s concern for land use had manifested itself in the widespread adoption of municipal zoning. The federal government's Department of Commerce under its secretary, Herbert Hoover, developed the Federal Standard and State Zoning Enabling Act in 1924. It was intended to serve as a model for state zoning legislation. It was widely adopted and, in fact, according to Boschken, "Public Control," p. 496, most state zoning legislation is still based on it. In 1926, the Supreme Court became involved in public regulation of private land use when it weighed and then upheld the basic constitutionality of land use zoning in the landmark *Village of Euclid* vs. *Ambler Realty* case. After this, federal interest in private land use diminished sharply. According to Windsor, "Political Economy," p. 396, "The federal judiciary systematically withdrew from this entire field and left review of the local land use policies to the state courts." Executive and legislative branches of the federal government followed suit in abandoning concern for public land use control. This lack of interest was broken only by occasional assertions by the federal government that the states were sovereign in this area. The states in turn delegated land use administration to the county and municipal governments. With regulation left to the lowest levels of government, zoning and planning anarchy was frequently the norm, policies varied widely from one municipality to the next, and within munic-

ipalities local zoning boards were frequently no match for land developers in power or legal skills.

13. Reilly, *Use of Land*, p. 34. Coke and Brown "Public Attitudes," detected a similar change of public consensus and a rejection of "traditional local government routines" in land use regulations.

14. Reilly, *Use of Land*, p. 34, also wrote that:

The new mood reflects a burgeoning sophistication on the part of citizens about the overall, long-term economic impact of development. Immediate economic gains from job creation, land purchases, and the construction of new facilities are being set against the public costs of schools, roads, water-treatment plants, sewers, and the services new residents require. . . . But the new attitude toward growth is not exclusively motivated by economics.

15. Coke and Brown, "Public Attitudes."

16. Plotkin, "Policy Fragmentation." Walker and Heiman, "Quiet Revolution," make essentially the same argument, i.e., that the emergence of land use control as a national issue can be traced directly to the changing nature of the building industry, and specifically to (1) the entry of large corporations into the field and (2) the increasing scale of land development.

17. Plotkin, "Policy Fragmentation," p. 419.

18. Most of the national leaders who advocated increased ordering of the development process through comprehensive planning attributed previous chaos to a pervasive individual selfishness and carelessness with land. According to Nixon, "Environmental Quality," the American had been too much concerned with his personal rights and too cavalier in his relationship to nature, too willing to treat our land as if it were an unlimited resource. The reasons for this, according to Nixon, were rooted in the way Americans felt what they did with land was their own business, and this attitude was a natural outgrowth of the pioneer spirit. The cause of this new interest on the part of the American public was, according to Nixon, a new clarity of vision. "Today we are coming to realize that our land is finite."

Earlier, Udall and Johnson had both attributed chaotic land development and pollution problems to a collective national carelessness. In *Quiet Crisis*, p. 177, Udall unfavorably compared the contemporary attitude toward land with that of the continent's former proprietors: "Our irreverent attitudes toward the land and our contempt for the Indians' stewardship concepts are nowhere more clearly revealed than in our penchant to pollute and litter and contaminate and blight once-attractive landscapes." If subdividers made quick profits from ill-conceived projects that ruined valley streams and obliterated open spaces, it was because "we" were not concerned enough with what was happening. Just as was the case with environmental degradation, rampant, ill-planned land development was, at its core, a problem of public attitude. The solution therefore would have to grow out of a change in popular attitude.

19. John F. Kennedy, quoted in Plotkin, "Policy Fragmentation," p. 414.

20. Udall, *Quiet Crisis*, p. 165.

21. National Commission on Urban Problems, *Building the American City*; Advisory Commission on Intergovernmental Relations, *Urban and Rural America.*

22. Healy and Rosenberg, *Land Use*, p. 6.

23. See ibid. for a detailed and perceptive discussion of these state-level planning initiatives. Reflecting (and undoubtedly encouraging) the "new spirit," the American Land Institute (ALI), a research and public relations wing of the American Bar Association, published its Model Land Development Code in 1976. It was intended to replace the U.S. Commerce Department's 1924 model for state land statutes. Unlike its predecessor, the ALI's model code stressed retention of much land use control at the state level.

24. Williams, "On from Mount Laurel."

25. Plotkin, "Policy Fragmentation," p. 422.

26. *Proceedings of the National Conference on Outdoor Recreation*, Sen. Doc. 151, 68 Cong. 1 sess. (1924) p. 125.

27. Russell Train, quoted in Fitch and Shanklin, *Bureau of Outdoor Recreation*, p. 5.

28. Conservation Foundation, *National Parks for the Future*, p. 87.

29. Plotkin, "Policy Fragmentation."

30. Conservation Foundation, *National Parks for the Future*, p. 62.

31. U.S. Department of the Interior, Bureau of Outdoor Recreation, *Recreation Imperative*, p. 346.

32. According to McPhee, "Profiles—George Hartzog," Hartzog saw upgrading the quality of water in the national wild and scenic rivers under his charge as one of his major responsibilities, and he knew that once Gateway National Recreation Area came into the System, one of his biggest concerns there would have to be reducing the pollution of its ambient waters.

33. 42 U.S.C. 4321 et seq. 1969.

34. U.S. Department of the Interior, Bureau of Outdoor Recreation, *Recreation Imperative*, p. 347.

35. Lee, *Family Tree*, p. 65.

36. Conservation Foundation, *National Parks for the Future*, pp. 88–89.

37. Nicholson, "What Is Wrong," p. 37.

38. Glazer and Moynihan, *Beyond the Melting Pot.*

39. Conservation Foundation, *National Parks for the Future*, p. 138.

40. U.S. Department of the Interior, Bureau of Outdoor Recreation, *New England Heritage.*

41. U.S. Department of the Interior, Heritage, Conservation and Recreation Service/National Park Service, *National Urban Recreation Study*, vol. I, p. 50.

42. Sax, "Helpless Giants," writes that the adverse impact of these activities on the parks has become a major issue only in recent years.

43. Ibid.

44. Related to the author by Thomas Harrison in an interview, March 6, 1981.

45. One source of power available to park superintendents was derived from their control of park access points. Since much development near the parks was aimed at servicing park visitors, the closing of a park entrance could have a serious impact on local commercial interests. In at least one instance, a superintendent's threat to close a park entrance was enough to convince local entrepreneurs that they should heed the superintendent's advice on development matters.

46. For a brief summary of NPCA interest in land development beyond the parks, see Rushing, "NPCA: Sixty Years."

47. Chadwick, "Cutting Glacier," p. 13.

48. Ibid.

49. Ibid., p. 17.

50. Conservation Foundation, *National Parks for the Future*, p. 46.

51. Everhardt, "A New Look," p. 54.

52. Shands, *Federal Resource Lands*.

53. U.S. Department of the Interior, National Park Service, *State of the Parks— 1980*.

54. Ibid., p. viii.

55. Ibid., p. 1, writes that:

For the first time, events occurring external to park boundaries began to be recognized as the cause of serious damage to park values. These adjacent activities included residential, commercial, industrial and road development; grazing; logging; agriculture; energy extraction and production; mining; recreation; and a myriad of others.

56. Conservation Foundation, *National Parks for the Future*, p. 74.

57. Lee, *Family Tree*, p. 43, writes that:

Partly because of the increasing difficulty of adding new Natural Areas to the System, the Service launched a Natural Landmarks Program in 1962. Its purpose was to recognize and encourage the preservation of significant natural lands by diverse owners, mostly nonfederal, including state or local governments, conservation organizations, and even private persons.

58. Nature Conservancy, *Preserving Our Natural Heritage*, vol. I, p. 285.

59. Comptroller General of the United States, *Federal Drive*, p. v. For a similar conclusion, see U.S. General Accounting Office, *Federal Land Acquisition*.

60. Foresta, *Open Space Policy*, found that although the provision of public open space is one of the most popular activities a government can undertake, condemnation of private land for parks can quickly lead to political trouble since

parks are popularly viewed as less essential than other things for which land is condemned, i.e., roads, schools, public offices, etc.

61. U.S. Department of the Interior, Heritage, Conservation and Recreation Service/National Park Service, *National Urban Recreation Study*, vol. I, p. 158, writes that, "Cape Cod National Seashore, authorized in 1961, was the pioneer greenline area in this country." According to one national park administrator, it was "an important prototype for the [greenline] idea." Cape Cod was similar to the [greenline] concept in the way its development plan called for a permanent mix of public and private land, and in the way it mixed respect for, and desire to, preserve a settled landscape with efforts to make it accessible to the public. For a more detailed description of the Cape Cod formula, see Nature Conservancy, *Preserving Our Natural Heritage*, vol. 1, pp. 30–31.

62. Mosher, "Congress Considers," p. 1658.

63. Ibid.

64. William Whalen, interview with the author, December 1, 1980.

65. U.S. Department of the Interior, National Park Service, various internal documents.

66. See U.S. Comptroller General, *Federal Drive*. Also see U.S. Department of the Interior, Heritage, Conservation and Recreation Service/National Park Service, *National Urban Recreation Study*, p. 46.

67. Sharkansky, *Public Administration*, p. 44.

68. Downs, *Inside Bureaucracy*, p. 18.

69. Jack Hauptman, interview with the author, September 25, 1980.

70. Orlans, "Political Uses," p. 31, found that such problem definition was not uncommon. He wrote that: "Once a particular profession becomes entrenched in an agency or institute it is not easily dislodged. Once this happens the organizations often concentrate on the methods and theories of one profession . . . rather than exploring whatever methods and theories are most pertinent to the problem at hand."

71. The value of easements of New Jersey farmland as discovered by Boyce, Kohlhase, and Plaut, "Estimating the Value," p. 239, are as follows:

County	Market value (per acre)	Current use value (per acre)	Development easement value (per acre)	Development easement value as a percent of market value
Burlington	$1,900	$ 400	$1,500	79
Cumberland-Salem	1,100	1,100	100	9
Hunterdon	1,500	300	1,200	80
Warren	1,700	300	1,400	82

72. U.S. Department of the Interior, Heritage, Conservation and Recreation Service/National Park Service, *National Urban Recreation Study*, p. 229.

73. National Park Service, "Scenic Easements," unsigned internal memorandum, p. 6.

74. Ibid., p. 6.

75. U.S. Comptroller General, *Federal Drive*, p. 23. See p. 25 of this report for a list of the advantages and disadvantages of less-than-fee acquisition as the U.S. General Accounting Office saw them.

76. National Park Service, "Scenic Easements," unsigned internal memorandum, pp. 8–9.

77. Ibid., p. 2.

78. U.S. Department of the Interior, Heritage, Conservation and Recreation Service/National Park Service, *National Urban Recreation Study*, vol. I, p. 117, defines the greenline park thus: "Greenline . . . areas are coherent resource areas containing a mix of public and private lands which are comprehensively planned, regulated and managed by an authority set up specifically to preserve its recreational aesthetic, ecological, historical and cultural values." See this work for a detailed discussion of the concept. Also see Kusler, *Public/Private Parks*, and Little, *Greenline Parks*.

79. U.S. Department of the Interior, Heritage, Conservation and Recreation Service/National Park Service, *National Urban Recreation Study*, vol. I, p. 120.

80. Ibid., p. 132.

81. Ibid., p. 64.

82. Siehl of the Congressional Research Service writes in his "Alternative Strategies," p. 8, that the greenline concept "is modelled after the experience gained in the organization and management of New York State's Adirondack Park and the national parks of England and Wales."

83. Goddard, "What the United States Can Learn," p. 36.

84. Ibid.

85. For a detailed treatment of the Adirondack park, see Liroff and Davis, *Protecting Open Space*.

86. Charles Little, who coined the term greenline park and who under that term did much to popularize the concept from his position on the staff of the Congressional Research Service, was one of those who felt that the Jackson bill would do more harm than good to the cause of rational land use planning. (Personal communication with the author, March 16, 1981.)

87. Plotkin, "Policy Fragmentation"; various interviews by the author.

88. Laura Beatty, interview with the author, September 19, 1980. When President Reagan abolished HCRS in 1981, federal responsibility for the pinelands planning effort was transferred to the Park Service, along with many other HCRS responsibilities and much of its staff.

89. *Congressional Record*, April 20, 1977, vol. 123, p. H11475.

90. The reserve was established as part of the 1978 parks omnibus bill, PL 95-625.

91. Others within the agency were as enthusiastic as Whalen but for a very different reason; they saw greenline parks as a defense against Congress's tendency to add inferior parks to the System for its own political purposes. According to one agency administrator, "the [greenline] approach is a bone we can throw a hungry congressman." In other words, if a powerful congressman was intent on getting a national park for his district, establishing this kind of unit was a way of appeasing him without the expense and management responsibility of a traditional national park.

92. S. 2306 95 Cong. 1 sess. (November 4, 1977).

93. The source on this and the following data is National Park Service, *Summary of Land Holdings, 1981*, internal agency document.

94. Some of the other new units with greenline features as cited by the U.S. Department of the Interior, et al., *National Urban Recreation Study*, vol. I, pp. 161–188, were:

Golden Gate National Recreation Area
Lower St. Croix National Scenic Riverway
Pictured Rocks National Seashore
Point Reyes National Seashore
Sleeping Bear Dunes National Lakeshore
Chattahoochee River National Recreation Area
Lowell National Cultural Park
Upper Delaware National Scenic and Recreational River

See this study for the specific provisions these units made for permanent inholdings.

95. Much land bordering on the parks, especially in the West, is under the management of the Forest Service and neither it nor its affiliated interest groups would look kindly on bills which conceded a measure of control over forest land to the Park Service. As for regulating the private land outside the parks, the federal government is, as touched on in the text, in a very weak constitutional position. Moreover, Congress has shown little willingness to circumvent this weakness through innovative legislation. See Sax, "Helpless Giants," for a discussion of both the legal grounds on which the Park Service might be able to have a say in land use beyond park boundaries and the reluctance of Congress to allow the agency to move too far in the direction of land use regulation.

96. U.S. Department of the Interior, Heritage, Conservation and Recreation Service/National Park Service, *National Urban Recreation Study*, p. 163.

97. Ibid.

98. Ibid., p. 104.

99. Coke and Brown, "Public Attitudes."

100. Rourke, *Bureaucracy*, p. 135.

101. Senator Hayakawa, quoted in Wilderness Society, *Land Use Planning Report*, September 29, 1980, p. 310.

102. Elkin, cited in Coke and Brown, "Public Attitudes," p. 97. This point is one of the key themes of Perin, *Everything in Its Place*.

103. The use of such inducements was not unique to the National Park Service or to land use planning. Elizar, "Fiscal Questions," p. 478, writes that "the federal government frequently gains [the] right to sit in on games previously considered exclusively local by contributing needed funds to them." Indeed the federal government tries to establish entirely new games with its funds in some instances, and municipal planning was just such a new game in many places.

104. U.S. Department of the Interior, Heritage, Conservation and Recreation Service/National Park Service, *National Urban Recreation Study*, p. 104.

105. Sax, "Helpless Giants," p. 244.

106. Ibid. This point was collaborated by Joel Pickelner, superintendent of Fire Island National Seashore, in an interview with the author, December 23, 1981.

107. Joel Pickelner, interview with the author, December 23, 1981.

108. Sax, "Helpless Giants," p. 244.

109. Gitelson, "Tocks Island," p. 42.

110. U.S. Department of the Interior, Heritage, Conservation and Recreation Service/National Park Service, *National Urban Recreation Study*, p. 126.

111. Ibid., p. 228, writes that "the dominant conventional wisdom" among park management personnel is that development easements do not work.

112. Related to the author by Kurt Gellerman, New Jersey canoe activist, in an interview with the author, October 30, 1980.

113. Another agency administrator expressed the fear that "regional planning will cost [the Park Service] people contact," that unlike the management of the national parks, where the agency's work is seen and appreciated by millions of people annually, regional planning efforts will go largely unnoticed by the public at large while arousing the ire of those whose freedom is affected.

114. Robert Herbst, interview with the author, April 7, 1981.

115. Ibid.

116. Ibid.

Chapter 8. Conclusion

1. Selznick, *TVA*, p. 4.

2. Pfeffer and Salancik, *External Control*, p. 78. See also McGowan, "Enacting the Environment."

3. Arrow, *Limits of Organization*, p. 52.

4. Downs, *Inside Bureaucracy*, p. 3. Sharkansky, *Public Administration*, pp. 40–41.

5. Holden, " 'Imperialism' in Bureaucracy," p. 950.

6. Simon, *Administrative Behavior*, p. 135; Schutzenberger, "Tentative Classification," p. 101.

7. Emery and Tryst, "Causal Texture."

8. Shankland, *Steve Mather*, p. 147.

9. Mishan, *Technology and Growth*.

10. *American Forests*, "Former Agency Directors Speak Out," p. 50.

11. Schell, "Reflections," p. 55.

12. For an eloquent expression of this view from an admittedly unusual source, see Badger's speech in Grahame, *Wind in the Willows*, pp. 77–79.

13. Undoubtedly there tends to be a fitting of world view to immediate circumstances. For a superintendent caught between the conflicting demands of environmentalists and those of concession holders with powerful political backing, a view which incorporates a geological perspective—and which therefore justifies moving at a glacial pace—is likely to be an attractive one. On the other hand, for an administrator in the agency's Office of Legislative Affairs, who is often swept along in the momentum of the fast-paced action of the House parks subcommittee, a view which sees time running out before the juggernaut of civilization justifies disregarding significance criteria and quickly adding to the Park System anything which remains unspoiled.

14. The literature of the environmental movement is full of contempt for park visitors. For example see Sax, *Mountains Without Handrails*, pp. 5–15.

15. See Lowenthal, "American Scene."

16. Sax, *Mountains Without Handrails*, pp. 47–59. Conservation Foundation, *National Parks for the Future*, pp. 31–45 argues similarly.

17. For a full treatment of Say's Law, see Sewell, *Say's Law*.

18. The Statue of Liberty is in Theme 9: Society and Social Conscience; Subtheme 9a: American Ways of Life; Facet 9a1: The Melting Pot.

19. See Caro, *Power Broker*, for the attitudes of New York's politicians, as well as those of the press, toward efforts by local interests to interfere with Robert Moses's parkway and mass recreation plans.

20. White and White, *Intellectual Versus the City*.

21. Warner and Havens, "Goal Displacement," p. 540, refer to this switching process, "where original objectives are supplemented by alternate ones," as goal diversion. Nienaber and Wildavsky, *Budgeting and Evaluation*, p. 10, make the point that "the advantage of multiple . . . objectives is that attention can be switched from one to the other as interests and opinions change without appearing to sacrifice the principle." Multiplicity of goals can also insure against the appearance of failure. If the ostensible aim of a policy has not been attained, the presence

of multiple goals will make it likely that one can save face by finding some goals which have been achieved, even if only incidentally, and by switching attention to these secondary successes. The recent widespread popularity of multipurpose planning (and the concomitant disdain for single-purpose planning) legitimizes and encourages this process of switching goals—which perhaps accounts for its acceptance as official doctrine by so many agencies.

22. Wamsley and Zald, "Political Economy," p. 65, write that:

where goals are clearly defined and subject to surveillance, an agency . . . may be left little room for choice or maneuver in goals, program objectives, and perhaps even means of task accomplishment. But if goals are ambiguous or multiple, an organization's elite may press for one definition or another and, within the bounds of political feasibility, allocate resources internally in pursuit of this choice.

Warner and Havens, "Goal Displacement," p. 543, make a similar point:

the intangibility of goals makes it possible for the organization to accommodate diverse and even inconsistent subgoals. The program can be 'all things to all people' at an abstract level, so that different people can feel satisfied that their interests are being served . . . the intangibility of goals allows the flexibility in interpretation of the claimed goals.

23. These two broad goals of use and preservation have been interpreted by students of the National Park Service as inherently contradictory and the need to reconcile the two has been seen as a great misfortune for the agency. Perhaps this is so, but the presence of the two goals also increased agency flexibility, allowing it over the years to stress one or the other as circumstances dictated.

24. Selznick, *Leadership*, p. 213.

25. Simon, Smithburg, and Thompson, *Public Administration*, p. 387.

26. Many policy analysts are aware of this point and have expressed it in different ways. For example, Pressman and Wildavsky, *Implementation*, p. 184, write that "the beginnings of an idea are, generally speaking, an insufficient measure of its capabilities or its scope." Downs, *Inside Bureaucracy*, p. 203, writes that "all creativity inherently involves experimenting on a trial and error basis. There is always the possibility that some attempts may fail."

27. Most organizations have policy thrusts and involvements that end in failure. One good example, close to the Park Service, was the move of its rival, the Forest Service, into the regulation of forests on private land in the 1930s. (See Robinson, *Forest Service*, pp. 11–13.) At the time, it looked like an astute and proper move. Accepted public interest was expanding from the management of public land to the productivity of privately held forests. This expansion created a new demand on the federal bureaucracy for the regulation of private forests, and this in turn created a fine opportunity for the agency which could meet the demand. Accordingly, the Forest Service lobbied for and received some power to oversee forest management on private land. The tide of positive government ran out after the Second World War, however. A countercurrent of reaction to the New Deal got underway and what looked like an opportunity before the New Deal came to look like a liability after it. When this happened, Raphael Zon, formerly a close associate of Pinchot

and one of the most outspoken advocates of Forest Service regulation of privately held forest land, reluctantly concluded that such efforts "no longer conformed to the realities of the situation," and the USFS would do well to turn its attention more exclusively to the management of the national forests. (Quoted in Schmaltz, "Raphael Zon," p. 91.) The USFS did recognize the realities of the situation and with congressional encouragement it wrote off most of its efforts to regulate private forests.

28. See Platt, *Open Space Decision Process*, for a detailed examination of the circumstances surrounding the establishment of the Indiana Dunes unit.

29. U.S. Department of the Interior, National Park Service, *Fire Island National Seashore General Management Plan*, discusses the natural attributes of the park. They are not unique; rather, Fire Island's flora, fauna, and ecology are similar to those of barrier islands and coastal areas both to the north and south of it along the Atlantic seaboard. What makes the island unique, according to most agency documents which address the subject, is the presence of such natural features in such close proximity to a major metropolitan area.

30. Plotkin, "Policy Fragmentation," p. 422.

31. Reilly, *Use of Land*, p. 34.

32. Plotkin, "Policy Fragmentation," p. 416.

Bibliography

Advisory Commission on Intergovernmental Relations. *Urban and Rural America: Policies for Future Growth,* U.S. GPO, Washington, D.C., 1968.

Albright, Horace M. and Frank J. Taylor *"Oh Ranger!": A Book About the National Parks,* Dodd, Mead, New York, 1947.

———. *Origins of National Park Service Administration of Historic Sites,* Eastern National Park and Monument Association, Philadelphia, 1971.

Altshuler, Alan, ed. *The Politics of the Federal Bureaucracy,* Dodd, Mead, New York, 1968.

———. "The Politics of Transportation Innovation," *Technology Review* May 1977, pp. 51–58.

American Forests. Vol. 82, no. 6, June 1976, pp. 26 et passim, "Former Directors Speak Out."

American Public Works Association. *History of Public Works in the United States, 1776–1976,* American Public Works Association, Chicago, 1976.

Arrow, Kenneth. *The Limits of Organization,* Free Press, New York, 1974.

Babcock, Richard F. *The Zoning Game: Municipal Practices and Policies,* University of Wisconsin Press, Madison, Wis., 1966.

Blau, Peter, and W. Richard Scott. *Formal Organizations,* Chandler, San Francisco, 1962.

Boschken, Herman L. "Public Control of Land Use: Are Existing Administrative Structures Appropriate?" *Public Administration Review* vol. 37, no. 5, September/October 1977, pp. 495–504.

Bower, Joseph L. "Descriptive Decision Theory from the 'Administrative' Viewpoint," pp. 103–148 in Raymond A. Bauer and Kenneth J. Gergen, eds., *The Study of Policy Formation,* Free Press, New York, 1968.

Boyce, David E., Janet Kohlhase, and Thomas Plaut. "Estimating the Value of Development Easements: Methods and Empirical Analysis," Appendix to *Saving the Garden: The Preservation of Farmland and Other Environmentally Valuable Landscapes Under Pressure of Urbanization,* Report to the National Science Foundation by the Regional Science Research Institute, Philadelphia, 1977.

Brower, David. "A New Decade and a Last Chance: How Bad Shall We Be?" *Sierra Club Bulletin* vol. 45, January 1960, pp. 3–4.

Browne, William P. "Organizational Maintenance, the Internal Operation of Interest Groups," *Public Administration Review* vol. 37, no. 1, January/February 1977, pp. 48–57.

————. *Politics, Programs and Bureaucrats,* Kennikat Press, Port Washington, N.Y., 1980.

Cahn, Robert. "State of the Park Service," *Audubon Magazine,* July 1980, pp. 20–23.

Caro, Robert. *The Power Broker: Robert Moses and the Fall of New York,* Knopf, New York, 1974.

Chadwick, Douglas H. "Cutting Glacier Down to Size," *National Parks and Conservation Magazine* vol. 54, no. 7, August 1980, pp. 13–17.

Chadwick, George F. *The Park and the Town: Public Landscapers in the Nineteenth and Twentieth Centuries,* Praeger, New York, 1966.

Clark, Peter B., and James Q. Wilson. "Incentive Systems: A Theory of Organizations," *Administrative Science Quarterly* vol. 6, no. 2, September 1961, pp. 129–166.

Clawson, Marion. *Land and Water for Recreation: Opportunities, Problems, and Policies,* Rand McNally, Chicago, 1963.

————, and Jack L. Knetsch. *Economics of Outdoor Recreation,* Johns Hopkins University Press for Resources for the Future, Baltimore, Md., 1966.

Clements, Frederick Edward. *Plant Physiology and Ecology,* H. Holt, New York, 1907.

Cleveland, Treadwell, Jr. "National Forests as Recreation Grounds." *Annals of Political and Social Science* vol. 35, March 1910, pp. 25–31.

Coke, James G., and Steven R. Brown. "Public Attitudes About Land Use Policy and Their Impact on State Policy-Makers," *Publius,* Winter 1976, pp. 97–134.

Comptroller General of the United States. *The Federal Drive to Acquire Private Lands Should Be Reassessed,* CED-80-14, December 14,

1979, prepared by the U.S. General Accounting Office, U.S. GPO, Washington, D.C., 1980.

Conservation Foundation. *Letter: A Monthly Report on Environmental Issues,* July–August 1975, Washington, D.C.

———. *National Parks for the Future,* Washington, D.C., 1972.

Coopers and Lybrand, Inc. "Management Improvement Project: Phase 1 Report," Washington, D.C., November 1977.

Darling, F. Fraser, and Noel D. Eichhorn. *Man and Nature in the National Parks,* Conservation Foundation, Washington, D.C., 1967.

Deutsch, Karl W. *The Nerves of Government,* Free Press, New York, 1966.

DeVoto, Bernard. "Let's Close the National Parks," *Harper's Magazine,* October 1953, pp. 49–52.

Domenici, Peter. "Can Congress Control Spending?" *Policy Review* vol. 19, Fall 1980, pp. 49–65.

Downs, Anthony. *Inside Bureaucracy,* Little, Brown, Boston, 1967.

Edelman, Jacob Murray. *The Symbolic Uses of Politics,* University of Illinois Press, Urbana, Ill., 1964.

Elazar, Daniel J. "Fiscal Questions and Political Answers in Intergovernmental Finance," *Public Administration Review* vol. 32, no. 5, September/October 1972, pp. 471–478.

———. "Citizenship and Public in Metropolitan America," *Publius* vol. 6, no. 2, Spring 1976, pp. 1–12.

Elson, Ruth Miller. *Guardians of Tradition: American Schoolbooks of the Nineteenth Century,* University of Nebraska Press, Lincoln, Neb., 1964.

Emery, F. E., and E. L. Trist. "The Causal Texture of Organizational Environments," *Human Relations* vol. 18, 1965, pp. 21–32.

Etzioni, Amitai. *Modern Organizations,* Prentice-Hall, Englewood Cliffs, N.J., 1964.

Everhardt, Gary. "A New Look at Our National Parks," *American Forests* vol. 82, no. 6, June 1976, pp. 25–26, 54–59.

Everhart, William C. *The National Park Service,* Praeger, New York, 1972.

Fenno, Richard F. *The Power of the Purse,* Little, Brown, Boston, 1966.

Fitch, Edwin M., and John F. Shanklin. *The Bureau of Outdoor Recreation,* Praeger, New York, 1970.

Foresta, Ronald A. *Open Space Policy: New Jersey's Green Acres Program,* Rutgers University Press, New Brunswick, N.J., 1981.

Foss, Phillip O., ed. *Recreation,* in the series *Conservation in the United States: A Documentary History,* Chelsea House Publishers, New York, 1971.

Gardon, J. S. "Time to Save the Vanishing Prairie," *National Parks and Conservation Magazine* vol. 49, no. 9, September 1975, pp. 4–9.

Gerth, Hans H., and C. Wright Mills. *From Max Weber: Essays in Sociology,* Oxford University Press, New York, 1968.

Gitelson, Alan R. "The Tocks Island Project: A Case Study of Participation and Interaction Patterns in an Intergovernmental Decision-Making System," *Publius,* Winter 1976, pp. 21–47.

Glazer, Barney G. *Organizational Scientist,* Bobbs-Merrill, Indianapolis, Ind., 1964.

Glazer, Nathan, and Daniel Patrick Moynihan. *Beyond the Melting Pot: The Negroes, Puerto Ricans, Jews, Italians and Irish of New York City,* M.I.T. Press, Cambridge, Mass., 1963.

Goddard, Maurice K. "What the United States Can Learn," pp. 35–43, in Henry Jarrett, ed., *Comparisons in Resource Management,* Johns Hopkins University Press for Resources for the Future, Baltimore, Md., 1961.

Grahame, Kenneth. *The Wind in the Willows,* Scribner's, New York, 1961.

Graves, Henry S. "A Crisis in National Recreation," *American Forestry* vol. 26, July 1920, pp. 391–400.

Green, Arnold W. *Recreation, Leisure and Politics,* McGraw-Hill, New York, 1964.

Guetzkow, Harold. "The Creative Person in Organizations," pp. 35–45, in Gary A. Steiner, ed., *The Creative Organization,* University of Chicago Press, Chicago, 1965.

Hall, Richard H. "Some Organizational Considerations in the Professional-Organizational Relationship," *Administrative Science Quarterly* vol. 12, December 1967, pp. 461–478.

Hays, Samuel P. *Conservation and the Gospel of Efficiency: The Progressive Conservation Movement 1890–1920,* Harvard University Press, Cambridge, Mass., 1959.

Healy, Robert G., and John S. Rosenberg. *Land Use and the States,* 2d ed., Johns Hopkins University Press for Resources for the Future, Baltimore, Md., 1979.

Heckscher, August. *Open Spaces: The Life of American Cities,* Harper and Row, New York, 1977.

Herbst, Robert. "Threats to the National Parks," *Courier: The National Park Service Newsletter,* July 1980, p. 5.

Hoch, Irving. "Economic Analysis of Wilderness Areas," chapter 6 in ORRRC, Study Report no. 3, *Wilderness and Recreation: Resources, Values and Problems,* U.S. GPO, Washington, D.C., 1962.

Holden, Matthew, Jr. " 'Imperialism' in Bureaucracy," *American Political Science Review* vol. 60, no. 4, December 1966, pp. 943–951.

Horn, Steven. *Unused Power: The Work of the Senate Committee on Appropriations,* Brookings Institution, Washington, D.C., 1970.

Huntington, Ellsworth. "The Conservation of Man," pp. 559–574 in A. E. Parkins, and J. R. Whitaker, eds., *Our Natural Resources and Their Conservation,* 2d ed., Wiley, New York, 1939.

Huth, Hans. *Nature and the American: Three Centuries of Changing Attitudes,* University of Nebraska Press, Lincoln, Neb., 1972.

Ickes, Harold L. "Keep It a Wilderness," *National Parks Bulletin* vol. 14, no. 66, pp. 9–13.

Ingram, Helen M., and Scott J. Ullery. "Policy Innovation and Institutional Fragmentation," *Policy Studies Journal* vol. 8, no. 6, Summer 1980, pp. 664–682.

Irland, Lloyd C. "Citizen Participation—A Tool for Conflict Management on the Public Lands," *Public Administration Review* vol. 35, no. 3, May/June 1975, pp. 263–269.

Ise, John. *Our National Park Policy: A Critical History,* Johns Hopkins University Press for Resources for the Future, Baltimore, Md., 1961.

Johnson, M. A. *Whitehead's Theory of Reality,* Beacon Press, Boston, 1952.

Kaufman, Herbert. *The Forest Ranger,* Johns Hopkins University Press for Resources for the Future, Baltimore, Md., 1960.

————. *The Limits of Organizational Change,* University of Alabama Press, University, Ala., 1971.

————. "The Natural History of Human Organizations," *Administration and Society,* August 1975, pp. 131–148.

————. *Are Government Organizations Immortal?* Brookings Institution, Washington, D.C., 1976.

Kusler, Jon A. *Public/Private Parks and Management of Private Lands for Park Protection,* University of Wisconsin, Institute for Environmental Studies, Madison, Wis., 1974.

Lee, Ronald F. *The Antiquities Act of 1906,* prepared for the Office of History and Historic Architecture, Eastern Service Center, U.S. National Park Service, Washington, D.C., 1970.

————. *The Origin and Evolution of the National Military Park Idea,* U.S. Department of the Interior, National Park Service, Office of Park Historic Preservation, Washington, D.C., 1973.

————. *Family Tree of the National Park System,* Eastern National Park and Monument Association, Philadelphia, 1974.

Leopold, Aldo. *A Sand County Almanac,* enlarged ed., Oxford University Press, New York, 1966.

Lieberman, Charles Joel. "Fiscal Policy Formation in Congress: The Legislative Budgetary Process," unpublished Ph.D. dissertation, University of California, Santa Barbara, Calif., 1976.

Lindblom, Charles E. "Policy Analysis," *American Economic Review* vol. 48, June 1958, pp. 298–312.

———. *The Policy-Making Process,* Prentice-Hall, Englewood Cliffs, N.J., 1968.

Lipset, Seymour Martin. *Agrarian Socialism: The Cooperative Commonwealth Federation in Saskatchewan,* University of California Press, Berkeley, Calif., 1950.

Liroff, Richard A., and G. Gordon Davis. *Protecting Open Space: Land Use Control in the Adirondack Park,* Ballinger, Cambridge, Mass., 1981.

Little, Charles E. *Greenline Parks: An Approach to Preserving Recreational Landscapes in Urban Areas,* A Report of the Environmental Policy Division, Congressional Research Service, Library of Congress, Washington, D.C., 1975.

Long, Norton. "Power and Administration," *Public Administration Review* vol. 9, Autumn 1949, pp. 257–264.

———. "Public Policy and Administration: The Goals of Rationality and Responsibility," *Public Administration Review* vol. 14, Spring 1954, pp. 22–31.

———. *The Polity,* Rand McNally, Chicago, 1962.

Louisiana Parks and Recreation Commission. *Louisiana Statewide Comprehensive Outdoor Recreation Plan, Supplement 1,* Baton Rouge, La., August 1966.

Lowenthal, David. "The American Scene," *Geographical Review* vol. 58, no. 1, 1968, pp. 61–88.

Lynch, Kevin. *Managing the Sense of a Region,* M.I.T. Press, Cambridge, Mass., 1976.

McConnell, Grant. "The Conservation Movement—Past and Present," *Western Political Quarterly* vol. 7, no. 1, March 1954, pp. 463–478.

McCool, Daniel. "The National Park Service: The Politics of Appropriations," unpublished master's thesis, University of Arizona, Tucson, Ariz., 1980.

McGowan, Robert P. "Enacting the Environment, Organizational Persistence and Change," *Public Administration Review* vol. 40, no. 1, January/February 1980, pp. 86–91.

McPhee, John. "Profiles—George Hartzog," *The New Yorker* vol. 47, September 11, 1971, pp. 45–48 et passim.

Marshall, Robert. "The Problem of Wilderness," *Scientific Monthly* vol. XXX, 1930, pp. 142–148.

Mazmanian, Daniel, and Jeanne Nienaber. *Can Organizations Change? Environmental Protection, Citizen Participation and the Corps of Engineers,* Brookings Institution, Washington, D.C., 1979.

Meltsner, Arnold J. "Political Feasibility and Policy Analysis," *Public Administration Review* vol. 32, no. 6, November/December 1972, pp. 859–867.

Merrium, Lawrence C., Jr. "The National Park System: Growth and Outlook," *National Parks and Conservation Magazine* vol. 46, December 1972, pp. 4–12.

Mishan, Edward J. *Technology and Growth: The Price We Pay,* Praeger, New York, 1970.

Moore, Nancy A. "The Public Administrator as Policy Advocate," *Public Administration Review* vol. 38, no. 5, September/October 1978, pp. 463–468.

Morris, Charles R. *The Cost of Good Intentions: New York City and the Liberal Experiment, 1960–1975,* W. W. Norton, New York, 1980.

Mosher, Lawrence. "Congress Considers a Novel Approach to Managing California's Big Sur," *National Journal,* October 4, 1980, p. 1658.

Muir, John. *Our National Parks,* Houghton Mifflin, Boston, 1901.

Nakamura, Robert T. *The Politics of Policy Implementations,* St. Martin's Press, New York, 1980.

Nash, Roderick. "The American Invention of National Parks," *American Quarterly* vol. 22, no. 3, Fall 1970, pp. 726–735.

———. *Wilderness and the American Mind, rev. ed.,* Yale University Press, New Haven, Conn., 1973.

———. *The American Environment: Readings in the History of Conservation,* 2d ed., Addison-Wesley, Reading, Mass., 1976.

Nathan, Richard P. "The New Federalism Versus the Emerging New Structuralism," *Publius* vol. 5, no. 3, Summer 1975, pp. 111–129.

National Commission on Urban Problems. *Building the American City,* U.S. GPO, Washington, D.C., 1969.

National Parks and Conservation Association. *Preserving Wilderness in Our National Parks,* Washington, D.C., 1971.

National Resources Planning Board. *Recreational Use of Land in the United States,* part 11 of the *Report on Land Planning.* U.S. GPO, Washington, D.C., 1938.

Nature Conservancy. *Preserving Our Natural Heritage,* vol. I *Federal Activities,* prepared for the Office of the Chief Scientist, U.S. National Park Service, U.S. GPO, Washington, D.C., 1975.

Nicholson, E. Max. "What Is Wrong With the National Park Movement?" pp. 32–38 in National Parks Centennial Commission, ed., *Second World Conference on National Parks,* International Union for the Conservation of Nature and Natural Resources, Morges, Switzerland, 1974.

Nienaber, Jeanne. "Differentials in Agency Power: Determining Innovative Capability in Water Resources Agencies," OWRT Project Completion Report, September 1980.

———, and Aaron B. Wildavsky. *The Budgeting and Evaluation of Federal Recreation Programs: Or, Money Doesn't Grow on Trees,* Basic Books, New York, 1973.

———, Helen Ingram, and Daniel McCool. " 'The Rich Get Richer' Phenomenon: Comparing Innovation in Six Federal Agencies," paper prepared for presentation at the 1976 Annual Meeting of the Midwest Political Science Association.

Novogrod, R. Joseph, Gladys O. Dimock, and Marshall E. Dimock. *Casebook in Public Administration,* Holt, Rinehart and Winston, New York, 1969.

Orlans, Harold. "The Political Uses of Social Research," *Annals of the American Academy of Political and Social Science* vol. 394, March 1971, pp. 28–35.

ORRRC. *Federal Agencies and Outdoor Recreation,* ORRRC Study Report no. 13, U.S. GPO, Washington, D.C., 1962.

———. *Outdoor Recreation for America,* U.S. GPO, Washington, D.C., 1962.

———. *Outdoor Recreation Literature: A Survey,* ORRRC Study Report no. 27, U.S. GPO, Washington, D.C., 1962.

Ostrom, Vincent. *Crisis in Public Administration,* University of Indiana Press, Bloomington, Ind., 1973.

Papageorgiou, Alexis. "Architectural Schemata for Outdoor Recreation Areas of Tomorrow," *Daedalus* vol. 96, no. 4, Fall 1967, pp. 1158–1171.

Parsons, Talcott. *Essays in Sociological Theory, Pure and Applied,* Free Press, Glencoe, Ill., 1949.

Paterson, Thomas T. *Glasgow Limited,* Cambridge University Press, London, 1960.

Peffer, E. Louise. *The Closing of the Public Domain: Disposal and Reservation Policies, 1900–1950,* Stanford University Press, Stanford, Calif., 1951.

Perin, Constance. *Everything in Its Place: Social Order and Land Use in America,* Princeton University Press, Princeton, N.J., 1977.

Pfeffer, Jeffrey, and Gerald R. Salancik. *The External Control of Organizations: A Resource Perspective,* Harper and Row, New York, 1978.

Pinchot, Gifford. *Breaking New Ground,* Harcourt Brace Jovanovich, New York, 1947.

Platt, Rutherford H. *The Open Space Decision Process: Spatial Allocation of Costs and Benefits,* Department of Geography, University of Chicago, Chicago, 1972.

Plotkin, Sidney. "Policy Fragmentation and Capitalist Reform: The Defeat of National Land-Use Policy," *Politics and Society* vol. 9, no. 4, 1980, pp. 409–445.

Pressman, Jeffrey L., and Aaron Wildavsky. *Implementation*, 2d ed., University of California Press, Berkeley, Calif., 1979.

Prophet, Edward C. "Recreational Resources," pp. 533–557 in A. E. Parkins and J. R. Whitaker, eds., *Our Natural Resources and Their Conservation*, 2d ed., Wiley, New York, 1939.

Redford, Emmette. *Ideal and Practice in Public Administration*, University of Alabama Press, University, Ala., 1958.

Rehfuss, John. *Public Administration as Political Process*, Scribner's, New York, 1973.

Reilly, William K. *The Use of Land; A Citizen's Guide to Urban Growth*, Crowell, New York, 1973.

Richardson, Elmo. *Dams, Parks and Politics: Resource Development and Preservation in the Truman-Eisenhower Era*, University Press of Kentucky, Lexington, Ky., 1973.

Robinson, Glen O. *The Forest Service: A Study in Public Land Management*, Johns Hopkins University Press for Resources for the Future, Baltimore, Md., 1975.

Rourke, Francis E. *Bureaucracy, Politics, and Public Policy*, Little, Brown, Boston, 1969.

Rowntree, R. A., D. E. Heath, and M. Voiland. "The United States National Park System," pp. 91–140 in J. G. Nelson, R. D. Needham, and D. L. Mann, eds., *International Experience with National Parks and Reserves*, Department of Geography Publication Series no. 12, University of Waterloo, Waterloo, Ontario, 1978.

Runte, Alfred. *National Parks: The American Experience*, University of Nebraska Press, Lincoln, Neb., 1979.

Rushing, Kathryn Karsten. "NPCA: Sixty Years of Idealism and Hard Work," *National Parks and Conservation Magazine* vol. 53, no. 5, May 1979, pp. 6–12.

Sax, Joseph L. "America's National Parks: Their Principles, Purposes and Prospects," *Natural History*, October 1976, pp. 59–87.

———. "Helpless Giants: The National Parks and the Regulation of Private Lands," *Michigan Law Review* vol. 75, no. 2, December 1976, pp. 239–274.

———. *Mountains Without Handrails: Reflections on the National Parks*, University of Michigan Press, Ann Arbor, Mich., 1980.

Sayre, Wallace S., and Herbert Kaufman. *Governing New York City*, W. W. Norton, New York, 1965.

Schell, Jonathan. "Reflections (Nuclear Arms—Part II)," *The New Yorker*, February 8, 1982, pp. 48 et passim.

Schick, Allen. "Congress and the 'Details' of Administration," *Public Administration Review* vol. 36, no. 5, September/October 1976, pp. 516–528.

———. *Congress and Money: Budgeting, Spending and Taxing,* Urban Institute, Washington, D.C. 1980.

Schiff, Ashley L. "Innovation and Administrative Decision Making: The Conservation of Land Resources," *Administrative Science Quarterly* vol. 11, June 1966, pp. 1–30.

———. "Outdoor Recreation Values in the Public Decision Process," *Natural Resources Journal* vol. 6, no. 4, October 1966, pp. 542–559.

Schmaltz, Norman J. "Raphael Zon: Forest Researcher, Part 2," *Journal of Forest History* vol. 24, April 1980, pp. 86–97.

Schrepfer, Susan. "Perspectives on Conservation: Sierra Club Strategies in Mineral King," *Journal of Forest History* vol. 20, October 1976, pp. 176–190.

———. "Conflict in Preservation: The Sierra Club, Save-the-Redwoods League and Redwood National Park," *Journal of Forest History* vol. 24, April 1980, pp. 60–77.

Schutzenberger, M. P. "A Tentative Classification of Goal-Seeking Behaviors," *Journal of Mental Sciences* vol. 100, 1954, pp. 97–102.

Seidman, Harold. *Politics, Position and Power: The Dynamics of Federal Organization,* 3d ed., Oxford University Press, New York, 1980.

Selznick, Phillip. "Foundations of the Theory of Organizations," *American Sociological Review* vol. 13, 1948, pp. 25–35.

———. *TVA and the Grass Roots,* University of California Press, Berkeley, Calif., 1949.

———. *Leadership in Administration,* Harper and Row, New York, 1957.

———. "Institutional Integrity and Precarious Values," in Alan Altshuler, ed., *The Politics of the Federal Bureaucracy,* 1968, pp. 213–218.

Shands, William E. *Federal Resource Lands and Their Neighbors,* Conservation Foundation, Washington, D.C., 1979.

Shankland, Robert. *Steve Mather of the National Parks,* Knopf, New York, 1951.

Sharkansky, Ira. *Public Administration: Policy-Making in Government Agencies,* 2d ed., Markham, Chicago, 1970.

Sharpe, Maitland S. "The National Parks and Young America," pp. 197–212 in Conservation Foundation, *National Parks for the Future,* 1972.

Shepsle, Kenneth A., and Barry R. Weingast. "Political Preferences for the Pork Barrel: A Generalization," *American Journal of Political Science* vol. 25, no. 1, February 1981, pp. 96–111.

Sherman, Harvey. *It All Depends: A Pragmatic Approach to Organization,* University of Alabama Press, University, Ala., 1966.

Siehl, George H. *Visitor Pressures on the National Parks,* Congressional Research Service Report E-160, 71-84 EP, Library of Congress, Washington, D.C., March 5, 1971.

Simmons, I. G. "National Parks in Developed Countries," pp. 393–407 in Andrew Warren and F. B. Goldsmith, eds., *Conservation in Practice,* Wiley, New York, 1974.

Simon, Donald E. "A Prospect for Parks," *Public Interest,* Summer 1976, pp. 27–39.

Simon, Herbert A. *Administrative Behavior: A Study of Decision-Making Processes in Administrative Organizations,* 3d ed., Free Press, New York, 1976.

————, Donald W. Smithburg, and Victor A. Thompson. *Public Administration,* Knopf, New York, 1950.

Smith, James N. "The Gateways: Parks for Whom?" pp. 213–236 in *National Parks for the Future,* Conservation Foundation, Washington, D.C., 1972.

Sowell, Thomas. *Say's Law, An Historical Analysis,* Princeton University Press, Princeton, N.J., 1972.

Starbuck, William H. "Organizational Growth and Development," pp. 451–533 in James G. March, ed., *Handbook of Organizations,* Rand McNally, Chicago, 1965.

Stegner, Wallace. "The Wilderness Idea," in David Brower, ed., *America's Living Heritage,* The Sierra Club, San Francisco, 1961, pp. 97–102.

Stratton, Owen, and Phillip Sirotkin. *The Echo Park Controversy,* University of Alabama Press, University, Ala. Published for the Inter-University Case Program, 1959.

Swain, Donald C. *Wilderness Defender, Horace M. Albright and Conservation,* University of Chicago Press, Chicago, 1970.

Tannenbaum, Edith Greenbaum. "The Preservation of Open Space in Seven New York Counties," unpublished Ph.D. dissertation, Columbia University, New York, 1965.

Teilhard De Chardin, Pierre. *The Phenomenon of Man,* Harper and Row, New York, 1961.

Tilden, Freeman. *The National Parks: What They Mean to You and Me,* Knopf, New York, 1951.

Udall, Stewart L. *The Quiet Crisis,* Holt, Rinehart and Winston, New York, 1963.

U.S. Army Corps of Engineers, North Atlantic Engineering District. "Delaware River Basin Report, Vol. IV," New York, 1960.

U.S. Congress, House Subcommittee on National Parks of the Committee of Interior and Insular Affairs. *Review of National Park Service Poli-*

cies, Serial no. 16, 88 Cong. 2 sess., U.S. GPO, Washington, D.C., 1964.

U.S. Department of Agriculture, Soil Conservation Service. "Recreation Memo-3, Supplement-3," Washington, D.C., April 23, 1964.

U.S. Department of the Interior. "Budget Justifications, F.Y. 1981: National Park Service," Washington, D.C., 1980.

U.S. Department of the Interior, Bureau of Outdoor Recreation. *New England Heritage.* U.S. GPO, Washington, D.C., 1968.

————. *The Recreation Imperative: The Nationwide Outdoor Recreation Plan,* July 1970. (The report was never published by the Department of the Interior. Instead it was published as a committee print of the Senate Committee on Interior and Insular Affairs, 93 Cong. 2 sess., U.S. GPO, Washington, D.C., 1974.)

U.S. Department of the Interior, Bureau of Outdoor Recreation, Lake Central Regional Office. *Water-Oriented Outdoor Recreation: Lake Erie Basin,* Chicago, March 1966.

U.S. Department of the Interior, Heritage, Conservation and Recreation Service/National Park Service. *National Urban Recreation Study: Technical Reports—Volume One,* U.S. GPO, Washington, D.C., undated (1978).

U.S. Department of the Interior, National Park Service. *Cuyahoga Valley National Recreation Area, General Management Plan,* U.S. GPO, Washington, D.C., July 1977.

————. *Fire Island National Seashore, Draft Master Plan,* U.S. GPO, Washington, D.C., March 1975.

————. *Fire Island National Seashore, Final General Management Plan,* U.S. GPO, Washington, D.C., 1978.

————. *Gateway: A Proposal,* U.S. GPO, Washington, D.C., undated (1969), commonly referred to as the Greenbook.

————. *Gateway National Recreation Area, General Management Plan, Final Environmental Statement,* U.S. GPO, Washington, D.C., August 1979.

————. *Historic Camden: Study of Alternatives,* Denver, Colo., July 1980.

————. *Index: National Park System and Related Areas as of June 30, 1979,* U.S. GPO, Washington, D.C., 1979.

————. *Management Policies,* Washington, D.C., 1978.

————. *Our Fourth Shore, Great Lakes Shoreline Recreation Area Survey,* U.S. GPO, Washington, D.C., 1959.

————. *Our Vanishing Shoreline,* U.S. GPO, Washington, D.C., 1955.

————. *Pacific Coast Recreation Area Survey,* U.S. GPO, Washington, D.C., 1959.

————. *Part One of the National Park System Plan: History,* U.S. GPO, Washington, D.C., 1972.

————. *Part Two of the National Park System Plan: Nature,* U.S. GPO, Washington, D.C., 1972.

————. *A Report on the Seashore Recreation Survey of the Atlantic and Gulf Coasts,* U.S. GPO, Washington, D.C., 1955.

————. *Santa Monica Mountains National Recreation Area, Draft Environmental Impact Statement and General Management Plan,* U.S. GPO, Washington, D.C., September 1980.

————. *State of the Parks—1980: A Report to the Congress,* U.S. GPO, Washington, D.C., May 1980.

U.S. General Accounting Office. *Federal Land Acquisition and Management Practices,* CED-81-135, September 11, 1981, U.S. GPO, Washington, D.C., 1981.

Utley, Robert. "Living History: How Far is Too Far?" *Courier: The National Park Service Newsletter,* August 1974, p. 2, U.S. Department of the Interior, National Park Service, Washington, D.C.

————. "Toward a New Preservation Ethic," *National Park Service Newsletter,* October 15, 1974, p. 7, U.S. Department of the Interior, National Park Service, Washington, D.C.

————. "A Preservation Ideal," *Historic Preservation,* April–June 1976, pp. 40–44.

Vogel, David. "The Public-Interest Movement and the American Reform Tradition," *Political Science Quarterly* vol. 95, no. 4, Winter 1980–81, pp. 607–627.

Vogler, David J. *The Politics of Congress,* 3d ed., Allyn and Bacon, Boston, 1980.

Walker, Richard A., and Michael K. Heiman. "Quiet Revolution for Whom?," *Annals of the Association of American Geographers* vol. 71, no. 1, March 1981, pp. 67–83.

Wamsley, Gary L., and Mayer N. Zald. "The Political Economy of Public Organizations," *Public Administration Review* vol. 31, no. 1, January/February 1973, pp. 62–73.

Warner, W. Keith, and A. Eugene Havens. "Goal Displacement and the Intangibility of Organizational Goals," *Administrative Science Quarterly* vol. 12, March 1968, pp. 539–555.

White, Morton Gabriel, and Lucia White. *The Intellectual Versus the City,* Harvard University Press and the M.I.T. Press, Cambridge, Mass., 1962.

Wildavsky, Aaron. "The Analysis of Issue Contexts in the Study of Decision-Making," *The Journal of Politics* vol. 24, November 1962, pp. 717–732.

———. *The Politics of the Budgetary Process,* Little, Brown, Boston, 1964.

———. "The Agency, Roles and Perspectives," pp. 78–88, in Edward V. Schneier, ed., *Policy Making in American Government,* Basic Books, New York, 1969.

Williams, Norman, Jr. "On from Mount Laurel: Guidelines on the 'Regional General Welfare'," pp. 79–103 in Jerome G. Rose, and Robert E. Rothman, eds., *After Mount Laurel: The New Suburban Zoning,* Center for Urban Policy Research, New Brunswick, N.J., 1977.

Williams, Raymond. *The Country and the City,* Oxford University Press, New York, 1973.

Wilson, James A. *Political Organizations,* Basic Books, New York, 1973.

Windsor, Duane. "The Political Economy of Land Use Control," *Public Administration Review* vol. 40, no. 4, July/August 1980, pp. 396–400.

Wirth, Conrad L. *Parks, Politics and the People,* University of Oklahoma Press, Norman, Okla., 1980.

Wisconsin Conservation Department. *A Comprehensive Plan for Wisconsin, Outdoor Recreation,* Madison, Wis., 1966.

Woll, Peter. *Public Policy,* Winthrop, Cambridge, Mass., 1974.

Yamada, Gordon T. "Improving Management Effectiveness in the Federal Government," *Public Administration Review* vol. 32, no. 6, November/December 1972, pp. 764–770.

Yard, Robert Sterling. "Historical Basis of National Park Standards," *National Parks Bulletin* vol. 10, no. 58, March 1930, p. 7.

Zimmerman, Joseph F. "Neighborhoods and Citizen Involvement," *Public Administration Review* vol. 38, no. 3, May–June, 1978, pp. 201–210.

Index

About the Author

Ronald A. Foresta received his doctorate in geography and planning from Rutgers University. He currently teaches geography at the University of Tennessee, where he specializes in the study of public land policy.

He was a Resources for the Future Gilbert White Fellow in 1981 and is the author of *Open Space Policy: New Jersey's Green Acres Program.*